PENGUIN
Jung

'*Jungle* is a bold, ambitious and truly wonderful history of
the world that shows the vital importance of tropical
forests to life on Earth' Peter Wohlleben, author of
The Hidden Life of Trees

'A fascinating story and a crucial revision of the momentous
importance of tropical forests to human history. Spanning
from our very evolution as a species, to the early stages
of globalization, to how we fill our kitchen cupboards today,
we all owe far more to jungles than we realize'
Lewis Dartnell, author of *Origins*

'Welcome to the "Jungle" – a breathtaking book showing
that tropical forests were key to our evolution, provide fossil
fuels for our modern carbon-hungry society and ultimately
must be protected and restored if we are to have a future.
This insightful and captivating book will ensure you never take
our jungles for granted ever again' Mark Maslin, author of
How to Save Our Planet

'There are many books on the history of trilobites and dinosaurs
and other animals, but so few on the history of plants. Here the
dynamic young scientist Patrick Roberts tackles the history of
the tropics, from the coal swamps of 300 million years ago,
through the co-evolutionary dance of dinosaurs and mammals
and flowers, to how our own human history has been shaped by
vegetation. As environments are changing rapidly around us
today, this is a timely, readable and highly relevant history that
celebrates the wonder and importance of jungles' Steve Brusatte,
author of *The Rise and Fall of the Dinosaurs*

'An enthralling jungle journey from the origins of life on
this planet to the present day. *Jungle* provides a brilliant new
perspective on our interaction with tropical forests, placing
them at the centre of human experience – and it delivers a
timely warning about our abuse of the environment'
David Abulafia, author of *The Great Sea*

'*Jungle* sweeps the reader into the primordial heart of the Earth, as if the crucible of life welcomed you to its sanctuary. Its revelations and stories will stir, rearrange and populate your mind for years to come. As a book, it is a joy: pure intellectual chocolate' Paul Hawken, editor of *Drawdown*

'Finally, a book on rainforests that does justice to their majesty and importance. Patrick Roberts skilfully and lucidly shows why tropical forests matter. He builds the case that people and tropical forests are intimately linked, whether you live in the rainforest or seemingly a world away. Those intricate links are more important than ever today, with ending deforestation playing a key role in solving the twin climate and biodiversity crises we face this century' Simon Lewis, co-author of *The Human Planet*

'Enormously ambitious, deeply researched and moves with great skill from ecology and evolution to history and politics' Michael Marshall, *New Scientist*

'Many European and American books and films imply that tropical forests are incapable of sustainably supporting large human societies. *Jungle* provides a superbly argued refutation of this long-held view . . . a thrilling reappraisal of our origins and our dependence on tropical forests' Charlie Pye-Smith, *Literary Review*

ABOUT THE AUTHOR

Dr Patrick Roberts is W2 Research Group Leader at the Max Planck Institute for the Science of Human History, Germany. He completed his PhD at Oxford University, has worked in jungles across the world and has received numerous prestigious awards, including a European Research Council Starter Grant of 1.5 million euros. He has written or co-authored over ninety peer-reviewed journal articles and his work has featured on the BBC, Channel 4 and in *The Times*, among others. He is the author of the academic book *Tropical Forest Prehistory, History and Modernity*, and this is his first book for a general readership.

Jungle

How Tropical Forests Shaped World History – and Us

PATRICK ROBERTS

PENGUIN BOOKS

UK | USA | Canada | Ireland | Australia
India | New Zealand | South Africa

Penguin Books is part of the Penguin Random House group of companies
whose addresses can be found at global.penguinrandomhouse.com.

First published by Viking 2021
Published in Penguin Books 2022

001

Copyright © Patrick Roberts, 2021

The moral right of the author has been asserted

Typeset by Jouve (UK), Milton Keynes
Printed and bound in Great Britain by Clays Ltd, Elcograf S.p.A.

The authorized representative in the EEA is Penguin Random House Ireland,
Morrison Chambers, 32 Nassau Street, Dublin D02 YH68

A CIP catalogue record for this book is available from the British Library

ISBN: 978-0-241-99078-0

www.greenpenguin.co.uk

For Rhys, Ida and Livia — that they may yet walk through their own 'jungles'.

Contents

List of illustrations

Integrated illustrations

Inset illustrations

1. Artist's reconstruction of a Carboniferous forest floor (Bob Nicholls)

2. A photogrammetric model of a preserved footprint from *Dromopus lacertoides* (Luke Meade/Meade et al., 2016)

3. Image of an angiosperm leaf recovered from the Cerrejón Formation in northern Colombia (Carlos Jaramillo)

4. Artist's reconstruction of the Neotropical forest of Cerrejón with Titanoboa in the foreground (Jason Bourque/Florida Museum of Natural History)

5. Artist's reconstruction of Jurassic forest environment (Bob Nicholls)

6. Fossil of mammaliaform *Vilevolodon diplomylos* (Zhe-Xi Luo/University of Chicago)

7. Gliding Jurassic mammaliaform (April Neander/University of Chicago)

8. *Eohippus*, or 'dawn horse', the earliest known ancestor of the modern horse (Daniel Eskridge/Alamy)

9. Artist's reconstruction of 'Ardi', *Ardipithecus ramidus* (Jay Matternes/Smithsonian)

10. Photographs of cutmarks on an otter bone (adapted from Wedage et al., 2019)

11. Preserved microscopic botanical remains from Kuk Swamp shown alongside their modern counterparts (courtesy of Tim Denham, Karimui, Papua New Guinea, 2007/Fullagar et al., 2006/JAS)

12. The UNESCO-protected Banaue rice terraces in Ifugao on the island of Luzon in the Philippines (Getty Images)

13. The ancient Classic Maya ceremonial centre of Tikal, Guatemala (Michael Godek/Getty Images)

Preface

Battling through the heart of the Brazilian Neotropics in July 2019, my student Victor Caetano Andrade and I fended off poisonous snakes, gigantic mosquitoes that can bite through the tough skin of crocodiles, and heat that made sweat cascade like waterfalls into our eyes. Over the past ten years, my fieldwork in the world's tropical forests had taken me from the leech-filled rainforests of Sri Lanka to the flammable dry forests of Australia and the volcano-skirted misty cloud forests of Mexico. But nothing had prepared me for the teeming life, stifling climate, and vast array of green that flanks the Amazon, the largest river in the world by volume. Just to reach our field site and host village required a 523-kilometre 'slow ferry' (*recreio*) journey that took a staggering thirty-six hours, followed by another two-hour ride on a small, open boat, which we undertook while loaded down by several awkward boxes of scientific equipment, a rudimentary toolkit for on-site maintenance, and whatever clothes the remaining space and willpower would allow us to carry. Over the course of his summer research, Victor would suffer a bout of dengue fever, a pus-producing infected boil the size of a small plum on his hand, and the failure of an aeroplane engine during landing at a local city. This all likely sounds fairly familiar to those of you, particularly in Europe and North America, who have experienced 'jungles' only in terms of the films and novels that are set there. From the trials of *Apocalypto* to Mowgli running with animals, most of us see tropical forests as a kind of *terra incognita*. They provide artistic metaphors and natural resources, but they are fundamentally different to our own ideas of 'home'. In fact, the overwhelming implication of most Euro-American books, series and blockbusters set in the 'jungle' is that tropical forests are fundamentally incapable of sustainably and safely supporting large human societies.

This notion informs not just popular, but also academic, thought.

Tropical forests are often ignored in discussions of our human story and the history of life on Earth. The dominant narrative of human evolution, for example, is that our hominin ancestors left behind the perils and slim pickings of the forest as soon as they were able, striding out on to open savannahs with novel tools ready to exploit the rich game opportunities that these environments would provide.[1] The search for the origins of our own species, *Homo sapiens*, and its rapid 'routes' of dispersal around the globe, has similarly focused on open grasslands or, alternatively, coastal settings.[2] Likewise, discussions of the origins of 'agriculture' or cities almost never involve tropical forests. We have deemed them 'unproductive' and assume that they have uniformly poor soils, deadly natural hazards, elusive animals, and extreme climates that would make it impossible to maintain the types of agriculture and cities we associate with our own, purportedly 'complex', human societies.[3] Given the seemingly inevitable devastation inflicted by industrial farming and urban populations on these environments today, we wonder how these habitats could ever have supported sweeping monoculture croplands, vast pastures and bustling metropolises. Instead, accounts of communities living in tropical forests tend to describe small, often supposedly 'isolated', groups, relying on hunting and gathering to survive.[4] These assumptions shape not only how we understand the history of tropical forests, but also how we go about trying to protect them. Traditional conservation approaches to tropical forests assume that humans are simply incapable of living in them sustainably and that the best way of protecting them is to treat them as intact ecological 'wildernesses',[5] with minimal human disturbance and presence. Even the word commonly used in English to refer to tropical forests – 'jungle', from the Hindi word *jangal*[6] – was originally meant to classify something outside of the realm of human settlement and home comforts.

Yet, we should be sceptical of the stories we have so often been told about tropical forests. During my life-changing expedition to the Amazon Basin, two specific moments – far less expected, but far more intense, than any of Victor's and my struggles under the canopy – stayed with me. These encounters not only highlight the longevity and intimacy of human interactions with these majestic environments, but also their ongoing significance to all of us – whether we live in the tropics or not. First. One morning, as we woke to the gentle shaking of the ageing *recreio* and the sounds of parakeets, Victor, a seasoned traveller in the Amazon and a Brazilian national, pointed at the tree-tops and said, 'We will see a village soon.' I tried to follow his finger, but could see no sign of people, houses, or even a clearing that might imply some kind of human presence – everything simply appeared wild and green. Then Victor pointed towards the *types* of vegetation that had suddenly begun to dominate the riverbanks. On closer inspection, in contrast to the previous mass of forest, a dense concentration of two particular types of plant became clear: *açai* berry-bearing palms and Brazil nut (or more correctly 'Amazon nut') trees. Long-term work as an ecologist and frequent visitor to settlements up and down the banks of the Amazon Basin meant that Victor knew that these plants acted as a living signpost for human populations. Sure enough, a village slowly emerged from the tangle of dense vegetation on the riverbank – one that was fittingly named '*Ponta da Castanha*' (or point of Amazon nut trees). As Victor and the local people, including our generous host Jucelino, well knew, today's river-shore settlements are almost always located on top of prehistoric occupation sites. Here, past human societies modified the fertility of the earth and the composition of vegetation over millennia. So much so, that these spots still attract local Amazonian food producers today. If Amazonian tropical forests were indeed 'untouched' and largely free of human history, how could I be standing in a location that had repeatedly seen settlement for thousands of years?

Second. At the end of our visit, we pushed off from the shore in a small motorboat. Riding just above the waterline and gripping our belongings and equipment in our laps for dear life, I looked up into the sky. What I saw was a fast-moving blanket of clouds, sailing above the Amazon rainforest and its human settlements, so close you could nearly touch it. It was in this moment that, perhaps for the first time, I truly appreciated the regional, continental and global significance of tropical forests. If these forests disappeared, the loss of water evaporating from billions upon billions of leaves would turn this blanket of cloud into a threadbare rag. What few of us realize is that tropical forests are globally responsible for a significant portion of the planet's terrestrial rainfall.[7] In the case of the Amazon Basin, if its tropical forests disappeared rain would decrease, not just locally, but also across vast swathes of the South American continent. Thanks to the network of ocean and atmosphere circulation systems that cross the globe, climates as far away as Europe could be impacted. These same forests are also often major carbon sinks, performing over one-third of the planet's photosynthesis and trapping and storing around a quarter of the world's land-based carbon. If they disappeared, this carbon would be released into the atmosphere. Meanwhile, less atmospheric CO_2 would be recaptured into the biosphere in their absence.[8] In the context of a growing urgency to address the role of emissions in human-induced climate change, I am sure you can imagine what that would mean for temperatures all around our planet.

These two experiences provide an entry point into the vital, but often remarkably neglected, place of tropical forests in human history and the functioning of the Earth that I hope to reveal to you in this book. Tropical forests are not 'green hells',[9] hostile to human habitation and useful only for distant, relentless extraction and clearance. They can be productively lived with and lived in (a fact that I believe casts strong doubts on the assumptions we often make about how human societies, economies and settlements should inevitably be organized). Indeed, given their importance to the entire planet, we are all going to have to find ways to do this sooner rather than later. As a multidisciplinary archaeologist working in the tropics, I have seen how the discoveries of my predecessors and colleagues have increasingly highlighted the adaptive flexibility of human societies living in these environments on a global scale, from our earliest ancestors to the inhabitants of some of the largest urban areas ever to have existed prior to industrialization. I have seen how biologists and ecologists have long emphasized the importance of these, the oldest land-based ecosystems on the planet, to the evolution and sustenance of the world's greatest concentration of plant and animal diversity.[10] And I have seen how Indigenous[11] peoples have long advocated the critical role of tropical forests for their economic and cultural survival, as well as the importance of their active stewardship to the wellbeing of these ecosystems.[12] In light of all this, it is remarkable that many of us still think of tropical forests as so inherently separate from human existence. We are happy to hear about Tarzan living among wild animals, to watch searches for 'lost' cities by directionally challenged explorers, and to let the dulcet tones of David Attenborough wash over us as he describes a fabulous exotic bird attempting to dazzle an unimpressed mate. Yet, the remoteness, biodiversity and increasingly dramatic disappearance of tropical forests, while fascinating and shocking, also feel disconnected from many of our day-to-day lives. Even when we call for their protection, we tend to call for humans to be moved out of them rather than searching for ways in which humans might live *alongside* them. The plights facing these environments today are undoubtedly poignant, but their apparent distance, isolation and exoticism all too often allow us to ultimately ignore them.

This book is a history of the world according to the tropics and their 'jungles'. It begins with a journey through the colourful cast of tropical forests that have inhabited our planet for nearly 400 million years (Chapters 1 and 2), from the first tree-like organisms emerging into a warm, wet world to forests that would be recognizable to those inhabiting (or visiting) the tropical portions of our planet today. We will see how, from their arrival on the Earth's surface, tropical forests have shaped – and been shaped by – barrages of climate change from above and crunching forces of tectonic plates from below. Tropical forests sculpt our planet's atmosphere, water cycle and soils, and have played a major role in the evolution of life on Earth. We will explore the role of these remarkable environments in hosting the first flowering plants and the earliest four-legged land-based creatures on Earth (Chapter 2), in impacting the evolution of the dinosaurs (Chapter 3), and in shaping the survival and evolution of many of the main ancestral lines of mammals that we flock to zoos to see today (Chapter 4). Tropical forests also played an essential role in human evolution. They were the leafy 'cradles' where the first hominins appeared in Africa, splitting off from the last common ancestor between other great apes and ourselves (Chapter 5). They were also one of the diverse environments in which our own species, *Homo sapiens*, emerged in Africa, before we went on to occupy nearly all the planet's continents between 300,000 and 12,000 years ago (Chapter 6).

Despite the remarkable position of tropical forests in these major evolutionary processes, many of us tend to see them as somewhat detached from our story: impossible to farm in the absence of deforestation, deserted, isolated islands, or unattractive and fragile in the face of urban developments. What I want to show you is that these preconceptions are often the product of Euro-American ideas of what 'agriculture' or 'cities' should be. Many of the crops and animals we rely on today were, in fact, first managed and domesticated in the tropics (Chapter 7). Many different human societies grappled with sustainable living on tropical islands, managing local resources alongside new crops and animals brought from elsewhere (Chapter 8). And tropical forests were also home to some of the largest, and also comparatively most successful, pre-industrial urban populations ever to have existed (Chapter 9). This, in turn, forces us to ask the question: given all this evidence for previous innovative human habitation and management of tropical forest environments around the world, why do we commonly think of them as 'empty' and 'vulnerable' to human presence today? The answer is to be found in the historical processes of the last 500 years, where Europe and the tropical world collided. Disease, warfare and murder ravaged the cities and villages of Indigenous populations (Chapter 10). New profit-driven approaches to tropical landscapes saw mining and monoculture plantations strip away forests, erode soils, and act as the sites of forcibly abducted and transported labour as part of the transatlantic slave trade (Chapter 11). The spread of imperial and capitalist forces around the tropics resulted in degraded landscapes, in wealth imbalances between the western half of Europe and northern North America and the tropical world, in racial discrimination and violence, in a marginalization of Indigenous knowledge, and in the onset of significant human-induced climate change, which continue to represent perhaps the most major ecological, social, political and economic challenges facing our societies in the twenty-first century (Chapters 11 and 12). They also led many of us to falsely conclude that these habitats cannot, and do not, support large numbers of people.

Although a number of readers may already care deeply about the environment and the decline of tropical forests, it is highly likely that, to some extent, you still feel a certain safe estrangement from some of the world's oldest land-based environments. Indeed, harsh working conditions, dense vegetation, and difficult navigation – as encountered by Victor and me on our Amazon expedition – have often hindered exploration of tropical forests and their histories. However, this book will look at how the latest scientific advancements, from laser scanning from the air to plant genetics in the lab, can take us through the canopy, to show us how these habitats actually penetrate every inch of our lives, wherever we are. Our kitchen cupboards are filled with groceries that owe their beginnings to tropical forests. Our travel to work depends on latex originally tapped from tropical trees. Our consumer decisions, from furniture to beauty products, continue to influence the nature and extent of tropical environments. Extinction, deforestation and fires in these environments can seem a world away from many of us as we hear about them in the news. But the loss of tropical forests, and the colonial legacies that shape their plight today, impact not just the lives and climates of people on the other side of the world, but also our own weather, politics, societies, and economies – from Manila to Munich, Colombo to Cardiff, and from Nairobi to New York. This book is an attempt to convince you that the history of tropical forests is your history too. But before we get ahead of ourselves, it is time to look back beyond 500 million years ago. Back when plants had not yet graced the land surfaces of the Earth, and things looked very, very different to how they do today.

1. Into the light – the beginning of the world as we know it

School tours of natural history museums around the globe frequently rush past the fossilized remains of early plants on the way to the reconstructed dinosaur skeletons or stuffed blue whales that often take the plaudits as the heavyweight 'stars' of evolution. Novels and films such as *Jurassic Park* have brought the work of palaeontologists, picking away at long-extinct bones to determine the evolution of fascinating creatures, into the public arena. But no such dramatization has emerged for the explorers of the world's ancient plants. This speaks to a more general apathy. As we go about our daily lives, we rarely spare a thought for the mosses growing on our pavements, the grasses covering our fields, the flowers occupying our gardens, and the trees that, if we're lucky, line our streets. We often take for granted that plants have always been on our planet and always will be. There are also fewer well-known documentaries that explore the emergence, evolution and conservation status of plants. Plants are considered traditionally less 'exciting' than other life forms. They are harder to empathize with than animals, with their emotional eyes and a semblance of what we might recognize as families. Plants also do not make noises audible to the human ear, apart from perhaps some accidental creaking in the wind, making it unlikely that they will feature in a viral YouTube video 'talking'. Yet, without plants, the world as we know it today would not exist.

Indeed, recent research makes it seem like we might have been too quick to dismiss our photosynthesizing friends. For example, we now know that plants actually share some features that make us so able to relate to other animals. Time-lapse cameras have revealed the amazing, dynamic way in which plants constantly grow and move in different directions to access light.[1] While they do not feel pain in a traditional sense, plants being consumed by caterpillars will release defence chemicals that make them less appealing to their attacker.[2]

Similarly, trees being eaten by giraffes in eastern Africa will release compounds into the air that warn other trees to protect their own leaves.[3] Such 'communication' has also been demonstrated in forests where neighbouring trees will try to route nutrients to their damaged or ailing companions, making use of mass fungal 'webs' that are some of the largest organisms in the world.[4] In an attempt to maximize growth, trees will also sometimes gamble on the shedding of their leaves in temperate winter environments, with daring individuals paying the potentially fatal penalty of frost if they keep their leaves for too long.[5] However, by far the most impressive part of these organisms is the way in which they engineer our world to make it hospitable to almost all other life. The sheer scale of this accomplishment is only truly apparent if we explore what the Earth was like *without* plants, which, thanks to recent scientific advances, we can now do.

Let's travel back to the early Cambrian period (between around 538.8 and 509.0 million years ago).[6] While it might seem an odd place to start in a book about tropical forests – being named after *Cambria*, the Latin word for Wales – this famous geological period witnessed a diversification of complex life, often termed the 'Cambrian explosion', that was to eventually produce the evolutionary lineages of all multi-cellular animals left on Earth today. Were we to dive into the significantly warmer oceans of the early Cambrian period, we would see many different species of trilobites (the characteristic woodlouse-like fossil arthropods many of you will be familiar with) and other invertebrates practising predation, scavenging and filter feeding as marine ecosystems began to take on a familiar form.[7] However, if we left this marine cornucopia and stepped out on to the land, a seemingly deserted, and frankly apocalyptic, scene would greet us. The continents would have been covered by dry, rocky landscapes, and the main type of life present would have been in the form of patchy films of microbes smeared across the surface. Beyond that, the only visible signs of life on land would have been the occasional slug-like mollusc precariously venturing out of the oceans to try and scratch a living on the slim pickings available.[8] During the early Cambrian there was no variety of terrestrial life at all, let alone something we might associate with a tropical forest. However, the world was certainly 'tropical' by common climate classification systems that define a tropical climate as one with monthly average temperatures of ~18°C or more.[9] In the Cambrian, the *global* average temperature was around 19°C, a staggering 5°C higher than the present day,[10] and it still remained at an average of ~18°C at the beginning of the Devonian (419.0 million years ago).[11]

It was into this 'tropical' world that the first land plants arrived to perform their greatest trick, with an alchemical skill we can only dream of. All of you will have had a high-school introduction to photosynthesis, the process by which most plants and certain bacteria can use the energy of the Sun to convert carbon dioxide (CO_2) in the air or oceans into sugars and oxygen (O_2). However, the fundamental nature of this process to life is easily forgotten. In the absence of land plants, even with some bacteria and aquatic algae performing photosynthesis, early Cambrian atmospheric CO_2 concentration was around 4,500 parts per million by volume. That's ten times higher than it is today.[12] Meanwhile oxygen made up only about 7% of the Earth's atmosphere,[13] around three times lower than today.[14] As a result, on our tour of the Cambrian's land surface we would require oxygen tanks to survive. Furthermore, in the absence of plants, there were no root systems to efficiently break down bedrock, and nutrients that are crucial to most food chains today remained locked away in the Earth's tough crust. We will now see how the broadly 'tropical' state of our planet between the Cambrian and the Devonian represented the perfect 'greenhouse' for the emergence of not only the first land-based plants, but also the first trees and forests. While these pioneering organisms actually put an end to widespread warm conditions, causing significant drops in CO_2, global cooling, and even the formation of polar glaciers, they also formed the building blocks for the first complex, land-based ecosystems that were to emerge around the remaining warm, wet regions of the equatorial tropics in the Carboniferous period (359.3 to 298.9 million years ago). These forests fixed soils, released nutrients, and stabilized the climate and the composition of the atmosphere. They also provided ideal homes for a growing variety of interacting animal life that would eventually lead to the amphibian, reptilian and mammalian lineages that surround us today. It is no exaggeration to say that they changed the face of the world for ever.

The process of photosynthesis, and the extension of roots into the very surface of the Earth, meant that, from their first arrival, plants had the potential to be significant movers and shakers in the context of our planet's atmosphere, climate and geology. Professor Timothy 'Tim' Lenton is Director of the Global Systems Institute at the University of Exeter and is interested in how the first plants on land might have influenced the various 'systems' (e.g. atmosphere, geosphere, hydrosphere and biosphere) that interact to shape the climate of Earth. As he puts it, 'There is a complex interplay between our oceans, the proportions of different gases in the air around us, and, crucially, what covers the land that affects how the Sun's energy is stored and moved around the world.' Earth systems scientists such as Tim rely on powerful computers to 'model' how changes to the amounts of different gases in our atmosphere, to the speed and direction of ocean currents, and to the distribution of different types of vegetation will alter the overall condition of the planet, and whether different parts of the Earth will be impacted in different ways by such changes. The outputs from what might be best described as an earth scientist's version of *Minecraft* can then ultimately be tested against available records of CO_2 and O_2 concentration and temperature from ancient sedimentary or ice core records, to see whether they 'work' and can effectively explain the state of the Earth at a given point in time.

One of the main uses of this climate 'sandbox' has been to make predictions about how twenty-first-century greenhouse gas emissions will impact global temperatures in the decades to come. However, it has also been used to explore how the arrival of the first plant life would have altered the state of the planet. Yet, in order to do this properly, these computer whizzes must turn to the more 'classic' approaches of exploring ancient plants, or 'palaeobotany', to answer the key question: when did the first true land plants (or 'Embryophyta') appear? It is generally accepted that the first land plants were non-vascular, or 'lower plants', meaning that they lacked complex structures, such as the transport network of xylem cells that carry water and phloem cells that convey nutrients from a plant's roots to the rest of its body. These plants evolved from aquatic algae and have long been thought to have resembled the modern liverworts, hornworts and mosses (or 'bryophytes') you will have likely seen covering large patches of ground, rocks and trees in your back garden. These spore-producing plants are, today, well-known for their ability to retain water and can also act as important sources of food for animals. However, suggestions that liverworts, in particular, were the earliest diverging 'sister' lineage of land plants[15] had meant that climate models often neglected the impacts of the earliest land plants on the Earth's surface and atmosphere.[16] This is because liverworts are 'lacking' a number of physical features such as

The liverwort *Marchantia polymorpha*. The earliest land plants are thought to have resembled living 'bryophytes' such as liverworts, hornworts and mosses. Liverworts, such as *Marchantia polymorpha*, have often been considered the oldest lineage of land plants, acting as a model system for a transition between aquatic algae and complex vascular land plants. However, the exact nature and physiological features of the earliest plants to colonize the land have remained unknown until recently.

stomata – the tiny holes in leaves that can be opened and closed to help plants control their uptake of CO_2 and loss of water during photosynthesis – that are found in more complex, 'vascular' land plants. As a result, they seemed to represent a perfect halfway house between 'primitive' aquatic algae and complex, terrestrial vascular plants.[17] Nevertheless, as you can imagine, directly testing this assumption by actually finding and studying the earliest of these plants has posed a major challenge to even the most enthusiastic of palaeobotanists.

The most direct evidence for the appearance of land plants comes from fossilized impressions of their bodies or the spores they used to reproduce that were left behind in ancient sediments. Following a series of high-pressure geological processes, these traces were eventually converted into rock, to be preserved to this day. The earliest embryophyte-like spores come from sediments in Argentina dating to around 470 million years ago,[18] during another geological period with a Welsh connection, the Ordovician (486.9 to 443.1 million years ago), named after an ancient group (the Ordovicii) that was subjugated by the Romans in what is now North Wales. More recently, five bryophyte clades have been argued to be represented by fossil imprints from a sinkhole excavated during the creation of the Douglas Dam, Tennessee, dating to ~460 million years ago.[19] This would be significantly earlier than one of the most widely accepted earliest fossils with a land plant body structure from Ireland that have a minimum age of ~427 million years ago.[20] Nevertheless, if you have ever touched a bryophyte while working in your garden or taking a walk in the woods, you will have noticed its fragility, and it will often simply crumble away in your hands. As a consequence, existing fossil discoveries are almost certainly under-estimates of the true age of the first land plants. Comparisons of modern and fossil plants can be used to estimate how long would have been needed for different 'morphologies', or body shapes, to have emerged in the past.[21] However, this is imprecise, as evolution can occur at markedly different rates. Furthermore, the incredible patchiness of the fossil record of plants, particularly during this early timeframe, means that there is frustratingly little to go on when attempting to understand when and how land plant lineages emerged and proliferated, as well as what they looked like and what implications they might have had for the planet.

Some scientists have therefore adopted different approaches to exploring the evolutionary history of land plants. For example, they have sampled the modern genetic data of different families of liverworts, mosses and other land plants to try and produce a kind of 'molecular clock'.[22] These 'clocks' work on the principle that mutations within the genetic or protein sequences of plants will have occurred at a constant rate that can be calculated. With this in mind, the modern genetic sequences of different plant groups can then be compared, with the amount of difference providing some kind of idea as to how long ago the different lineages formed and separated from each other.[23] This work has allowed scientists to determine that vascular plants, or 'tracheophytes', which include the trees, shrubs, flowers, and ferns and fern allies that characterize the green plant life most visible to us today, indeed emerged *after* the liverworts and mosses.[24] More significantly, it has also been demonstrated that liverworts are not actually the earliest land plant lineage, calling into question the relevance of their physical features for characterizing the earliest plants to colonize the land.[25] Nevertheless, the earliest evolutionary history of plants still remained hazy and it has been incredibly challenging to estimate a robust date for the first land plants. Given the complexity of genetic mutations and differences in the rate of change between populations, species, and even plant families, dates based on modern genetic data alone are notoriously unreliable, providing a rough ballpark figure at best.[26] If the date for the first land plants, and insights into their biological features, cannot be determined, then working out when and if they had impacts on the Earth's climate obviously becomes a major challenge.

Fortunately, in 2018, scientists decided to combine the strengths and weaknesses of the different fossil and genetic approaches to the origins of land plants to develop a more accurate and comprehensive model for their appearance. This team, headed by scholars at the University of Bristol, combined the genetic and morphological estimates of rates of plant evolution, with the available 'fixed points' provided by actual remains in the fossil record. As Dr Silvia Pressel, head of the Algae, Fungi and Plants Division at the Natural History Museum of London, and co-author of the study, states, this work allowed 'initially rough evolutionary models to be tested and refined using the evidence that we can hold in our hands'. Moreover, by repeating these models thousands and thousands of times the researchers could work out just how likely it was that they represented reality. All their analyses converged to indicate that the first land plants actually appeared about 500 million years ago, slap bang in the middle of the Cambrian period, just after the evolution of multicellular animal species had started to take off underwater.[27] Even more significantly, biomolecular analysis of different plant lineages undertaken by a separate multidisciplinary team of palaeo- and neo-botanists indicated that the ancestor of land plants was probably more complex than assumed so far, already possessing root-like filaments and stomata on its surfaces.[28] These features would have allowed them to process more soil and more carbon dioxide than had previously been thought.[29]

Silvia's work has also added a further dimension to this emerging picture. You might have read about how plants and fungi combine to produce a close 'symbiotic' relationship (known as mycorrhizae). The fungi, interlocking with the roots of a plant, provide more water and soil nutrients to the plant, helping them to break down bedrock and soils, and even protect them from pathogens.[30] Meanwhile, the plants provide the fungi with sugars for sustenance from photosynthesis. We now know that most land plants make use of this win-win partnership. About 90% of plant species today are supported by fungal helpers, and the mapping of species of fungi and bacteria between the roots of trees across the world has come to be known as the 'wood wide web',[31] enabling flows of nutrients across entire ecosystems.[32] This relationship was already thought to be ancient because of the presence of fungal structures, including little tree-like structures referred to as arbuscules, in the cells of 400-million-year-old fossil plants.[33] However, some bryophytes, namely mosses, lack these interactions, leaving the question open as to whether the earliest plants on land had also already established fungal partnerships. By studying the

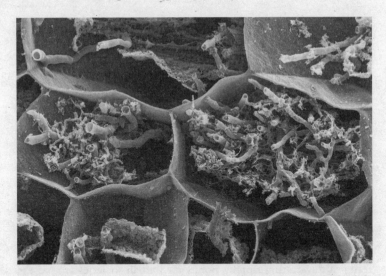

Fungal structures seen within a modern liverwort. These mycorrhizal fungi help the plant to absorb water and nutrients from the soil, and the relationship between them and plants likely extends back to the first plants to grow on land.

diversity and function of fungi across the bryophyte groups, and by sequencing the modern genomes of mycorrhizal fungi, researchers like Silvia have confirmed that this 'subterranean cooperation between plant and fungus extends back to the roots of plant evolution on land', as she puts it,[34] and likely involved a greater diversity of fungi than previously thought.[35] Together, then, it is becoming increasingly clear that the first land plants emerged earlier, and, alongside their fungal aides, had a more significant impact on the planet than has often been considered.

Returning to the computer studio, back in 2012, Tim's team had used their simulation-based approach to make two significant arguments about the first land plants. First, they performed experiments that showed that modern mosses, even with limited root systems, would significantly boost the weathering of rocks to nearly the same degree as more complex 'higher' vascular plants. They do so by releasing acids from their tissues on to the surrounding land surface. The weathering of rocks traps CO_2 from the atmosphere in acidic solutions with ions (bicarbonate, calcium), while at the same time CO_2 is absorbed from the atmosphere by photosynthesis. Applying these new weathering estimates, which would only be significantly increased by the findings of the 2018 study and the discovery that, unlike mosses, the earliest land plants would have already established a joint venture with fungi, enables an estimation of the planet's CO_2 concentrations and temperatures during the Ordovician period. Although these computer models are inevitably imprecise to some degree, assuming that these non-vascular land plants had colonized just 15% of the currently vegetated parts of the Earth's surface between 475 and 460 million years ago (plants make up around 32% of the Earth's surface today[36]), they suggested a staggering *halving* of atmospheric CO_2 and a global cooling of at least 4 to 5°C.[37] The modern aim to limit global warming to 1.5°C in order to avoid a series of devastating tipping points for human societies around the world[38] should put the scale of this planetary change into some perspective. Taking an even earlier date of plant colonization means that changes could perhaps have begun during the late Cambrian ~490 million years ago. Significantly, corresponding climatic alterations are, at least, clearly identifiable in available records from the Ordovician period,[39] indicating that these processes did indeed have immense implications for the beginnings of complex life on Earth.

Firstly, the reconstructed cooling was enough to trigger an 'ice age' by the time of the Ordovician–Silurian geological boundary 443.1 million years ago. The Silurian is yet another name derived from Wales and, although the chilly conditions at the start of this period may be more in keeping with popular ideas of a trip to the Brecon Beacons, the real reason so many of these periods have names deriving from this country is that a lot of formative work in the discipline of geology was done by British people in Britain, with the names of classic blocks of time first identified on these isles frequently being imposed upon the rest of the world. The drop in temperatures at the Ordovician–Silurian boundary was so extreme that it was to be one of the coldest periods in the last half a billion years of Earth history[40] (or over 1,500 times longer than we have existed on the planet!). Secondly, enhanced weathering would have released significant nutrients, and specifically the element phosphorus, from the previously barren rocky land cover. Phosphorus is critical to the growth of organisms. However, too much phosphorus can be a bad thing, particularly in aquatic settings, and, today, phosphorus pollution can stimulate a rapid growth of toxin-producing algal blooms. Back in the Ordovician, the washing of newly-weathered phosphorus into the oceans would have had a similar impact. Although these 'blooms' would have supported more and more marine life, this would also have gradually sucked the limited oxygen supply from the water. Ultimately, suffocation and a mass extinction of marine life characterizes the end of the Ordovician. While earlier dates for the first plants ~500 million years ago may indicate that this process was a relatively slow one, a similar cooling and mass marine extinction is actually also documented in the middle of the Cambrian, which could hint at a more immediate role for these new green organisms. Regardless, the latest research is clear in showing that even the most primitive of land plants could, and did, dramatically alter the climate, atmosphere, and surface of the Earth. These plants survived their self-induced Ice Age, huddling in the warmer equatorial areas of the planet. There they waited, ready to grow and change the world to an even greater extent. While the oceans initially suffered, the land was bracing itself for an unprecedented expansion of life.

Although the first plants certainly affected the Earth's systems, it was only with the arrival of the most complex of vascular plants, the trees, that these changes began to stick. The earliest 'higher' vascular (or 'tracheophyte') fossil body plant form, with xylem, phloem, and complex, invasive rooting systems that fit with our more traditional school-based definitions of 'plants', is dated to around 420 million years ago in Canada,[41] nearly 80 million years after the first plants colonized the land, and represents the beginning of the plethora of complex green plant forms we are familiar with today. Shortly afterwards, during the Devonian (419.0 to 359.3 million years ago), the first trees appeared on the planet. Trees have certain evolutionary benefits over other plants that enabled their rapid success. Unsurprisingly, their sturdy structure enables them to grow taller and gain better access to light for photosynthesis. More photosynthesis meant more body mass and their root systems could search deeper and more widely for nutrients, with weathering from roots and their fungal collaborators increasing the rate of soil formation and greatly expanding the area in which subsequent plants could take root. More photosynthesis also increased the amount of CO_2 absorbed from the atmosphere.[42] As much of this CO_2 became stored in robust lignin-based and, later, woody trunks, resistant to decomposition after death, it also became locked away from the atmosphere, increasing the relative proportion of O_2 in the air that remained. The true giants of vegetation cover, and its impact on the Earth's systems, had arrived.

Until recently, the story of our planet's trees was thought to have begun at a stoneworks near Albany, New York State. Here, at Riverside Quarry near Gilboa, workers extracting stone needed for a nearby dam project in the 1920s discovered hundreds of log-like shapes, one after the other, sticking out upright from the rock they were chipping away. New York State's first female palaeontologist, Professor Winifred Goldring (1888–1971), was called to the scene and her research on these amazing finds[43] shaped discussions of ancient forests for nearly a century. The trunks were identified as *Eospermatopteris*, a plant very similar to modern tree ferns. While it was considered to have a broadly 'tree-like' (what palaeobotanists call an 'arborescent') structure, *Eospermatopteris* lacked proper leaves. Instead, photosynthesis occurred on fronds (like those of palms) inserted into almost-vertical branches. The trunks of *Eospermatopteris*, dating to around 380 million years ago, were also not composed of wood but rather consisted of a tough, hollow lignin structure. Later work in the 2000s identified a woody 'aneurophytalean' vine at Gilboa, which, with mosses, apparently climbed the *Eospermatopteris* tree forms. While *Eospermatopteris* has subsequently often been heralded as the first anatomical 'tree',[44] the aneurophytalean vines were actually considered a better biological example of an early ancestor of the woody, seed-bearing trees of today[45] – or a 'progymnosperm' – that is, until 2019, when another fierce competitor for the title of the oldest 'tree' emerged.

The findings at Riverside had stimulated further palaeobotanical investigations of the rock formations on the east coast of North America, which were eventually to happen across another major ancient fossil. Again in New York, and again in a quarry (albeit this time abandoned), scientists uncovered an incredible record of a different type of ancient arboreal life-form in the region. Dating 2–3 million years older than Riverside, the new Cairo site produced the same hollow *Eospermatopteris* and the same woody aneurophytalean vines as Gilboa. In addition, the scientists found so-called 'lycopsids' (distant relatives of the still living *Lycopodium* genus of clubmosses, commonly known as 'ground pines') and liverworts. However, most amazingly, these Devonian sediments also preserved the remarkable root and body structure of a member of the *Archaeopteris* genus. Although known from previous studies, the Cairo record clearly demonstrated that, unlike *Eospermatopteris*, *Archaeopteris* also already had a clear set of seed plant ('spermatophyte') features much earlier than previously thought.[46] This included a vast, deep root system indistinguishable from modern seed plants, a large, upright form, thick woody trunk, varied spore types, and flat green leaves for the efficient capture of sunlight and absorbance of CO_2.[47] As a result, it is considered to represent an ancestral lineage from which all seed-producing plants, which dominate the Earth's forests today, evolved. Furthermore, while *Eospermatopteris* was largely limited to flooded, lowland swamp habitats, the root systems of *Archaeopteris*, and the ability of its sophisticated leaves to function in dry systems, enabled these trees to expand much more widely across the Earth's surface, with the potential for more wide-scale planetary impacts.[48] By ~388–359 million years ago, fossil records suggest that *Archaeopteris* was not only a dominant member of many forests in North America,[49] but also in Morocco[50] and China,[51] among other places.

Whichever group of organisms best represents the first trees, together, the Gilboa and Cairo fossil sites provide a remarkable snapshot of the world's first forest ecosystems. These new leafy kids on the block began to further reconfigure the Earth's surface and atmosphere, and not necessarily to their own advantage. These novel Herculean plants were the world's first 'geo-engineers'.[52] True, deep, wide root systems converted significant amounts of rock into plentiful, supportive soils. As the roots reached outwards, they continued to form ever larger bonds with fungi and microbes that characterize the dynamic floors and soils of modern forests.[53] Calcium, magnesium, sodium, iron, aluminium and potassium were mobilized from the Earth's crust into new, thick, nutrient-, and even clay-, rich soil ecosystems and mineral carbonates.[54] Once again, however, not all these minerals remained on land. As at the end of the Ordovician, they made their way into the Earth's seas, driving an expansion of life.[55] More rapid and wide-scale weathering trapped more CO_2 in solution, further reducing CO_2 in the atmosphere,[56] alongside more efficient photosynthesis by extensive leaf systems. This new plant-based shock to the planetary system once more sent the world into a drastic phase of cooling, even more significant than before. Polar glaciers pushed right up to the lower latitudes.[57] Numerous extinctions occurred both in the oceans and on land at the end of the Devonian. This cooling also resulted in the widespread decline of many of the first forests that had dared to stake their claim on our planet.

Nevertheless, the pioneering biological and geological success of these new, tree-based configurations meant that the world was now forced to contend with something of an 'inevitability of forests'.[58] By the time of the Carboniferous period, 359.3–298.9 million years ago, a time when the supercontinent of Pangaea was forming, glaciers were contracting, and much of the continental landmasses of present-day Europe and America existed around the warm, wet equatorial regions,[59] forests were well and truly back. Woody trees with thick bark, alongside their decking of vines and climbers, expanded in moist, swamp-like habitats across these latitudes. Indeed, the name of this geological period comes from all the 'carbon' that became trapped in the ground following the deaths of these new, more resistant trees.[60] The thick layers of peat that formed from these extensive forests, in the absence of microbes and fungi that could digest tough wood fibres, were buried deep under the surface. There, millions of years of heat and intense pressure turned them into the Carboniferous coalfields, which geologists around the world could use to visibly recognize a distinct period of time. In an ironic twist of fate, these same coalfields, formed from forest habitats that had begun to turn the Earth's surface into a more lush and comfortable setting, were rediscovered as a key fuel source for the industrial revolution of the eighteenth and nineteenth centuries. The human-induced release of CO_2 back into the atmosphere has reversed a process that began hundreds of millions of years prior, and threatens not only the forest descendants of these Carboniferous ecosystems, but also the habitability of various parts of the Earth's surface for plants and animals (including ourselves) alike.

Various ancient trees (known as lycophytes) lived in these first Carboniferous forest environments, extending up to between forty and fifty metres high and up to two metres in diameter.[61] Sweeping vines climbed these gigantic arborescent versions of modern-day clubmosses, while horsetails added density to the extensive, humid, tropical swamp 'coal forests' that, by ~300 million years ago, made up the majority of the now plentiful forest cover visible in fossil records dating to this period. The remaining third consisted of progymnosperms and even true seed-producing trees, such as the now-extinct giant 'cordaites' that may have been ancestral to later conifers, ginkgos and cycads, which broke free of the wetlands and adventurously struck out into well-drained, drier, upland conditions.[62] True seed-producing plants (spermatophytes) emerged in the Devonian, but it was within the warm conditions of the early Carboniferous that they expanded across latitudinal zones.[63] The appearance of fossil evidence for vast mats of roots from 305 million years ago[64] shows the earth-shattering capabilities of Carboniferous forests – producing ample soils they could anchor into both horizontally and vertically. The resulting active, nutrient-rich soils were teeming with fungi, bacteria, and also now invertebrates (mainly insects) that were highly effective at breaking down organic matter, and would not have been out of place on the ground-level of twenty-first-century tropical forests.[65] Photosynthesis continued to absorb and trap significant amounts of CO_2, leading to atmospheric levels less than five times that of pre-industrial levels. At certain times, O_2 was at an even higher abundance in the atmosphere than it is today.[66] The shift in the atmosphere during the Devonian and into the

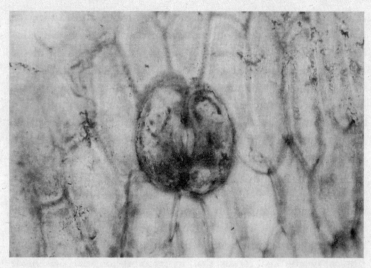

Image of a 400-million-year-old stoma from the Rhynie Chert fossil plant *Aglao-phyton majus.*

Carboniferous was so significant that even the plants themselves likely had to adapt. While we have already seen that these pores have inferred origins as far back as the Cambrian with the first land plants,[67] there appears to have been an increase in frequency and surface area covered by stomata on plants as the Devonian period wore on.[68] It has been suggested that the proliferation of stomata, and a growing function in gas regulation, was an adaptation to declining CO_2 and environmental conditions at this time,[69] though this remains debated.[70] Regardless, overall mild global climatic and atmospheric conditions were now ripe for the expansion of life and, as we will see, a proliferation of familiar land-based organisms. Forests growing in the tropical latitudes of the time were the key catalyst for this process, which is only now beginning to be undone by human hands.

In contrast to our tour of the desolate landscapes of the Cambrian, if we strode through the new forests of the Carboniferous we would feel significantly more at home. Walking across the Earth's surface we would now see organisms and ecosystems that look like the plants and forests we know today. Compared to the chatter and squawking of diverse Amazonian rainforests, or even the sounds of British woodlands, Devonian and Carboniferous forests would have been eerily silent. No herbivores yet existed that could process the new cellulose- and lignin-rich plants that had begun to populate the planet, leaving them largely unmolested to grow upwards and outwards. Nevertheless, the rapid formation of soils, thanks to the lofty trees, now provided moist, warm homes for millipedes and centipedes, consuming rotting plant matter or hunting other tiny organisms trying to make their way in this very new world. The earliest insects appeared in Devonian and Carboniferous forests,[71] some flying through the humid swamp forests on wingspans as large as ~60cm by the end of the latter period thanks to soaring atmospheric oxygen.[72] Predatory spiders and scorpions had likewise begun to call these lush forests their homes, while worms and snails likely slithered along the ground, although their soft body tissues have not preserved well enough as fossils for us to be certain. Ultimately, if we were to dig away at the ground of the Carboniferous forests, the communities of creepy-crawlies we would uncover would not have been significantly different to those seen around the world today.

Turning our heads up from the floor, if we looked closely enough at the dense, dark forest sub-canopy, we might begin to see how these earliest forests, while relatively quiet, also hosted some of the first major step-changes in land-based animal communities. To observe this most clearly, it is best if we travel to Hamstead in Birmingham in the United Kingdom, and the 'Jewel of the Midlands'. Like Wales, it might seem a peculiar location to pick when discussing the origins of tropical forest ecosystems on Earth. However, the coal layers which attracted industrialists to Birmingham in the eighteenth and nineteenth centuries provide an ideal place to look at the forests of the Carboniferous – a time when Birmingham would actually have been located near to our planet's equator. In 2016, scientists presented an incredibly significant find in the context of early tetrapod (four-legged animal) evolution.[73] The discovery concerned twenty red sandstone slabs that had been sitting neglected at the Lapworth Museum of Geology since their discovery by a schoolteacher in the early twentieth century. Using state-of-the-art scanning approaches, scientists took around 100 photographs at a range of heights and orientations (a methodology known as photogrammetry) around the specimens to build 3D models of the marks left behind on the slabs. Re-examining these blocks after over a century, lead author of the resulting study Luke Meade notes how the results 'opened a window on to Birmingham at the end of the Carboniferous'. Analysing the 3D models, what they found were a series of footprints left by increasingly active animal life on the muddy, forested floodplains during this period of Earth's history.

Most numerous were amphibians or amphibian-like creatures. These were the first four-legged animals to emerge from the oceans and inherit the Earth – making short trips on to land as early as the Late Devonian. The warm, humid swamp forests of the Carboniferous period, and their regulation of the planet's atmosphere, enabled amphibians to diversify in terrestrial habitats. They became the top land predators, some reaching several metres in length, consuming the available creepy-crawlies on land and fish in the swamps and rivers. Yet, like frogs and newts today, they were always forced to return to the water to lay their eggs, limiting the degree to which they could successfully colonize the land. The Birmingham footprints, however, reveal two other vertebrate groups that were beginning to walk, reproduce and form permanent land-based communities. Rare tracks of large pelycosaurs, which may have looked a bit like Komodo dragons, represent one of the first members of an ancient lineage that would eventually produce mammals.[74] Meanwhile, smaller tracks show the appearance of the sauropsid reptiles.[75] Initially small and lizard-like, their production of amniotic eggs, whose surrounding fluid could resist drying out on land, represented the beginning of a reptilian dominance of Earth that would include creatures as diverse as the dinosaurs, crocodiles, turtles, and, later, birds. Much of what first springs to mind when we think of animal life thus began in these warm, wet equatorial forests.

The earliest forests of the tropics completely redesigned the Earth's surface, providing nutrients, shelter, stable soils and atmospheric conditions for lively terrestrial ecosystems that contained insects and various animals living at different trophic positions and exploring different adaptations. Within these new types of environment appeared all the main terrestrial groups of animals that we know today, captured in time by the Birmingham footprints. Tropical forests were thus the first complex land-based ecosystems on the planet. They also had a lasting impact on the Earth's atmosphere and climate. Absorbing significant amounts of CO_2 through a now expanded factory line of photosynthesis and aiding the cycling of water into the atmosphere and through the ground thanks to evaporation from larger leaves and effective root systems, they made themselves critical to life on Earth. In essence they developed an entirely new planetary order that was to host a diversity of life on land from that point onwards. Tropical forests were vital to the air breathed in by new, terrestrial organisms and the climates that they experienced, and, as we will see in Chapter 12, they remain crucial to the functioning of the various systems of our planet today. This is something we should perhaps consider more often as they rapidly disappear from its surface. Yet this does not mean that they stood unmoving. In fact, they had only just started evolving. The 3D scans of the Birmingham sandstone reveal much more than just footprints. They reveal the raindrops that fell in the humid Carboniferous 'coal forests' but also the mudcracks that formed in drier periods that were becoming increasingly common in basins of rivers ~300 million years ago.[76] Tropical forests were going to need to change if they were to survive, as were the creatures that inhabited them. And change they did, impacting and hosting some of the most dramatic periods in planetary history that were to come.

2. A tropical world

Tropical forests have been on Earth for over 1,000 times longer than we have, so it is easy to act as though they've just always been here – timeless, primordial guardians of our planet. Their great age is even often actively used by conservation efforts, which contrast unchanged and uninhabited forests with the dangerous, destructive 'progress' of twenty-first-century industry and development. Even how we commonly present or talk about tropical forests in popular culture works to sweep away any ideas of transformation or diversity. Novelists and film producers are all too happy to describe or capture the same tangles of vines, the same dense undergrowth, and same dark, humid settings to represent a stock perilous 'tropical' environment while, as we have seen, the colloquially used word for tropical forests in English, 'jungle', simply refers to a generalizable 'wilderness'. If these perceptions of static, homogeneous forests were correct, we would still be living in a world decked in the clubmoss- and fern-dominated forests of the Devonian and Carboniferous that we met in Chapter 1. Today, however, tropical forests, as a whole, are home to over half the Earth's plant and animal species.[1] This is not because they are all the same and it is certainly not because they have remained unaltered for 300 million years. Instead, over the course of their history, they have proven themselves to be just as dynamic, vibrant and geographically variable as the rest of the Earth's environments.

What is a tropical forest? As we have already seen from Chapter 1, there are two broad ways of thinking about tropical forests. One focuses on the climatic necessities known to be required for the growth of certain tropical forest types (2,000mm of annual rainfall, minimum temperatures of 18°C, and no significant dry season for tropical rainforests, for example[2]). These definitions can be incredibly useful in 'deep-time' when exploring periods in the Earth's history, like the Devonian and Carboniferous, when much of the world has been what we would conventionally consider 'tropical' in a climatic sense, and forests reliant on warm and wet conditions, once present, could extend well beyond the equator. They also provide us with a way of looking at the surroundings of the first plants and trees to have colonized the Earth (Chapter 1). Definitions focusing on necessary climatic parameters are also useful in acknowledging that the forests of the past may not have had exact modern counterparts. For example, we will see in this chapter how terms such as 'mega-thermal moist forests'[3] have been used to explore how even the most unlikely of regions could have hosted hot, wet, frost-free forest ecosystems of a type that we might expect solely to prevail in the tropics today. The movement of the Earth's surface (and thus the 'palaeolatitudes' of different continental landmasses at different points in time) and changes in the planet's climate systems[4] add yet another layer of drama to the winding road of 'tropical' forest origins and evolution.

The second way of defining 'tropical forests', and the one that will be formally used when explicitly using the term in this book from now on, is 'any forest located between the Tropic of Capricorn and Tropic of Cancer'. Although the tropical latitudes were also often the warmest, by not focusing on the climatic requirements of a particular forest 'type' this definition allows us to explore the vast variety of the tropical forests we co-exist with at present. From the wet lowland evergreen rainforests we know so well from documentaries on the Amazon and Congo River Basins, to the seasonally dry and thorny tropical forests in places such as Central America and Madagascar, and the snow-tipped branches of montane tropical forests that disappear into the heavens in regions such as New Guinea and the Andes.[5] Although rainforests receive a lot of attention for the incredibly diverse insect communities they host,[6] as well as their great apes and forest elephants,[7] more seasonal tropical forests are home to equally rare, though perhaps less conventionally rousing, specialized species of plants[8] and animals, such as lemurs, fruit bats and the world's most endangered tortoise.[9] Crucially, this definition also allows us to investigate varied forest ecosystems that inhabited the tropical latitudes at different points in time and how they led to the origins of the tropical forest ecosystems that exist there in the present. We are able to chart a course, from the equatorial Carboniferous 'coal forests' we met in Chapter 1 to the first closed canopy Neotropical rainforests that represent the origins of ecosystems that have made the Amazon Basin one of the most renowned homes to tropical forests today.

Long-term tropical forest evolution, and its significance for the emergence and diversification of life on Earth, is a story of change. The first complex tropical forests, emerging under warm, moist conditions at the equator, had already played a role in the permanent alteration of the Earth's systems and the appearance of some of the first land-based animal life during the Carboniferous. But this was by no means the final word. To truly understand how and how far tropical forests have impacted the various evocative life-forms that have graced our planet in the last half a billion years, and to truly determine the degree to which they managed to extend their roots into the functioning of the Earth's atmosphere, climate, and nutrient cycles, we must now investigate how they themselves went on to change. This chronicle is one of ongoing shifts in global climate and, significantly, the crunching process of continental drift and plate tectonics. It is key to the evolution of all plant life and to its increasingly permanent seat at the table of atmospheric, geological and climatic change. So, without further ado, let's begin our tour of tropical forest evolution and dynamism. While the Devonian and Carboniferous both represented major ecological accomplishment for arboreal life-forms, tropical forests, not for the last time, were about to grow headlong into catastrophe.

Given the variety of life that we met beneath their canopy in Chapter 1, the Carboniferous 'coal forests' that covered the continental landmasses at the equator could already be considered a highly successful evolutionary experiment by ~300 million years ago. Nevertheless, the final coalescence of the world's continents into a single supercontinent, Pangaea, ~330 million years ago began a slow shift towards a warmer and, more significantly, drier planet, that would completely overhaul the vegetation communities of the tropics, and beyond. So severe was the resulting 'rainforest collapse', as it has become known, that it is argued to have led to one of just two mass extinction events ever recorded for plants in the fossil record[10] (by comparison animals have faced five mass extinction events, and may soon be heading for their sixth . . .[11]). Tree and tree-like species that had begun to dominate the earliest ancient forests, such as clubmosses, giant horsetails, and the immense cordaites we met in Chapter 1, gradually disappeared from the fossil record, with the number of tree families slashed in half by the time of the Permian (298.9–251.9 million years ago). Not only did species change but, by the start of the Triassic period (251.9 million years ago), forests in all forms had almost entirely disappeared from the face of the Earth. Amazingly, this was the only point in time, from the evolution of the first tree until today, that forests have been practically absent from our planet's surface.[12]

As these forests had already inserted themselves into the Earth's systems, knock-on effects were inevitable. Anchoring trees were suddenly missing, causing a thinning of life-giving soils, soil erosion and the movement of once reliable rivers, and a return to a bleak surface with the exception of a few isolated 'islands' of vegetation.[13] Furthermore, although the planet was drying anyway,[14] a reduction of forests, greenery, and the evaporation of water from leaf surfaces, almost certainly compounded this increasing global aridity.[15] Any water that did make its way onto or into the ground was no longer captured by dense root systems. Instead, it flowed away, carrying valuable nutrients from the Earth's crust into rivers and, eventually, thriving Permian marine ecosystems. At the end of the Permian, explosive volcanic activity in Siberia spat dust and poisonous gas into the atmosphere for hundreds of thousands of years, choking photosynthesis and the diversity of four-legged creatures that had begun to expand within and beyond forests. Over half the families of animal life on land went extinct.[16] However, while these dramatic eruptions at the end of the Permian rightly get attention in most popular accounts of this period – for nearly wiping out all life on Earth – the forests had actually already encountered their apocalypse. Although plants faced a further reckoning at the end of the Permian and the Early Triassic, with an additional slashing in number of the remaining plant families,[17] much of the damage to the planet's greenery had been done. Without widespread forests, animal life continued to face frequent flooding and a relatively barren planetary surface, even as it recovered from its brush with volcanic hell. The first real attempts by forests to make their way in the world seemed to have ended in failure.

As is so often the case in evolution, however, adversity can lead to innovation. Out of disaster sprouted the beginnings of all plant and forest life as we know it today. Nowadays we divide seed-producing (spermatophyte) plants into two broad categories, the gymnosperms and the angiosperms. The first of these groups can actually trace its origins back to the 'pro-gymnosperms' that we met in Chapter 1, and were the first beneficiaries of the Carboniferous 'rainforest collapse'. Their name comes from the Greek word 'gymnospermos', meaning naked seed, and refers to the fact that the seeds they rely on for reproduction are unenclosed prior to fertilization. While they had already begun to dominate ecosystems during the Carboniferous, they were often found on forest floors – in and among the earliest cast of tree-forms. However, Pangaean drying and increasing CO_2 concentrations caused problems for other plant life, creaking open a window in the canopy for gymnosperms, such as cycads and ginkgoales (made famous by the ginkgo tree or *Ginkgo biloba* which is the only surviving member of this group), to rise to the top. Plants reproducing via seed have the advantage that their seeds can lie dormant, enabling them to outlast dry spells and more seasonal conditions.[18] When forests did finally begin to return to Earth as the Triassic period (251.9 to 201.4 million years ago) wore on it was these seed-producing gymnosperms, this time including the familiar conifers, that now dominated the forest over-storeys[19] in the tropics and beyond. The results were ecosystems that are still represented by today's temperate forests, from northern Siberia to their southern outposts in Chile and New Zealand.[20] Meanwhile, spore-bearing plants were left to pick themselves up on the forest floors, and the ancient 'progymnosperms' such as *Archaeopteris* were consigned to the Carboniferous fossil record.

Given the cold temperatures they inhabit today, you might be thinking that the boreal forests of Russia are not really what you would consider 'tropical', and you would be entirely entitled to do so. Indeed, while the gymnosperm cycads are still represented widely in the tropics, and these gymnosperm-dominated forests extended across the tropics for much of the Mesozoic era (Triassic to Cretaceous period), to find the true beginnings of the types of forests that inhabit the tropics in the twenty-first century we must turn to the evolutionary history of the other group of plant seed-producers, the angiosperms (or flowering plants). To do so, there is no better guide than Dr Carlos Jaramillo, appropriately based at the Smithsonian Tropical Research Institute in Panama. With his wide-brimmed hat, geological pick and mud-covered trousers, Carlos is one of the finest explorers of ancient flowering plants and tropical forests around. Undertaking long-term fieldwork and excavations in the Neotropics, he and his team are on the lookout for all types of fossil plant evidence, from minuscule pollen grains to huge petrified trees, in order to piece together the deep-time history of tropical ecosystems. While he cannot resist the occasional analysis of giant turtle and crocodile fossils, which they also often find in their Central and South American excavations, he tells me that his heart truly lies with the enchanting, and under-appreciated, evolution of tropical plants.

The history of tropical forest vegetation is one of radical and fascinating change. While they might be ancient, they have never been short of excitement or dynamism, particularly across the last 200 million years of their evolution. As Carlos puts it, '120 million years ago the entire tropics were home to just a single flowering plant, yet today they are almost completely composed of some of the greatest concentrations of flowering plant diversity anywhere in the world.' In fact, the origins, expansion and diversification of the most disparate group of land plants, the group that yields the plants that today sit in our vases and manicured gardens, dominate celebration and mourning rites, and make up the majority of plants exploited by humans for food, medicine and fibres, are intimately linked to the evolution and emergence of the varied tropical forest habitats as we know them. So how did these enormous ecological changes happen? To find out, we have to look at the planetary drama out of which the angiosperms emerged. During the Jurassic period (201.4 to 143.1 million years ago), the super-continent of Pangaea began to experience a tectonic divorce, as sub-surface tensions bubbled up into the Earth's crust to unleash the irresistible forces of plate tectonics. First, Pangaea broke in two, forming a northern Laurasian and southern Gondwanan landmass. Then, by 140 million years ago, Gondwana was itself beginning to split into Africa, South America, India, Antarctica and Australia.

It is into this fiery Cretaceous (143.1 to 66.0 million years ago) break-up that the angiosperms entered the frame. After a cooler interlude in the Triassic that favoured conifer expansion, this world-making process released masses of CO_2 into the atmosphere as a product of volcanic eruptions, stimulated by the grinding apart of the Earth's crust. Temperatures and humidity rose, creating the ideal 'greenhouse' for the first angiosperms that were able to ride their way into a new world order on the backs of free-roaming continents. Based on the majority of existing genetic and fossil evidence,[21] these flowering plants likely first emerged around 140–120 million years ago in the tropical latitudes,[22] although some researchers have also suggested that mid-latitudes would have perhaps provided more suit-able, stable settings for angiosperm emergence during the Early Cretaceous (143.1 to 100.5 million years ago)[23] and others still argue that the angiosperms may have begun to emerge as early as the pre-ceding Jurassic.[24] Regardless, analysis of ancient angiosperm lineages certainly makes it clear that they first appeared as part of a warm, wet forest undergrowth,[25] lying low under gymnosperm canopies that were gradually losing diversity. Meanwhile, genetic evidence shows that around ~100 million years ago many orders of angiosperm that are widespread in tropical forests today diverged as Gondwana broke up and global temperatures became warmer and wetter.[26] Between 100 and 70 million years ago angiosperms spread across the tropics, experiencing rising species diversity and a growing domination over all other plant groups, particularly in warm wet land-based environ-ments. One final dramatic proliferation ~60 million years ago saw angiosperms spread extensively across the majority of the lower and middle latitudes.[27] This proliferation of angiosperms has also been implicated in alterations to the atmosphere and the water cycle. They have higher numbers of veins in their leaves and can therefore per-form water evaporation (transpiration) and photosynthesis more effectively. The release of more water back into the atmosphere from these novel leafy surfaces has been linked to the development of wet-ter and less seasonal climates.[28]

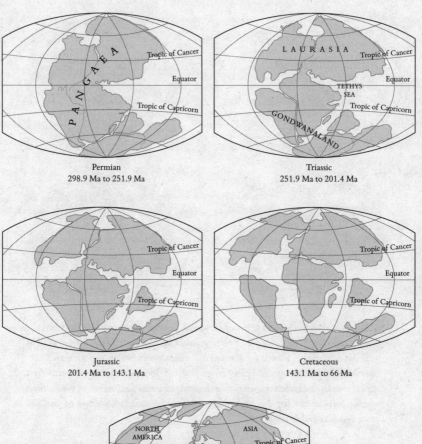

Permian
298.9 Ma to 251.9 Ma

Triassic
251.9 Ma to 201.4 Ma

Jurassic
201.4 Ma to 143.1 Ma

Cretaceous
143.1 Ma to 66 Ma

Present Day

The Pangaea super-continent and the process of its break-up over time. The ages of the geological periods are given according to the latest 'Geologic Time Series' of Gradstein et al.[29]

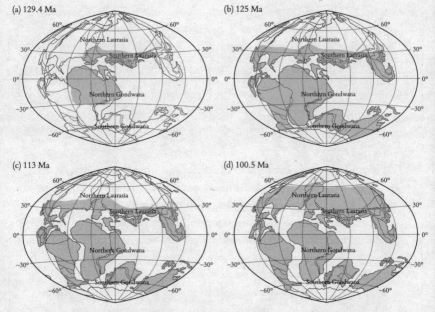

(a) 129.4 Ma

(b) 125 Ma

(c) 113 Ma

(d) 100.5 Ma

A tentative map of the Cretaceous latitudinal expansion of angiosperms from the tropics and sub-tropics based on pollen assemblages evaluated by Coiro et al.,[30] as well as additional evidence for Cretaceous angiosperm pollen in southern Laurasia. Note that this image refers to records of approximate angiosperm first appearance in the pollen record rather than angiosperm dominance of ecosystems which, particularly for the higher latitudes, only occurred around 60 million years ago as noted in the text.

The scale of the tectonic journeys of the angiosperms is visible when we focus in on the family that includes some of the true tropical rainforest giants that stand today (some as high as 80–100 metres), including major tropical timber trees which are exploited globally, the Dipterocarpaceae. Evidence from fossils, modern genetics, and comparisons of body shapes all indicate an origin for this group pre-120 million years ago on the landmass of Gondwana, which included Africa, South America, Antarctica, Australia and the Indian subcontinent. Some members of this family then rafted on top of the plate comprising India and Sri Lanka as it separated from Gondwana and eventually smashed into Laurasia (a plate including the rest of Asia), completing their global voyage by ~45 million years ago.[31] Aided by the forces of geology, and perhaps also pollinating insects,[32] the angiosperms accompanied global 'greenhouse' conditions that prevailed from the Cretaceous (143.1 to 66.0 million years ago) right up until the middle Eocene (48.1 to 37.7 million years ago). As early as the Late Cretaceous (100.5 to 66.0 million years ago), these flowering plants had also already started to become key members of tropical forest ecosystems, and fossils of broad, woody angiosperm trees and fruit-producing climbers have been found in Nigeria.[33] However, clear fossil records of the earliest angiosperm-dominated forest ecosystems, with a collection of species and structure (i.e. a number of stratified layers or 'storeys' within closed forests) that we expect from tropical forests today, only really began to appear in the Paleocene (66.0 to 56.0 million years ago), when temperatures started to rise and environments began to recover following the extra-terrestrial collision and volcanic winter that drew a dramatic line under the Cretaceous and dinosaurian dominance. To see what these new, angiosperm-dominated tropical forests looked like, and how they related to the diverse ecosystems we know from the tropics today, we must now follow Carlos and his team into yet another mine in the Americas . . .

The Cerrejón coalmine in northern Colombia is one of the largest open-pit coalmines in the world. So vast is the expanse of coal in the region that it has been recognized as its own geological entity, the Cerrejón Formation. As we have already seen in Chapter 1, thick beds of coal are something of a 'smoking gun' for detectives in search of ancient tropical forests, and these are no exception. The mine has yielded an unprecedented window into the communities of plants that formed in the tropics 60 to 58 million years ago, just after the dramatic Cretaceous–Paleogene boundary (66.0 million years ago) extinction event that wiped non-avian dinosaurs off the face of the planet. The Cerrejón coalmine does not just represent a treasure trove for fossil fuel companies, but also for palaeobotanists, and to date it has produced over 2,000 plant 'mega fossils', some nearly twice as big as an average human, which include ancient leaves, flowers, seeds and fruits. Today, when we think of tropical forests, and particularly rainforests, in the 'Neotropics' (i.e. the tropical portions of the

Americas), we imagine or, in the case of Victor and myself, experience a certain level of plant diversity, the presence of specific angiosperm families, and high numbers of plant-eating insects, high temperatures and plentiful rainfall. Unsurprisingly, these traits are immensely difficult to find all together in one spot in the fossil record, making it exceedingly challenging for us to properly identify the origins of this evocative twenty-first-century tropical forest community. Cerrejón coalmine is perhaps *the* major exception in this regard and represents a paradise for researchers like Carlos.

Carlos and his team have discovered that the plants found at Cerrejón represented many of the families that dominate tropical rainforests in this part of the world today. When analysing the fossils under the microscope the team could identify characteristic plants, including palms, legumes, and even the ancient relatives of the avocado, and show that angiosperms had truly come to dominate these new, post-dinosaurian forests,[34] with multi-layered rainforest canopies providing the stage for angiosperms to diversify and experiment with the amazingly different forms we see in contemporary tropical rainforests. Amazingly, when studying the microscopic (e.g. fossil pollen) and macroscopic (e.g. fossils of leaves) remains preserved at Cerrejón, it is possible to find about 60–80% of all the main groups of plants found in twenty-first-century Neotropical rainforests. While slightly less diverse, the likeness of this ancient forest to those seen in Central and South America today is remarkable given the geological processes (e.g. uplift of the Andes, which occurred between 55 and 40 million years ago) and changes in climate (fluctuating between warmer and cooler than present) that were to subsequently occur. Effectively, at Cerrejón we can start to see the origins of some of the modern tropical forests that we can still encounter around the world today, 50 million years before our first hominin ancestors would appear. These plant communities, while changing subtly and waxing and waning in spread, were to stand tall throughout major upheavals and remain around us, spanning a significant portion of the Americas including, of course, the Amazon Basin. After experiments of varying degrees of success, a new type of tropical forest had now established itself on Earth, and it was here to stay.

Carlos and his team did not stop there, however. The remarkable preservation within the Cerrejón Formation even allowed them to analyse the leaf structure of these ancient forests. The leaves had the same forms as those found in modern Neotropical rainforests, big with smooth edges, indicating similarly hot and wet conditions. Indeed, comparison with modern leaf shapes suggested a past rainfall greater than 2,500 millimetres per year and an average annual temperature of greater than 29°C, about 2°C higher than that of today. In fact, the leaves were so complete, even after a remarkable 60 million years, that Carlos and his colleagues could analyse the bite marks of insects that had feasted on these tropical bounties. This damage showed that a variety of plant-eating insects now bustled around tropical forests, though, like the plants, they were less diverse than the menagerie of organisms found on the trunks and floors of the Amazon rainforests today.[35] Finally, the available fossils also show the beginnings of a classic Neotropical animal community, including the remains of a now extinct giant snake, the terrifyingly named 'Titanoboa'. Related to living boa constrictors and anacondas, this monster would have been right at home in Luis Llosa's 90s Amazon-based horror film *Anaconda*. As in the case of the large leaves, the significant size of this snake (nearly 13 metres in length) suggests that the Cerrejón forest was warmer than the tropical rainforests of modern South America.[36] This is because snakes are 'ectothermic', relying primarily on environmental sources of heat to regulate their bodily functions.

Cerrejón was not the only tropical forest in town.[37] By the Paleocene–Eocene boundary (56.0 million years ago), tropical forests, with characteristic plant families and structures that we still see in these forests today, spanned across the tropics. Not only that, but similar 'megathermal' forests reached the greatest extent they have ever managed in the history of our planet. Although incredibly diverse, and not always 'tropical' in the context of the definition I provided at the start of this chapter, whether we set down in North America, Africa, South America, Australia, Asia or Europe, warm 'mega-thermal' forest ecosystems would have greeted us, reinforcing a global move towards warmer and wetter conditions as they further inserted themselves into the Earth's system. Mangrove-like forests even adorned the now frosty coasts of the United Kingdom, while so-called 'boreotropical' forests (the name for forest communities with similarities to those seen in the tropics today that extended well into northern and southern latitudes) reached as far south as Tasmania and as far north as Alaska.[38] Importantly, unlike the Carboniferous forests, the structure and species of the new forests in and around the equator, like that at Cerrejón, would have appeared much more familiar to those of you who live in, or visit, the tropics today. The distribution and evolution of various tropical angiosperm plants continued apace, and certain plant families were able to move rapidly between the tropical continents via land-bridges.[39] However, the ongoing processes of tectonic separation throughout the Eocene (56.0–33.9 million years ago) meant that tropical forest communities began to take on their own regional 'flavours'. By the end of this period, and the final division of Antarctica, South America and Australasia, the Earth's continental make-up was pretty much decided. Species which had previously surfed on moving geological plates now became largely trapped.[40] The result was the gradual, independent evolutionary formation of the unique forest ecosystems we know from different parts of the tropics today.

The process of continental separation, as well as isolating tropical forests in different parts of the globe, also stimulated the development of new climate systems. The moving apart of landmasses opened up new spaces for vast ocean circulation systems that extended across the planet. Cold water flowed from the north polar region into the Atlantic, cooling and drying the northern hemisphere thanks to a reduction in evaporation, while Antarctica became separate from warmer equatorial waters.[41] From ~47 million years ago, the formation and rise of many of the world's great mountain ranges exacerbated this trend. They began to force the 'mega-thermal' forests back towards their current distribution as a belt around the Earth's equator.[42] While trees retreated, new players began to expand from ~20 million years ago: players which had crept on to the vegetation scene during the time of the dinosaurs.[43] With a shape, root system and physiology adapted to resisting droughts and cold snaps, grasses provided a small, but collectively mighty, challenge to forests during times of disturbance.[44] And thus, present-day ecological battle-lines were formed within the tropics. On the one side, tropical and sub-tropical forest ecosystems, with their vast amounts of angiosperm diversity; on the other, novel grassland ecosystems. These two formations have been locked in an eternal skirmish ever since, waxing and waning in the face of periodic global climate swings.

10 million years ago the modern tropical forest toolbox was largely complete, with characteristic plant species, continental location, and the broad global conditions we associate with them today, all in place. This did not mean that they remained idle, however. Although their exact impact remains hotly debated, changes in precipitation, temperature and global CO_2 continued to drastically impact the extent and structure of tropical forests. During the Quaternary (2.58 million years ago until present) these shifts have been primarily driven by cycles in the distribution of the Sun's energy around the Earth, linked to our planet's changing orbit and tilt.[45] Broadly, over the last 1 million years these 'Milankovitch' cycles (named after the Serbian scientist who discovered them) have resulted in an Ice Age, and temporary dramatic expansion of the polar ice-caps, approximately every 100,000 years. Under Ice Age conditions, sea ice advances in the North Atlantic which, through circulation systems in the ocean and air, can also cool the tropics. This cooling in turn leads to movements and weakening in the operation of key climate systems, such as the Inter-Tropical Convergence Zone (where north-eastern and south-eastern trade winds meet around the equator), and the Indian summer monsoon system, that bring rainfall to many parts of the tropics today, resulting in drier conditions around the equator.[46] Nevertheless, these broad solar models do not always explain all the variation in the Earth's climate over the last 10 million years, in terms of both extremes and frequency. In fact, temperature, precipitation and CO_2 at a given time and place also remained dependent on changing relationships between how heat and water circulated between the oceans and the atmosphere as well as, crucially, land cover, including the greatest of earth systems engineers, the plants and forests themselves.

Taking a broad-brush approach, we can identify some major climate- and atmosphere-driven changes in the varying fortunes of tropical forests and their grassland competitors over the past 10 million years. Starting in the Miocene (23.04 to 5.33 million years ago), warm global temperatures and high CO_2 concentrations had seen 'megathermal' forests once again expand across much of Africa and Eurasia, reaching a maximum extent between around 21 and 14 million years ago.[47] From 10 million years ago, however, as CO_2 concentrations declined, profound changes in vegetation occurred. Grasses practising a new, never-seen-before form of photosynthesis (named 'C4' after the number of carbons in the first sugar produced by the plant during photosynthesis), which was more efficient at absorbing CO_2 and photosynthesizing in drier conditions and which dominates tropical savannah habitats as we imagine them today,[48] joined existing 'C3' grasses[49] in the fight back against warm, wet forests which saw the latter removed from Europe and Central Asia.[50] Ongoing, overall cooling and drying that had begun in the Eocene continued into the Pliocene (5.33 to 2.58 million years ago), dealing another, gradual, but lasting, blow to non-tropical 'mega-thermal' forests as cooler temperate adapted vegetation expanded in the northern and southern hemispheres and they retreated, regrouping near to the equator. This was a new dawn. The age of tropical grasslands had begun, while other types of seasonally dry tropical forests (including 'deciduous' plants that shed their leaves seasonally) also began to emerge and diversify across the sub-tropics and tropics.[51] Continued drying also saw deserts form in the most water-deprived areas of Africa, South America and Asia, resulting in the layer cake of environmental variation we see as we travel from north to south across the globe today.

While overall planetary cooling continued, from a mean annual surface temperature around 2°C warmer than today at the start of the Pliocene to as low as −6°C cooler than today during parts of the Pleistocene,[52] the greatest characteristic of the Pleistocene epoch (2.58 million years ago to 0.01 million years ago) was actually the intensification of the impact of Milankovitch cycles. In the early Pleistocene (2.58 to 0.77 million years ago) glacial cycles were more frequent (around every 41,000 years) but less extreme, while the middle and late Pleistocene saw the less frequent, but more significant,100,000-year swings in global climatic states. In a tropical context, Pleistocene glacial cycles, and a drying of the tropics, have been argued to have led to an expansion of 'savannah corridors' across different parts of Africa[53] and Asia,[54] and perhaps even reaching between the two continents.[55] Growing larger as glacial shifts became more extreme into the middle and late Pleistocene, these corridors would have provided ideal, homogeneous grassland conveyor belts for medium to large sized animals which, as opposed to the usual dense tropical forest that had covered these regions, enabled them to traverse vast areas. In Southeast Asia, for example, some argue that such a 'corridor' extended from southern China in the north down beyond the Wallace Line that runs through Indonesia, aided by the formation of a land-bridge between mainland and island Southeast Asia which appeared as growing ice-caps sucked up water from the oceans.[56] The end of wide-ranging savannah ecosystems was, nonetheless, inevitable, due to the fact that, although glacial ice ages were longer and more extreme in this period, they were not indefinite. The return of warmer, wetter, 'inter-glacial' conditions broke up savannah extents, replacing them with classic tropical rainforest environments – hostile to many of the large-bodied grazing animals that had moved in in their absence.

The last great glacial period, or Last Glacial Maximum (LGM), occurred between 26,000 and 21,000 years ago. Greater numbers of ancient archives of past climate and environment that relate to this more recent timeframe allow us to determine the exact effect such a period might have had on tropical forests around the world in more specific detail than during earlier glacials. Although changes were not as extreme as in temperate areas, staggering drops in temperature of as much as 4–8°C have been suggested for different parts of the tropics during the LGM (see summary in[57]). Given that our current rates of emissions predict a rise of tropical temperatures anywhere between 2 and 7°C from between 1980 and 1999 to between 2080 and 2099, this should highlight both the extreme nature of the LGM but also the disastrous biological consequences potentially awaiting us if we do not act.[58] The LGM also made the tropics more arid. In South America and Africa, for example, it may have resulted in a rapid retreat of tropical rainforest to just a few small 'refugia' within the usually vast Amazon and Congo Basin forests.[59] Glacial declines in atmospheric CO_2 may also have changed the structure of tropical forests, leading to smaller, more efficient leaves, open canopies, and increasing forest floor growth, as well as the increasing presence of mosaics of forests, woodlands and grasslands. Whatever the cause, sub-tropical and tropical forest retreat 26–21,000 years ago has been recorded in Africa, Southeast Asia, South Asia, New Guinea, Australia and South America. The rate of forest re-expansion varied by region and was often complicated by other factors including, in Southeast Asia, the impact of rising sea levels on rainfall patterns.[60]

Better dated, and finer-grained records of climatic and environmental change for our current geological period, the Holocene (11,700 years ago), show us that the broad changes in tropical forest structure and extent noted above for time periods where resolution is limited to 1,000 years, or even more, miss finer-scale variability that would have confronted tropical forest environments. Study of the changing operation of climatic systems today, and in the recent historically recorded past, has revealed that parts of the tropics can rapidly switch between cold and dry, and warm and wet conditions. For example, variations in how winds and sea surface temperatures interact over the Pacific Ocean (as part of the El Niño Southern Oscillation climate system), have led to extreme, and sometimes un-anticipated, droughts and flooding in the recent past.[61] Although the broader planetary changes seen on geological timescales undoubtedly influenced evolutionary processes, it is these short-term shifts that would have shaped the experiences of individual organisms and communities. Feedbacks between vegetation and climate systems are also important for these high-intensity processes. This is acutely visible when we take the dramatic example of the 'Green Sahara'. Stretching from the Red Sea in the east to the Atlantic Ocean in the west, the Sahara is today the world's largest non-polar desert, encompassing much of North Africa. Between 15,000 and 5,000 years ago, however, conditions were very different, and this gigantic field of dunes was covered in vegetation and water. Climate models have demonstrated that alterations to the orbit of the Earth are not enough to produce this effect alone. In fact, it is only when increasing precipitation, *plus* feedbacks from expanding forest and plant life, are factored into the computer software that earth scientists can reproduce a state of a vast oasis, with vegetation likely increasing water in the atmosphere through evaporation from their green surfaces, while also stopping the usual build-up of dust.[62] Certainly, by the Holocene, it is evident that sub-tropical and tropical forests – which enable increases in regional air moisture, stabilize landscapes to avoid run-off, act as a buffer against flooding, and absorb solar energy – were critical players in the translation of climate system changes into on-the-ground environmental dynamics. This is something we will see later in the book, as our own impact on these remarkable environments, and thus the planet's climate, comes to the fore.

As we have seen, the history of tropical forests on Earth is one of upheaval, not timelessness. Extreme geological processes saw the first tropical trial runs begin to disappear from ~300 million years ago to a point, ~250 million years ago, when there were basically no forests on Earth. This seeming calamity paved the way for new forests, eventually composed of a new form of pioneering, flowering plant life that came to dominate global vegetation between 140 and 60 million years ago. By around 60 million years ago, tropical forests that closely resembled their twenty-first-century descendants had appeared in the tropics (and even pushed out beyond into a warm world as 'mega-thermal' forests). But this was not the end. From 10 million years ago they were pushed back by new grassy adversaries to roughly their current equatorial distribution. Still the world was not done with them. Ongoing cycles in the availability of the Sun's energy and changes in climate circulation systems buffeted these sub-tropical and tropical forests, causing them to wax and wane, open and close, ever since. However, tropical forests were never just victims of external change. As geology and climate shaped and moved them, they slowly but surely inserted their green tendrils into the Earth's systems. While the Sun dominated planetary climate change, this new vegetation cover decided how this solar energy manifested on the ground, whether it made local temperature and precipitation changes more dramatic, or buffered local landscapes against them. Furthermore, they controlled the amount of CO_2 in the atmosphere and how it was distributed, providing some of the most significant land-based storage areas or 'sinks' of carbon on Earth.

This dynamism, both in terms of evolutionary change and the increasing role they played in climate systems, gives us a new appreciation of their significance for life on Earth. Tropical forests, broadly considered, were the sites of the most significant evolutionary events in the evolution of plants: the rise and diversification of the gymnosperms and angiosperms. They were also changing their distribution and structure as some of the most famous cast members of the animal kingdom came, evolved and went from the planetary stage. At the beginning, this involved insects and amphibians, but by the end it included the appearance of our species, its migration across the global tropics, and its formation of a vast variety of social, economic and political forms that have varied radically until the twenty-first century. Throughout the rest of this book we will meet the hugely diverse types of tropical forest – from the seasonally dry forests of Mexico and Guatemala to the often frosty forests of New Guinea, from the isolated island forests of Wallacea and the Pacific Islands to African tropical forest-savannah mosaics – met by our species and that are the ultimate products of this vibrant evolutionary history. Before turning specifically to *Homo sapiens* and its closest relatives, however, the significance of tropical forest change to our world as we know it is vividly visible when we look at two of the most evocative groups of animal life to walk the earth. The next two chapters explore how the dynamics of tropical forest, from the Permian to the Pleistocene, described above, influenced the evolution and fates of, first, the dinosaurs (Chapter 3) and, later, their eventual inheritors, the mammals. Captivated by the fossils of these fascinating creatures, we have perhaps too rarely explored, or given credit to, the leafy, canopied companions that escorted them along on their biological journeys. It is to these stories that we now turn.

3. 'Gondwanan' forests and the dinosaurs

I will always remember visiting the National Museum in Cardiff, the capital city of Wales, as a young boy with my family. Unlike many museums, which value academic silence and regular processions around their secure glass cases, this museum created a cinematic experience for its visitors – dramatically using sound and atmospheric visual displays to form the most vivid picture possible of our planet's history. The continental plate exhibit was accompanied by a darkness that focused your eyes on to the flowing red-orange lava of volcanoes and your ears on to the loud, crunching sounds of earthquakes. The exhibit of some of the earliest humans to reach a then icy Wales involved a trip into a cave-like chamber to meet moving, trumpeting mammoths. Most stunning of all, however, was the hall of the dinosaurs. Here, fossil casts of giant reptiles emerged out of shadows, illuminated by well-placed light and a chorusing of roars and brays. This room has stayed with me as one of my earliest, and perhaps also scariest, encounters with the evolution of life on Earth. However, looking back now, something major was missing from this drama. In fact, it has been missing from almost all the museum exhibits of dinosaurs I have encountered since. While we are staring at monstrous jaws, reconstructions of colourful, and now feathered or hairy,[1] skin, or artistic impressions of vicious hunting scenes, we always seem to forget one of the most key parts of dinosaur existence – their environment.

~240 million years ago, the dinosaurs began as a predominantly carnivorous group of initially small 'dinosaurmorph' reptiles. They bucked the trend of other sprawling and shuffling Triassic life and stood upright – adopting a very different way of life and movement.[2] Indeed, when many of you think of dinosaurs, your thoughts will quickly turn to hulking bipedal carnivores such as *Spinosaurus* and *Tyrannosaurus* that have haunted dreams since Hollywood pushed them on to our screens in the early 1990s. Yet these gigantic killers ultimately needed something to eat. From the Jurassic (201.4 to 143.1 million years ago) period and into the Cretaceous (143.1 to 66.0 million years ago), the dinosaurs began to form some of the largest land-based food chains by mass that have ever existed. Herbivorous dinosaurs accounted for around 95% of vertebrate biomass in some ecosystems,[3] and included the long-necked, long-tailed quadrupedal 'Sauropod' dinosaurs such as *Argentinosaurus*, *Diplodocus*, and the aptly named *Supersaurus* and *Patagotitan*, which were the largest organisms to ever walk on solid ground – with a weight maximum of 70 tonnes these sauropods would have dwarfed even the largest of African elephants, recorded at 10 tonnes.[4] Researchers estimate that the biggest sauropods would have needed to consume a staggering 200 kilograms of vegetation every day.[5] For reference, that's nearly 700 cans of baked beans. Amazingly, we still rarely consider the key role that plants must have played in the evolution and sustenance of dinosaur ecosystems, or the combined impact that herds of herbivorous dinosaurs must have had on ancient vegetation – with plants being portrayed as a convenient camouflage or as passive backdrops rather than a crucial part of life.

This is perhaps even more remarkable when we think about the spectacular geological and environmental processes that happened from the Permian (298.9 to 251.9 million years ago) to the Cretaceous period (143.1 to 66.0 million years ago). Major changes in vegetation

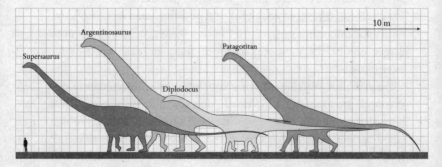

Comparison of selected giant sauropods based on conservative estimates of their body size.

accompanied the collapse of Carboniferous rainforests, the rise of the coniferous and cycad gymnosperms and, later, the flowering angiosperms. Warm, wet forests at times blanketed the Earth before shrinking to more restricted distributions. Meanwhile, the plates of the Earth kept drifting, as first Pangaea and then Gondwanaland were rent asunder to form the current continental layout of the tropics. Throughout, new forms of plants were emerging, moving and evolving in novel, increasingly isolated, ecosystems in the tropics. Today, we accept that climate and environmental change associated with such geological and biological shifts would have massive implications for animal life (and in this chapter, we will take a look at the science which enabled us to be more confident in thinking that). Nevertheless, while often used for dramatic effect in films, the vegetation of the dinosaurs has frequently been cut out of their evolutionary story, from their emergence and diversification through cataclysmic disaster and persistence as birds today. Is this due to the poor preservation of fossil plants? Or simply a result of plants lacking the media appeal of the reptiles they supported? Either way, it's high time we inserted plants into the history of these 'terrible lizards'. Today leading researchers are using state-of-the-art investigation of reptile (including dinosaur) fossils, fossil pollen and wood records of past environments, and, incredibly, even the occasional preservation of the very last meals of particular dinosaurs, to do precisely that.

It might be hard to imagine, but the first 'true' dinosaurs, defined by a series of particular features seen in their fossilized skulls, hips and limbs, did not immediately dominate the Earth when they appeared in the Middle to Late Triassic period ~240 to 230 million years ago. Scampering alongside large amphibians, other diversifying reptiles, and tiny primitive ancestors of mammals, the dinosaurs were just one of a number of organisms trying to survive on the surface of a planet that had been through the wringer of extreme change. Although things looked bleak for animal life on land following the massive plant extinction event at the end of the Carboniferous and the volcanic-induced extinction of animals at the end of the Permian, a number of researchers, including Dr Emma Dunne, believe that it was in these cataclysmic shifts that the dinosaurs grasped their chance to rise to the top of the animal kingdom. The changes to the Earth's surface that began with the collapse of the Carboniferous rainforest, whose canopies had coated practically all exposed equatorial land ~310 million years ago, were to gradually, but fundamentally, alter the course of evolution. In time, they were to pave the way for all the main groups of land-based animals we know today. The first key beneficiaries were, however, the reptiles and, more specifically, the dinosaurs.

Emma, a vertebrate palaeobiologist at the University of Birmingham, has been interested in fossils for as long as she can remember, though she sees them in a different way to many people. Perhaps unsurprisingly there is often a fascination in individual fossils, with people wanting to find the first type of a certain animal or particular body shape. However, a lot of information can be missed if these remarkable traces of life are treated as single objects. Emma prefers to think of them together as 'parts of a living, breathing ecosystem'. In this way she can try to reconstruct not just what they looked like, but also the type of environment they might have lived in. This is essential if we are to go from a stationary bone to determining what these animals might have eaten, what climatic and environmental conditions might have surrounded them, and what impact they themselves might have had on the world around them. Emma is particularly interested in how land animal communities changed between the Carboniferous and the early Mesozoic – an age that includes the Triassic, Jurassic and Cretaceous periods – taking us right up to the dawn of the dinosaurs. To look at this change, she and her colleagues have tried to find a way to properly measure the number of different types (or 'diversity') of animals that were living in different parts of the world, as well as the world as a whole, at a given point in time. This, in turn, allows them to investigate how and why this diversity might have changed across space and time. This is not as easy as it might sound, however. There are huge biases in the fossil record from this time period. Most of our knowledge, like that of the ancient forests we have already encountered, comes from particular sites, mainly in North America and Europe. This is a product of the longevity of geological and palaeontological study in Euro-American societies, as well as the availability of funding for such enterprises in different social, political and economic situations – something that, as we will see in Chapters 10 and 11, is often linked to historical global imbalances in wealth distributions.

To try to get around this problem, Emma and her team made use of a remarkable tool that is actually available to every single one of us – the 'Paleobiology Database'.[6] This database is a free-to-use record of fossil discoveries around the world, from the Proterozoic (2,500 to 538.8 million years ago) to the Holocene. I can guarantee that the beautiful software and easy access data will give you hours of distraction should you let it – allowing you to discover long-extinct creatures and the locations in which they have been found. Playing around with its functions, I went on an evening-long search for the find-spots of some of the most iconic creatures we commonly hear about from palaeontologists, from the gigantic Megalodon shark to *Triceratops*. Most importantly, however, it represents part of a growing push to make scientific data freely available or 'open access', so that they can be used and combined to answer larger, planetary-scale questions. In this case, as palaeontologists repeatedly update the database, adding the location and age of all finds from across the world, some of the more glaring geographical biases can be increasingly overcome. Instead of simply counting species that there are fossils for, larger databases such as this enable the application of novel statistical approaches so that we can get as complete a picture as is possible of the *entire* animal fossil record, spanning the earliest tetrapods (four-legged land animals) to the first dinosaurs, and beyond.

The results of Emma and her team's research[7] are fascinating. They demonstrate that the collapse of the tropical Carboniferous 'coal forests' had a dramatic impact on land animal diversity. Those affected most were the large, diverse amphibian communities, whose footprints were preserved in the rainforests of Birmingham ~310 million years ago. Although early dark, wet and warm forest ecosystems had provided a haven for animals that needed water to lay eggs that were vulnerable to drying out, more arid conditions initiated the dramatic decline in these habitats. What happened next is critical. The small, reptilian footprints, which were very much in the minority in Carboniferous Birmingham, suddenly proliferated, growing larger. Those tetrapod species that did survive the rainforest collapse began to disperse more freely across the planet, colonizing a number of new environments and moving further away from the tropics. Their laying of eggs, resistant to drying, unleashed a massive potential for expansion and gave them a number of advantages over the amphibians as forests continued to decline through the Permian and into the Triassic. In summary, then, the rise of the reptiles and, in turn, the dinosaurs was evidently, at first, actually aided by the disappearance of tropical forests. The changes in these ecosystems re-shuffled the hands of the different animal groups and set out a new field of ecological competition.

Dinosaurs have often been portrayed as 'tropical' animals, existing within a framework of stable, warm climates from the Triassic to the Cretaceous. Yet from the above it seems that they could both be affected by and benefit from climatic and environmental instability. This is certainly true when we look at the early evolution and expansion of the first 'true' dinosaurs from 230 million years ago. Although dinosaurs rapidly divided into three distinct successful lineages (Theropoda, Sauropodomorpha and Ornithischia) and had achieved a remarkable global distribution by 206 million years ago, it has remained something of a mystery as to why it took nearly 30 million years from their first appearance for them to enter the tropics and complete their global conquest. Furthermore, even when they did arrive, they did so in a tentative, rather than sweeping, manner.[8] The early theropod dinosaurs, which were later to include the infamous *Velociraptor*, made it to the tropics and sub-tropics by the Late Triassic. This clade (a group of organisms considered to all be the descendants of a single common ancestor) was characterized by animals with hollow bones and three-toed limbs, surviving in the form of modern birds, and was initially dominated by predatory carnivores hunting small animals scurrying along the ground. By contrast, the other two main dinosaur clades, the increasingly large herbivorous sauropodomorphs and ornithischians, were restricted to the higher latitudes, being somehow 'kept out' of the tropics until the Jurassic period

Artist's reconstruction of seasonal arid Triassic landscapes that acted as a backdrop for the origins of the dinosaurs. A steady growing dominance of conifer (gymnosperm) canopies can be seen in the background.

(201.4 million years ago). Although this division has often been explained by high humidity in the tropics causing bad fossil preservation or potential fights for survival with crocodile-like reptiles that already inhabited the equatorial regions, a growing collection of fossils and the uncanny division of dinosaur types by latitude suggests that a deeper, underlying geographical trend was at play.[9]

To try to unravel this problem, researchers with an interest in past climates have studied preserved woody charcoal, pollen remains, and the geochemistry of river sediments in New Mexico that have also preserved a variety of early dinosaur fossils.[10] The results demonstrated that although the world was still generally warm and dry during the Late Triassic, it was increasingly affected by frequent, rapid and extreme shifts in atmospheric CO_2 that led to sudden temperature spikes, increasing fluctuations in rainfall, and, most significantly, persistent wildfires that scorched wide areas of ground. The resulting climatic unpredictability in the tropics that we touched upon in Chapter 2 may have placed particularly strong stressors on animal communities living at the equator. As a result, these regions remained under the rule of other giant reptiles (pseudosuchian archosaur communities that bear greater resemblance to crocodiles and alligators), with some theropod dinosaurs eventually arriving and going about their business capturing prey, fishing or foraging. Meanwhile, the increasingly large-bodied, fast-growing sauropodomorph herbivores, requiring much more reliable food sources, could not break into lower latitudes until conditions stabilized during the Jurassic period. Even the ferocious dinosaurs, then, were shaped, from their earliest appearance and evolution, by the state of tropical environments. These habitats at once locked out some of the newer, more experimental forms, while also providing a productive home to some of the earliest types of carnivorous dinosaur that picked their way through the bleak world of the Triassic.

In 2009 Isabel Valdivia Berry and her husband, Erico Otilio Berry, were exploring the locality of Los Molles, Neuquén Province, Argentina. Many tourists come to Los Molles for its spectacular mountain views, hiking and hot springs. It is also exploited by multinational companies as one of the largest shale and gas reserves in South America. However, Isabel was more interested in the other treasures that the surrounding rocks are known to offer, at least to those who know where to look. The Los Molles Formation dates to the Early to Middle Jurassic (201.4 to 161.5 million years ago) and its rocky outcrops provide vivid snapshots into the marine and river delta ecosystems that were forming on the tropical coasts of the Gondwanan landmass, a considerable portion of which, including much of South America, was primarily located within the tropics during this time period. Ichthyosaurs, which resemble a cross between a dolphin and a crocodile and whose name aptly means 'fish lizards', ruled the Early Jurassic seas and are found in abundance at Los Molles. However, the Berrys, on their winter expedition to the region, happened upon a very different type of fossil. Poking out of the rock face was a near complete skull and significant portion of the body of a dinosaur which, as they would discover, had never been seen or described before.

After Isabel brought her startling find to the Museo 'Prof. Dr Juan A. Olsacher' of Zapala, a team of palaeontologists, headed by Dr Leonardo Salgado of the National University of the Río Negro–Conicet region, set off to perform detailed excavation at the find site, recovering

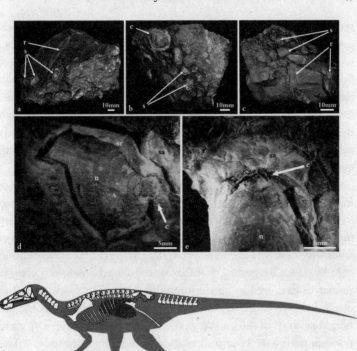

Images of the gut contents of *Isaberrysaura mollensis* gen. et sp. nov. discovered in the Early Jurassic Los Molles Formation in Argentina. Panels a to c show seeds of cycads (c) and other seeds (s) in relation to the dinosaur's rib (r). Panels d and e show details of the cycad seeds: their sarcotesta (sa), sclerotesta (sc), coronula (c), nucellus (n). Finally, the drawing shows the location of the discovered gut contents within the reconstructed skeleton of *Isaberrysaura mollensis*.

further fragmentary material. Following years of painstaking research, Leonardo and his team eventually published the find as *Isaberrysaura mollensis* in 2017.[11] More recently, this dinosaur has even been recognized as one of the earliest stegosaurs[12] – large, heavily built herbivores that were clad with rows of plates and had tails tipped with spikes. Beyond its evolutionary position, however, *Isaberrysaura* is significant for two main reasons. First, it represents overdue recognition of the remarkable women, from the famous nineteenth-century palaeontologist Mary Anning onwards, who have made important contributions to palaeontology, with 'Isaberry' dedicating the find to Isabel herself. Second, when the palaeontologists analysed the specimen, they realized that within the skeleton, where the stomach should have been, there was, astonishingly, a mass of mineralized food. As Leonardo puts it, 'we always ask ourselves what dinosaurs ate. Usually we have to rely on assumptions based on the shape of teeth. However, here was our chance to go one better.'

What the team found provides a new, more intimate picture of increasingly large dinosaurs and their increasingly green surroundings in the tropical latitudes of the Jurassic. Two types of seed were found in the gut of *Isaberrysaura*. One set could not be identified, but the larger seeds were clearly members of the cycad group. The seeds were almost entirely intact, suggesting that they were gobbled down, in a manner similar to that used by modern reptiles to eat their food, instead of being chewed. Furthermore, as cycads are toxic to most animals, if *Isaberrysaura* survived this culinary experience then the microbes inside its stomach must have been well adapted to digesting these plants. This one in a million find – while providing a vanishingly brief single meal snapshot into the dietary behaviour of dinosaurs – gives us a new appreciation of the close relationship that diversifying dinosaurs had with novel emerging plant life in the Late Triassic and Early Jurassic. Not only were herbivores consuming gymnosperms, the new green arrivals on Earth, but, Leonardo argues, they may also have had a role in their increasing success. As we have seen, he found that the seeds left in the stomach were almost complete. This means that when they passed out of the *Isaberrysaura* they could go on to be fertilized and grow. In essence, these early stegosaurs were potentially some of the first wide-ranging seed dispersers, perhaps even contributing to the health of these rapidly expanding gymnosperm plant communities.

At the end of the Triassic 201.4 million years ago, there was yet another mass extinction event on land. Global drying, an increasing instability and seasonality of climate, and the deadly accompaniment of volcanic eruptions killed many large amphibians and all reptiles with the exception of the dinosaurs and the ancestors of those that are alive today, including crocodiles, turtles and lizards. It also caused a major turnover in seed-producing plants.[13] With the rise of the Jurassic, however, the Earth's climate stabilized. The arid conditions of the Triassic were replaced by wetter and warmer ($\sim$3.5°C warmer than today) conditions that encouraged the expansion of lush vegetation, including tropical forests. While conifers continued to dominate, as they had done from the Middle to Late Triassic, these new conditions enabled the diversification of gymnosperm plants. Study under a microscope of the fossil pollen found from the same geological formation in which *Isaberrysaura* was preserved shows that it belonged to a mix of conifers (such as the family Araucariaceae, which now extended across the entire planet), cycads and non-seed-producing ferns. Together these plants would have provided a combination of rich open and forest environments for increasingly diverse dinosaur species, not just in the tropics but also beyond.[14] Ginkgo trees began to dominate at warmer temperate latitudes, with pine-based boreal forests extending into the cooler environments. Although atmospheric CO_2 levels stabilized in the Jurassic period, as was seen for the end of the Triassic, generally high CO_2[15] could still lead to frequent, regionally variable high temperature events and rainfall variability.[16] Nevertheless, overall increasing stability, and a lush gymnosperm vegetation relied upon by ever-growing herbivorous dinosaurs, provided some semblance of steadiness, and dinosaurs rapidly filled every single ecosystem and niche on an increasingly mixed landscape that was full of vibrant green life. The Jurassic was thus a time when not just dinosaurs, but also their gymnosperm food sources, came to rule the Earth.

The relationship between dinosaurs and gymnosperms may even have been so intimate as to represent an example of 'co-evolution'. The sauropods – dinosaurs with massive necks, huge tails and an overall tendency towards gigantism – flourished by the end of the Jurassic. Studies of the shape and size of their teeth, as well as reconstructions of their jaw movements and skull dimensions, show that they were herbivores.[17] The vast size of these animals and their rapid growth rate, when compared to known modern large herbivores, shows that they must have been bulk feeders, with a paraphernalia of gut bacteria that helped them to digest their leafy meals through fermentation.[18] As with large stegosaurs (later relatives of *Isaberry-saura*), they did not chew their food. Instead, they swallowed it whole.[19] Studies of the fossil plant remains found in the same contexts as sauropods have provided some insights into the kinds of plants that were likely on the menu of these giant vegans.[20] When all this information is combined, it seems clear that these proliferating long-necked dinosaurs relied on the bulk feeding of gymnosperms, including conifers such as araucaria and ginkgo trees, and ferns in a variety of different settings – ranging from humid forest edges to more open areas.[21] Subtle variations in the shape of their teeth suggest that they diversified, so that while some may have focused on treetops, others swept more mobile necks around ground level to feed.[22] This was not just a one-way street, however, and, like the large herbivores of the modern tropics,[23] the sauropods likely 'gave back' to their ecological settings. Their wide-ranging, crashing movements would have benefited gymnosperms, as their seeds were dispersed without damage, and plants such as cycads may have attracted colour-seeing dinosaurs with their brightly coloured seeds.[24] It has even been suggested that the cones on conifer trees, particularly the so-called 'monkey puzzle' Araucariacae, evolved their shape as a way to survive and benefit from dinosaur herbivory![25] Although the strength of this relationship has been hotly debated and an exact link between gymnosperm success and dinosaur abundance seems unlikely,[26] it is undeniable that the varied, enormous herbivorous dinosaurs of the Jurassic, as well as their predators, benefited from an expansion and diversification of lush gymnosperm vegetation. It is also intriguing that, in North America at least, sauropod diversity eventually decreased, while stegosaurs all but disappeared, at a time of gymnosperm decline in the Early-middle Cretaceous (143.1 to ~100 million years ago). In fact, it is at around this time that an even more dramatic link between dinosaurian green 'fingers' and plant evolution has been proposed. This time, centring particularly on the tropics.

Professor Paul Barrett has the job many of us would have wanted as children (and, let's face it, probably still want). His office in the Natural History Museum in London is located just down the hall from one of the largest and most famous collections of dinosaurs anywhere in the world. Even better, every day he gets paid to walk down the hall to study these bones, which include the first skeleton of an *Iguanodon* known to science, used as part of some of the first academic descriptions of dinosaurs as a biological group, and a part of the first *Tyrannosaurus rex* skeleton ever discovered. Nevertheless, like the other researchers in this chapter, Paul is not just interested in dinosaurs for dinosaurs' sake. He wants to know how they influenced life on Earth, and how they were a part of broad processes that have shaped the environmental and evolutionary history of our planet with a lasting legacy to this day. As many of us have only ever seen a dinosaur as a collection of bones, it can be hard for us to think about them as living, active beings, particularly when we often imagine them as ultimately unsuccessful and consigned to the static fossil record by an asteroid. Furthermore, in museum collections that include dinosaurs with machete-like teeth, the more gentle, herbivorous beasts of this earlier time are not often thought of as major players. Yet Paul's work over the last twenty years has looked into some of the most intriguing hypotheses of how populations of herbivorous dinosaurs impacted plant evolution, and how, in turn, the dinosaurs were affected by changing climate, environment and plant communities.

In the 1970s, the American palaeontologist Robert Bakker presented the intriguing idea, in the world-leading scientific journal *Nature*, that 'Dinosaurs invented flowers'.[27] Looking at broad changes in the fossil record, he noticed that there was an apparent link between a significant change in dinosaur feeding behaviour and the origin and spread of the flowering angiosperm plants, at the expense of gymnosperms, during the Cretaceous period (143.1 to 66.0 million years ago). In the Jurassic, a dominance of long-necked sauropods feeding high in the canopy allowed slow-growing gymnosperms to proliferate from the floor up. By the Early Cretaceous, ornithischian browsers that preferred to remain lower down for their meals, such as the tank-like ankylosaurs that had wrecking balls for tails and *Iguanodon* that were some of the most numerous dinosaurs across Europe, massively increased pressure on plants growing in the ground floor of vegetation communities. The resulting high disturbance was hugely problematic

for the lethargic gymnosperms, leaving the fast-growing angiosperms to radiate out from the tropics and attain dominance of the
plant world. Whether it was a change in behaviour, new, sophisticated jaw types developed by certain herbivorous dinosaurs to
specialize in angiosperm consumption,[28] or simply the scale of disturbance caused by hulking Cretaceous dinosaur communities
dominated by animals larger than 1 tonne, it appeared that the global
spread of angiosperms, and flowers, from the equator to the poles,
may have, remarkably, been a dinosaurian achievement.

'While it is a wonderful idea, I will unfortunately have to stop
your readers there,' says Paul. He tells me that when we review fossils, dinosaur feeding behaviour, and plant evolution in detail, it is
clear that the dinosaurs did not 'invent' the angiosperms. Nor can
they be given sole credit for their global expansion.[29] First, when
preservation and geographical biases in the dinosaur fossil record and
limited understanding as to how different dinosaur fossils relate to
each other are taken into account, it is clear that the sauropods did not
decline everywhere in the Cretaceous. In fact, they even flourished in
areas such as tropical Gondwana where the giant, 60-tonne, long-
necked and long-tailed *Patagotitan* roamed the forests of Argentina
late into this period. Nor did the ornithischians and their 'low browsing' tendency succeed everywhere, remaining largely absent from
Gondwana and the Southeast Asian portion of Laurasia. Second,
there is actually little difference in the spatial distribution of major
herbivore types between the Jurassic and Early Cretaceous. Third, as
noted above, not all sauropods browsed in the canopy and, alongside
stegosaurs and other herbivores, many long-necked giants would
have actually already placed major pressures on vegetation growing
under one metre tall in the Jurassic period, prior to the advent of
angiosperms. Finally, the major expansion of dinosaurs, such as the
iguanodonts, with special jaws occurred a considerable time *after* the
origins of flowering plants.[30]

Nevertheless, all is perhaps not lost in the search for dinosaur 'gardening', and there may be some indications that a more 'diffuse'
co-evolution, as Paul puts it, existed. The disturbance of ground vegetation by increasingly large dinosaur ecosystems, or 'dinoturbation',

evidently increased across America, Africa and Asia from the Early Jurassic, reaching its maximum during the Cretaceous.[31] This undoubtedly provided some assistance to fast-growing angiosperms and potentially drove their diversification by the Late Cretaceous. It also remains possible that the origins and rapid expansion of one major group of dinosaurs, the hadrosaurs, or 'duck-billed' dinosaurs, which occurred from the Early Cretaceous and exploded in number in the Late Cretaceous, both helped, and were dependent on, the planetary colonization of angiosperms.[32] In particular, it has been suggested that novel innovations in hadrosaur teeth differentiated them markedly from other reptiles, with a tooth complexity that rivalled that of modern herbivorous mammals and that would have allowed them to eat a diversity of plant matter.[33] Perhaps the starkest example of a direct relationship between dinosaurs and angiosperms exists in the form of *Kunbarrasaurus*, an ankylosaur fossil from the Early Cretaceous of Australia, whose 'last meal' included a variety of angiosperm fruits,[34] meaning that we know dinosaurs did eat these plants. Indeed, it has been suggested that the lack of mechanical defences against herbivores among the earliest angiosperms, such as thorns and spines, show that these plants actually wanted to be eaten and dispersed by bulk-feeding dinosaurs.

In fact, our record of dinosaur fossils and angiosperm origins and diversification is simply not yet complete enough to rule out a close relationship between dinosaurs and angiosperms. The origins and expansion of this new type of plant certainly benefited from increased levels of Cretaceous CO_2 as well as the expansion of their insect pollinators. It would not be surprising if dinosaurs also added their considerable weight to this process, especially given evidence for the role of large animals in seed dispersal and habitat modifications across the tropics, and beyond, today.[35] There also seems no doubt that dinosaurs benefited from the varied ecosystems that angiosperms were to dominate by the end of the Cretaceous. Spreading out from tropical forests, angiosperms accompanied dinosaurs all over the planet, into more arid, riverine environments and even as far north as the Arctic Circle in Russia,[36] potentially providing dinosaurs with more varied ecosystems in which to lay their eggs, reproduce, and

Hadrosaurids developed highly complex jaw and teeth morphologies that would have allowed them to consume a diversity of plant matter, including, perhaps, expanding angiosperms.

feed.[37] Larger and larger food chains, which supported the giant carnivorous tyrannosaurs by the end of the Cretaceous, suggest that the rise of the angiosperms and the loss of dominance of gymnosperms did not hinder the dominant land animals of the time, but rather drove them to bigger, wider and greater success. Popular imagery, and even conservation presentations, tend to link dinosaurs to 'ancient' Gondwanan cycads and conifer trees. Yet it is clear that the angiosperms, which diversified rapidly in Cretaceous 'mega-thermal' forests and went on to dominate the majority of modern environments, grew alongside dinosaurs and also sustained them right up until the end. Then, 66 million years ago, an asteroid hit Mexico and brought their green relationship to an end. Or did it . . . ?

Portrayals of a dramatic 'extinction' of dinosaurs often lead us to associate them with failure. Not only do we forget that by existing between 230 and 66 million years ago they would be one of the most successful animal groups ever to have roamed the planet (around 23 times longer than our own hominin clade, and 547 times longer than our own species), but we also commonly neglect the fact that they actually still live, or rather fly, around us today. Birds are dinosaurs. They share a huge number of clear, unique anatomical features with the theropod dinosaurs that we now know are their closest ancient relatives. We have the fossils, and the appreciation that dinosaurs had feathers, or at least feather-like structures, to show that the ancestors of birds metaphorically flew out of the end-Cretaceous doomsday brought on by an extra-terrestrial impact.[38] And, as we all know, birds are argued by some estimates to be the most successful terrestrial vertebrates on Earth, with a staggering 18,000 species distributed over practically the entire planet.[39] Birds also maintain an incredibly close relationship with seed-producing plants, including within the tropics. Tropical forests are home to a vast variety of birds today: from the fantastic bowerbird of Australia and Papua New Guinea, which clears itself a stage within the forest to dance and prance for potential mates, to the Harpy Eagle of the Amazon Basin, which is one of the largest eagles worldwide and terrorizes monkey communities in the canopy. Across the world, tropical forests, including both gymnosperms and angiosperms, continue to provide green shelters for some of the greatest diversity of living dinosaurs.

Whether or not Jurassic or Cretaceous dinosaurs co-evolved with gymnosperms or angiosperms, birds undoubtedly play some of the most important roles in keeping flowering plant populations alive and kicking around the world today. Plants actively seek to attract the colour vision of birds by producing bright fruits so that these wide-ranging fliers will eat and spread their offspring to new horizons. So important are birds to the spreading of seeds that, in the tropics, there is hardly a forest ecosystem that could remain healthy without them. Again and again, biologists find that birds ensure the reproduction and genetic diversity of tree populations, from the tropical montane forests of the Andes[40] to the lowland evergreen rainforests of the Amazon Basin,[41] from the dry evergreen forests of India[42] to the island tropical forests of the Pacific.[43] Not only that, but it has been shown that birds are some of the most important agents in the recovery of disturbed tropical forests. Modern agriculture, ranching and infrastructure not only cause dramatic deforestation but, by disturbing resident birds, also threaten any hope of a new beginning.[44] It seems likely that non-bird dinosaurs had significant relationships with plants and forest communities that were dynamically changing on an increasingly fractious tropical 'Gondwanan' surface. Their descendants certainly did and still do. However, the dinosaurs, while spectacular, are not the only major type of animal to have both been buffeted by forests and to have gnashed them back. We now turn to another key group of animal life with a deep-rooted, symbiotic relationship with tropical forests. One that is perhaps a little closer to home . . .

4.   'Tree-houses' for the first mammals

Mammals are everywhere. They stock our supermarkets with meat and dairy products. They fill our houses with barking, meowing and squeaking. They carry us trekking over rough landscapes or on a day out at the beach. They are some of the most frequently used symbols for brands, making appearances on logos ranging from football badges (the wolf of Roma) to boxes of cereal (Tony the Tiger). They dominate some of the most iconic ecosystems presented to us in documentaries or recreated for us in zoos. And, in the form of the trumpeting African elephant and the hulking blue whale, members of the *Mammalia* represent the most giant living animals on both land and sea. They are found in every single land-based environment on planet Earth. They are also, of course, us, comprising the ~7.8 billion of the world's still-growing human population. In this bustling 'Age of Mammals',[1] we rarely take pause to ask, 'How did we get here?' In fact, while we take our current mammalian success for granted, for two-thirds of the existence of mammals on planet Earth, from ~210 million years ago to 66 million years ago, they were never larger than a badger,[2] cowering beneath the mighty dinosaurs. Not only that, but it is mind-boggling to think that *all* modern mammals – from the Etruscan pygmy shrew that weighs in at ~2–3 grams[3] to the blue whale at 190,000 kilograms[4] – are actually related to just a single common and tiny ancestor from a time when the planet was ruled by reptiles.

Approximately 66 million years ago a giant asteroid crashed into the coast of Yucatán in Mexico, leaving a crater with a diameter of 180 kilometres – the centre of which lies directly under the modern-day town of Chicxulub. The impression left on our planet by this vast extra-terrestrial object was not just geological. Its impact incinerated much of North America, threw soot and ash into the atmosphere, and likely triggered massive volcanic eruptions in the

Deccan Plateau of India that induced a global winter.[5] With sunlight reduced to around half previous levels and lashings of acid rain caused by the released volcanic gases mixing with water in the air, the Earth's surface and the life it sustained were sent spiralling into disarray. Many species of plants, birds, insects, marine life and, perhaps most significantly, *all* the non-bird dinosaurs became extinct, more than 75% of all species on Earth at the time.[6] Although some mammal species were also wiped out in this 'end of days' event, traditionally the fire and brimstone at the end of the Cretaceous period has been seen as a window of opportunity for our evolutionary line.[7] With no dinosaurs left to molest us, we could fill landscapes that were left empty by disaster, take our pick of remaining food sources, and expand and diversify in a march of seemingly inevitable progress until the present day. However, while such a narrative is attractively simple, it certainly does not tell the whole story.

Fantastic fossil finds are now beginning to show us that mammals, and their closest, but now extinct, 'mammaliaform' (literally meaning 'mammal-shaped') relatives were actually already experimenting and expanding *before* the dinosaurs went extinct. These little, but increasingly varied, critters did not just scurry below these fearsome reptiles, but also climbed above and even flew around them – first in Jurassic gymnosperm-dominated forests and, later, in Cretaceous forests that were increasingly invaded by flowering (angiosperm) plants. Furthermore, the asteroid-induced mass extinction at the Cretaceous–Paleogene (commonly known as the K-Pg) boundary was for mammals only the beginning in our brushes with climatic and environmental change. During the following Paleocene (66.0 to 56.0 million years ago) mammals accompanied the expansion of angiosperm-dominated 'mega-thermal' forests across an increasingly fragmented planet by the end of the period. A climatic record from the Arctic has shown that mean atmospheric temperatures rose from 16–18°C to a balmy 25°C at the Paleocene–Eocene boundary (56 million years ago).[8] As we saw in Chapter 2, tectonic shifts continued to rearrange the world's continents towards their current orientation. In particular, in the middle Eocene, Australia and Antarctica split apart, creating a deep-water passage for the flow of cold polar water towards the equator.[9] The new patterns of ocean circulation and their relationship to the atmosphere saw swings between warmer and cooler states in the Eocene (56.0 to 33.9 million years ago) and Oligocene (33.9 to 23.04 million years ago).[10] They also resulted in the global expansion of grasslands and retreat of 'mega-thermal' forests as the planet broadly became cooler, drier and ever more seasonal heading through the Miocene (23.04 to 5.33 million years ago).[11] The latest science is showing that the geology- and climate-induced ebbs and flows of warm, wet forests over this 45-million-year span played a major role in shaping mammalian distribution, diversity and evolution, leaving us with the modern ecosystems we gaze in awe at today. Not just that, but without forests growing in the tropics of the Jurassic, the 'Age of Mammals' may never have even begun.

In north-eastern China, the provinces of Liaoning, Inner Mongolia and Hubei meet in one of the most important regions for palaeontologists trying to track the world-changing transition of life on Earth as it swung from the realm of the dinosaurs to the dominion of mammals. Here, past volcanic activity and flooding have seen fine particles of ash and water-borne mud seal in soft body parts and entire skeletons across millions of years[12] as part of what have become some of the world's most-celebrated and most well-preserved set of fossil beds. This part of north-eastern China has yielded one of the first dinosaurs ever to be found with intact feathers, so complete that palaeontologists could even identify cells that give the feathers of birds today their colour.[13] While not part of the direct lineage that eventually led to modern birds, this fossil, *Sinosauropteryx*, nonetheless highlights the biological innovations of some avian dinosaurs in the Cretaceous period that allowed them to hang on beyond the K-Pg boundary.[14] Even more importantly, this region has also preserved one of the most significant fossils in the evolutionary history of the group that was to replace the dinosaurs as the new heavyweights of the animal kingdom. In 2011, Professor Zhe-Xi Luo, of the University of Chicago, and his colleagues uncovered the remains of an animal they named *Juramaia sinensis*, which they dated, using radiometric dating methods, to ~165 to 160 million years ago and the

Artist's reconstruction of *Juramaia sinensis*, the 'Jurassic mother' of all placental mammals.

Middle to Late Jurassic period.[15] A time when this part of the world would have been located at the boundary of the tropics.[16] These methods compare the amount of naturally abundant radioactive isotopes (in this case uranium-235 or -238) to the elements they are known to decay into (lead-207 and -206) within minerals such as zircon trapped in the fossil layers of interest.[17] As the rate of decay can be calculated, these measurements can be used to estimate how long ago the fossil was buried. Based on its age, and the characteristics of the fossil bones, the team gave it a name that denotes it as the 'Jurassic mother' of all true placental mammals.[18] 'Despite its minute stature, it gave birth to almost all of the mammalian diversity we see around us today,' says Luo.

The word 'mammal' literally means 'breast' in Latin, giving immediate insight into the new biological features that separated these animals from their dinosaurian overlords. Although monotremes (like the duck-billed platypus) lay eggs which are later hatched and raised in a pouch, the majority of mammals are defined by the fact that they give birth to live offspring which they feed by milk. Back in the Jurassic and Cretaceous periods, this would have represented a major evolutionary adaptation, allowing mammals to develop their young from the comfort of safe burrows. Like their extant relatives, these early mammals would also have been warm-blooded and have had fur, enabling them to hunt at night when temperatures were low and most predators were asleep.[19] Unsurprisingly, however, it can be incredibly difficult to spot the first appearance of these traits in the fossil record. Soft tissues rarely, if ever, preserve over these vast timescales. This has meant that scientists have looked to sturdier, more resistant skulls for clues. Comparison of fossils with modern mammals demonstrates what set the earliest mammals apart from their reptilian counterparts: the separation of a newly united jaw bone from ear bones that moved backwards in the skull. A more stable jaw supported teeth that interlocked like a jigsaw, slicing and dicing through food. While reptiles tend to swallow their food whole, using their teeth simply to lock on and tear at prey, this new adaptation enabled a 'pre-processing' of food before it reached the stomach, ultimately releasing more nutrients.[20] Meanwhile, separate ear bones, located away from the jaw, meant better hearing (essential when active at night), and also provided space for mammalian brains to grow outwards and backwards.[21] The earliest 'mammaliaform' (or mammal-shaped) creatures in which these features have been found are known as the morganucodonts. Fossils uncovered from the Late Triassic across Europe, Africa, America and Asia show this group was starting to scamper around just after the arrival of the first dinosaurs.[22]

New and varied mammaliaform creatures continued to make an appearance throughout the Jurassic and Cretaceous periods. Palaeontologists have been frantically searching for the point at which these mammal-like creatures became the 'true' ancestors of most mammals alive today. In *Juramaia*, Zhe-Xi thinks we have it. *Juramaia* had a minuscule body length of between 7 and 10 centimetres. We can actually see its soft tissues, including hair, thanks to impressions left in its imprisoning rock. Importantly, detailed comparison of *Juramaia*'s paw bones and teeth with other extinct and living mammals demonstrates that it was alive shortly after the evolutionary division between all of today's placental mammals (mammals that protect and feed their young in a womb using a placenta until fully developed) and marsupial mammals (mammals like kangaroos that release their embryos early to be developed in a pouch). *Juramaia* which, like us, is part of the placental line, shows that this split must have occurred

~170 million years ago. This is in line with molecular 'clock' estimates which compare genetic diversity in modern mammals, as well
as assumed rates of mutation, to estimate the age of their last common ancestor,[23] although more recent finds by members of the same
team may ultimately push the beginnings of this process back still
earlier.[24] More incredible fossils, again from north-eastern China,
show that the split between placental and marsupial mammals was
certainly complete by the Early Cretaceous, with the earliest marsupial, the tiny pointed-nosed *Sinodelphys*,[25] and, until *Juramaia*, the
earliest placental mammal *Eomaia* ('dawn mother') dating to 125 million years ago.[26] From *Juramaia*, and from north-eastern China,
we can therefore see the building blocks needed to reconstruct the
evolutionary history of the majority of mammals that exist on the
planet today.

Significantly, the rich fossil beds of north-eastern China have also
provided insights into the interaction between some of the earliest of
our mammalian ancestors, and their closest 'mammal-shaped' relatives, and diverse sub-tropical and tropical forest ecologies. Until
recently, scientists had assumed that the earliest mammals were all
small, all ground-dwelling, and all nocturnal, with variation in
mammals occurring only *after* most dinosaur communities were
blasted into oblivion.[27] Nevertheless, the study of plant fossils found
alongside Jurassic and Cretaceous mammals, as well as analysis of the
shape and size of their arms and legs that provides insights into the
stresses and uses they were put to, has revealed that early mammals
and mammaliaforms went through two major periods of ecological
'radiation' *prior* to the K-Pg boundary: one in the Jurassic and one in
the Cretaceous.[28] In the Jurassic period (201.4 to 143.1 million years
ago), weighing in at 15–17 grams (just a little more than a 'AAA' bat-
tery), *Juramaia* would not have been particularly conspicuous, and
study of its teeth reveals that it mainly ate insects.[29] Nevertheless, its
strong forearms suggest that, rather than cowering in fear on forest
floors, it ascended trees to live up at eye-level with the dinosaurs.[30]
Indeed, living in trees has been shown to have significant benefits for
mammals in terms of life expectancy and protection from ground-
based threats.[31] Moving away from *Juramaia*, fossil finds of Jurassic

A reconstruction of *Repenomamus* snacking on a young *Psittacosaurus* dinosaur.

mammals have now revealed that they were practising a variety of arboreal,[32] ground-based, burrowing, and even aquatic[33] modes of movement. Another Jurassic fossil from north-eastern China has even revealed the existence of 'gliding' mammaliaforms in the Jurassic ~160 million years ago.[34] Remarkable preservation has enabled scientists to observe the wing-like membrane of these animals which would have enabled them to fly between trees. Not only that, but through comparison with modern gliders, which are all herbivores (with some supplementing their diet with insects), and a detailed analysis of the shape of their teeth, Luo and his team have argued that these Jurassic mammaliaforms had developed a new dietary adaptation.[35] One that involved the consumption of young leaves, cones, and the tender tissues of seed ferns and various lush gymnosperm forest vegetation that dominated Jurassic landscapes.[36] As Luo puts it, 'The forests of the Jurassic provided the perfect "tree houses" for the earliest mammals, not just for protection, but also for food as part of increasingly diverse mammalian ecological niches.'

This diversity continued into the Cretaceous period. In the Early Cretaceous, findings of fossils of the badger-like *Repenomamus*, again in north-eastern China, and dated to ~125 to 123 million years ago, overturned ideas of what early mammals were capable of. Not only were the teeth of this mammal adapted to carnivory, but it has also even been found with a dinosaur preserved in its stomach![37] The Late

Cretaceous saw a further, second major radiation in mammal and mammaliaform diversity. A number of the early mammalian groups went extinct between ~125 and 80 million years ago, with those surviving being primarily small and adapted to eating insects.[38] However, from ~80 million years ago, changes in the teeth, jaws and body sizes of eutherian ancestors of placental mammals, metatherian marsupial ancestors, and their close relatives indicate the development of varied dietary specializations.[39] So great was the diversity of different Late Cretaceous mammaliaforms, that it has been found to be similar to the renowned diversity of small-bodied mammal communities that live in tropical forests today.[40] Again, this mammalian ecological radiation seems to have links to tropical forests of the time; this time, forests that were being increasingly encroached on by flowering angiosperms, with their fleshy fruits, nutrient-rich leaves, seeds and tubers. Indeed, one of the earliest dated fossil angiosperm families (called 'Archaefructaceae' or 'ancient fruit') comes from the same part of north-eastern China that we have explored throughout this chapter.[41] The teeth of some Cretaceous mammaliaforms, such as the rodent-like 'multi-tuberculates' (so named because of the many tubercles (or 'cusp bumps') on their teeth), seem to have had a specific adaptation to the consumption of complex angiosperm vegetation, with sharper, blade-like teeth for slicing at the front, and bumpy teeth for crushing at the back.[42] Meanwhile, the teeth of other mammaliaforms, including our placental ancestors, were well tuned in to eating insect populations that would have also blossomed as the pollen couriers for new flowers.[43] Although the significance of the role of mammals in the initial Cretaceous origins and expansion of these new types of plants remains debated, the creation of an intimate relationship with these innovating plant communities was perhaps the best thing the mammals ever did. Not only did it give them a number of sturdy companions with which to survive one of the worst extinction events ever seen on Earth, but, coming out on the other side, it also guided them into a new era of global domination.

The amazing fossil finds from the Jurassic and Cretaceous show that diverse mammalian pioneers were flying around the heads of dinosaurs, scurrying between their feet, swimming in their rivers, and some were even beginning to turn the traditional predator–prey relationship on its head. This growing roster of mammals and their relatives seems to have quite literally been energized by gymno-sperm-dominated Jurassic forests as well as Cretaceous forests with an increasing representation of angiosperm communities. Neverthe-less, while our earliest mammalian forerunners were certainly more vibrant than is often appreciated, their main break still clearly came with the fall of the dinosaurs. To watch how mammalian understud-ies became the stars of the show, we must leave China and travel to North America. Here, at the Corral Bluffs in the Denver Basin, Col-orado, we can find one of just a handful of localities in the world with a fossil record that brackets the disastrous K-Pg impact. The Corral Bluffs are a single, continuous, exposed outcrop that runs for 27 square kilometres on either side of Highway 94. Thousands of motor-ists fly past this part of the windswept plains of western North America without even realizing its significance. Yet, 66 million years ago, this location was in the direct firing line of the pulse of heat emanating from the massive Chicxulub impact. Frozen in time, fos-sils found within the Bluffs span ecosystems that were wandering obliviously into disaster during the last 100,000 years of the Creta-ceous and, crucially, those that emerged from the rubble in the first 1 million years of a new, green Paleocene era.

'The fossil bonanza that has emerged from the Bluffs makes it one of the best sites, if not the best site, for studying the "comeback" tour of life after the lethal asteroid collision,' says Dr Tyler Lyson, a palae-ontologist at the Denver Museum of Nature and Science. Although ancient-fossil detectives are usually on the lookout for skeletons pok-ing out of sediments, this time the treasure could only be found by looking inside inconspicuous 'concretions'. To the untrained eye, these might look to all intents and purposes like solid rocks. How-ever, as the team began to split these concretions open back in the lab, they discovered the biological secrets that lay within. Specifi-cally, they found the incredibly well-preserved remains of turtles, crocodiles, plants and, most significantly, a rapid (at least in the con-text of geological time) succession of mammal forms that show the way in which these animals strutted to the top of the Paleocene.[44] The biggest mammals that escaped the extinction event weighed no more than half a kilogram. But just 100,000 years down the line their descendants weighed 6 kilograms. 200,000 years later and mammals had reached 20 kilograms – far heavier than any pre-K-Pg mammals. By 700,000 years after the disaster, mammals weighed over 45 kilograms,[45] that's more than an adult hyena or chimpanzee. Traditionally, the unrelenting arrival of ever larger mammals so soon after the global apocalypse at the end of the Cretaceous has been put down to a simple absence of dinosaurs. Basically, without the 'terrible lizards', our ancestors would no longer have had any reason to hide in the shadows. However, to settle for such a simple explanation would be to do the dynamic early mammals and their vibrant leafy companions a disservice.

The asteroid impact, and the reduction in access to the Sun's energy that followed, undoubtedly caused ecological disruption and localized loss of vegetation diversity.[46] However, many plant families found as Cretaceous fossils exist today and seem to have had a special capacity to adapt to rapid, global environmental change.[47] The angiosperms were here to stay, and were not about to stop expanding now. Fossil pollen from the Corral Bluffs record shows us that the increase in mammal body size occurred at the same time as plants were getting more diverse and more nutritious in a warming world where tropical climatic conditions prevailed across much of the planet.[48] While ferns were the first to emerge out of the North American ashes, it was the angiosperms, including palms, growing taller and more nutritious than ever before under the warming Paleocene conditions, that soon dominated the landscape. 300,000 years after impact, the largest mammal around was *Carsioptychus*, a distant ancestor of all modern hoofed mammals or 'ungulates'.[49] This animal had large, flat pre-molar teeth, with bizarre fold-like shapes, that were ideally adapted to the munching of hard objects, like the nuts of trees from the walnut family, which appears in the fossil plant record at the same time. 400,000 years later, and the largest mammal in the Corral Bluffs record occurs alongside the sweeping entrance of beans into the fossil plant records of North and South America.[50] Their

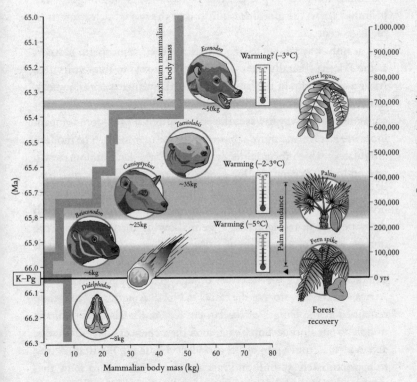

Schematic showing the development of mammal body size at the Corral Bluffs site in Colorado, USA, and their relationship with angiosperm plants and temperature. Timescale on the left refers to million years ago. Timescale on the right refers to years after the K-Pg boundary.

energy-filled leaves and protein-rich seed pods provided ideal sustenance for growing mammal bodies. The close relationship between increasingly large mammals and their tropical angiosperm companions can be seen in the Corral Bluffs fossil record through the ever more popular trend towards vegetarianism. The close bond between mammals and tropical plants seen in Colorado is something that can be found across the global fossil record of the warm, dense forests of the Paleocene.

The only way was up for mammals and their angiosperm partners as the Eocene (literally 'new dawn') broke at 56.0 million years ago. After a cooler blip in the late Paleocene, the planet warmed rapidly, spurred on by tectonic plate movements, and 'mega-thermal' forests expanded all the way towards the polar regions.[51] More forests meant more vegetation and more opportunities for mammals. The modern 'ungulates', which include the majority of large land mammals still around today, expanded. One group, the perissodactyls (which bear their weight on one of their five toes), including the ancestors of horses, tapirs and rhinoceroses, were particularly successful, with their special gut fermentation system allowing them to process the fibrous leaves of dense forests.[52] Another group, the artiodactyls (which bear their weight equally on two of their five toes), also emerged, including the ancestors of pigs, goats, hippos and cattle in Africa and the ancestors of the camel in North America.[53] This group remained at the fringes of the largely forested landscapes until the middle to late Eocene. Some even took their chances on the high seas, and evidence from a hippo-like artiodactyl found in Pakistan, dating to approximately 50 million years ago, has been used to show that modern whales and dolphins derive from early land-dwelling artiodactyls that gradually went on to adapt to ocean life through the Eocene.[54] In South America, marsupials, armadillos and now-extinct ungulate forms diversified. Early forms of carnivores with paws, rodents with claws, elephant-related 'proboscideans', bats and arboreal lemur-like primates had also all appeared on Earth by the middle Eocene, completing the sweep of most of the modern mammal groups across ecosystems that would have seemed remarkably similar to those of the modern world. In fact, the main difference to today would have been the size of these mammals, and many of these early forms would have looked like miniaturized models of their twenty-first-century relatives, with small sizes acting as a coping mechanism against the heat and dense forests that prevailed up to the middle Eocene.

The 'Messel Pit', a disused quarry that was previously mined for oily shale, just south-east of Frankfurt am Main in Germany, provides a remarkably vivid picture of these flourishing, cosmopolitan middle Eocene mammal ecosystems.[55] Here, slap bang in the middle of modern Europe, a fair distance from the tropics, palaeontologists have discovered a wet, warm forest ecosystem where early rodents, hedgehogs and even marsupials were menaced by early hyena relatives. Where a semi-aquatic otter-like mammal took dips in swampy wetlands. And where early relatives of horses and an ancestor of rhinoceroses made increasingly larger strides. In the trees themselves, early primates consumed fruits and the small cat-like carnivore, *Paroodectes*, stalked along the branches of tropical trees like canarium, cashew and nutmeg, which covered central Germany at this time.[56] Taking leave of the ground, early bats flew around the heads of these other flourishing mammals. Study of the small mammals at Messel shows that a 'rich' community existed that would not be out of place in the canopy of a multi-layered tropical rainforest today, sustained by angiosperm fruits and seeds. Undoubtedly, this cornucopia of early mammals found across the major continental landmasses at the time was being encouraged and supported by the spread of angiosperms to almost every corner of the globe. But this isn't simply a story of take, take, take on the part of the new and increasingly diverse mammals; it may be that their green companions also benefited.

Indeed, it is not just mammals that underwent major changes in shape and diversity between the Cretaceous and middle Eocene, but also the angiosperm plants themselves. First and foremost among these changes was the development of the fruits that many modern plants use to attract primates, bats and birds so that they will disperse their seeds, carrying their potential offspring to new parts of the landscape, often many kilometres away. This strategy is particularly important in tropical forests, where around 70–94% of woody species produce fleshy fruits.[57] Fruits have been preserved in the angiosperm fossil record back to the Late Cretaceous, but, significantly, angiosperm creativity in terms of seed size and fruit types gradually increased, peaking in the early Eocene 56.0 to 48.1 million years ago, just at a time when most modern mammalian orders were arriving on the scene.[58] Some trees, like the walnut family mentioned at Corral Bluffs, even shifted from winged seeds that were good for flying, to more robust seeds that provided better packaging for being couriered by animals.[59] Although the extent of the involvement of the earliest mammals in tropical angiosperm evolution and dispersal is disputed, it remains highly likely that at least some of the earliest mammals and their relatives, like the multituberculates, had a symbiotic relationship with the plants that were sustaining ever-increasing mammalian dominance.

The period from the late Eocene (37.7 to 33.9 million years ago)/Oligocene (33.9 to 23.04 million years ago) through to the Miocene (23.04 to 5.33 million years ago) saw the beginning of significant global swings between cooling and drying and warming and wetting as the continents and increasingly active flows of water between the world's oceans and hemispheres began to assume their present-day positions.[60] The late Eocene witnessed a dawning of seasonality in planetary climate conditions and the gradual retreat of warm, wet forests back towards their current tropical positions. To some extent, this perhaps heralded a decline in the role of these environments in mammalian evolution beyond the tropics. However, these broad vegetation changes forced mammals who had, to this point, been flourishing in 'mega-thermal forests' to adapt. This process left us with some of the most iconic mammals around us today, in both tropical and temperate regions. To take the first example, horses are not just important for transport, or for their cultural significance from Euro-American films to a day at the races, but also because their evolution provides one of the clearest perspectives on how the role of 'mega-thermal' forests has changed in mammalian evolution over the last 35 million years. As palaeoecologist Professor Gina Semprebon, of Bay Path University in Massachusetts, an expert in early horse evolution, states, 'We might not recognize it today, but our hooved friends, from their appearance in the North American Eocene, through to the

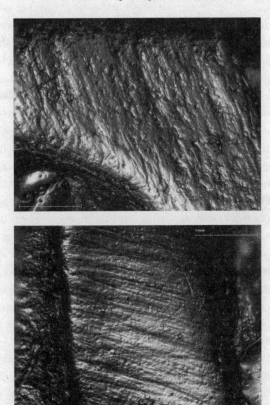

Microwear images of extinct horse teeth. Top: an image from the tooth surface of *Archaeohippus*, an early Oligocene horse. An increased number of 'pits' on the tooth surface indicate browsing on soft fruits within forests. Bottom: *Acritohippus*, a large Miocene horse. An increased number of 'scratches' on the tooth surface indicate a heavy reliance on the grazing of grasses.

present, have actually undergone radical changes in their body shape, size, and preferred diet that track global shifts in environments across much of the Earth.'

The teeth of animals, including our own, have much to say about their lifestyle and environment. For herbivores, specialized grazers can have high, ridged, 'hypsodont' teeth, which extend past the gums and provide greater resistance to gritty diets of coarse, fibre-filled material like grass. Meanwhile, browsers, such as wild goats and deer, tend to have flatter, lower-crowned, shorter teeth for crushing and grinding. Not only that, but the types of food eaten leave character-istic scratches, or 'microwear', on the tooth surface that can be observed through a microscope.[61] Gina is a specialist in using these approaches to study how the ecology of the horse has changed over time. The first horses, represented by the short-faced *Hyracotherium* and, later, *Eohippus* (or 'dawn horse') were no bigger than a relatively large dog such as a Labrador – a far cry from the long-legged, gallop-ing beasts we admire today. Horses are also generally thought of as grazers. However, the teeth of these early horses show only the faint-est traces of ridges that were to become so important later on in their evolution. Otherwise, they were generally low and flat with a rounded chewing surface – well suited for eating soft, lush leaves, fruits, nuts and plant shoots. You would not know it today, but the first horses were right at home in the dense 'mega-thermal' forests of the early to middle Eocene. Small-bodied, and focused on picking off bits of nutritious angiosperm vegetation, they made the most of expanding forest floors, from North America to Europe.

In the late Eocene and the early stages of the Oligocene (33.9 to 23.04 million years ago), things got drier and sandy grasslands began to expand more widely. The new types of horse, or 'equids', changed in response. 40 million years ago, the so-called *Mesohippus* (or 'middle horse') appeared in North America. Now that forests could not always be relied on for hiding away from growing predator threats, the longer and leggier *Mesohippus* could run faster and more widely in the expanding open areas. Still relatively small (only 60 centimetres at shoulder height), this new horse form had grinding cheek teeth with sharp crests that allowed it to break down some of the earliest gritty grasses that were emerging at this time, something also visible in Gina's microwear work.[62] The stage was now set for the rise of modern horses, and, from the Miocene to the Pleistocene appearance of the modern genus *Equus*, horses got bigger.[63] Their feet became increasingly focused on the single toe formation used by horses today, and their teeth increasingly tended towards high, ridged, 'hyp-sodont' cheek teeth that were ideal for dealing with the expanding C4

Period	~Millions of years ago			
Eocene	50	Eohippus		
Oligocene	28	Mesohippus		
Miocene	20	Merychippus		
Pliocene	5	Pliohippus		
Pleistocene	1.8	Equus		

The evolution of body size and tooth dimensions among horses from the Eocene through to the Pleistocene. A clear shift can be seen from the small *Eohippus*, with low, flat teeth and rounded chewing service adapted for consuming fruits within forest environments, to the long-legged modern *Equus* and its high teeth with sharp crests adapted for grazing.

grasslands of this time period. While horses were clearly able to switch between both leaves and grasses right up to the Pleistocene, and indeed still do on the Central Asian steppes, the overall trend towards sprinting, grassland-adapted animals is evident in the mouths of these animals. Continental land-bridges played a major role in this 'Great Transformation',[64] enabling new horse forms to repeatedly pass between America and Europe, and, later, Asia to make the most of expanding grasslands.

However, these changes did not just influence the mammal communities left outside the tropics. Many of us, all over the world, flock to zoos to see the peculiarly shaped giraffe. Appearing in the Miocene, a number of giraffids that roamed Africa and Eurasia, including the diminutive *Giraffokeryx* in India, show a pretty speedy tendency towards longer-neck evolution by 14 to 10 million years ago.[65] Landbridges between continents provided an ideal means to range widely as well as a safety net, to move to new areas should food run scarce in one region. The modern genus *Giraffa* seems to have evolved somewhere in East and South Asia before entering Africa ~7 million years ago. Although ongoing climate change then caused *Giraffa* to be extirpated from Asia, it found its long-term home in the sub-tropics and tropics of Africa where the modern-day giraffe eventually emerged 1 million years ago. The main evolutionary driver behind the unique giraffe features we know and love today is thought to be another retreat of 'mega-thermal' forests which occurred rapidly following the expansion of arid-adapted C_4 grasslands in India and Africa from 12 million years ago.[66] However, unlike other previously forest-adapted animals, such as the horses, which shifted to grasses, the giraffes stubbornly persisted in browsing whatever islands of trees were left. They developed long necks, like the dinosaurian sauropods, to access nutritious evergreen leaves in dry, deciduous woodland and grassland ecosystems in the tropics, and teeth and digestive systems that could cope with tough foliage and the toxins that some of these trees produce. One of the most key members of sweeping eastern African tropical savannah ecosystems known today formed among the chaos of the retreat of warm, wet forests to their current tropical realm.

Bats provide another example. Today, bats make up one-fifth of all living mammal species and are one of the most diverse of the mammal orders. The earliest bats of the Paleocene and Eocene evolved powered flight and their amazing echolocation senses in order to hunt the blooming insects in a warming, flower-dominated world. Yet it was only between the late Eocene and Miocene that some bats, including the group of 'mega-bats' (Pteropodidae) that still fill the skies and flock around tasty fruit trees in tropical Asia, started along an evolutionary path towards specialized fruit consumption. Changes to the skull, teeth and jaw occurred among different diversifying lineages of increasingly specialized bats that fed on first soft and then harder and more fibre-rich fruits.[67] This shift occurred at a time when 'mega-thermal' forests were contracting, leaving behind woodland and forest patches that were becoming less reliable for larger fruit-eating mammals. Flying creatures, such as birds and bats, however, were ideally placed to link up these patches.[68] The angiosperm plants themselves seem to have recognized this, and smaller seeds and fruits from the Oligocene among many tree families may be associated with the attraction of these smaller-bodied but more mobile consumers. Today, these adaptations have left fruit bats, giant or otherwise, as crucial players in seed dispersal in many tropical forests.[69]

Watching cows graze on pastures, wildebeest flow across savannahs, and whales dive among the deep blue, it is hard for us to imagine that our earliest mammalian forebears primarily lived and flourished in dense, warm, wet forest environments. From the first experiments among the dinosaurs to their rise and domination following the K-Pg global disaster, mammals were tropical creatures, living in trees, gliding between branches, eating insects or increasingly lush vegetation. The close relationship between mammals and the expanding angiosperms seems to have been a key part of their survival, as well as their spread and diversification around the planet. It may even be that the mammals, in turn, helped these new, innovative, nutritious plants to succeed. From the peak warm period at the middle Eocene, however, and despite some brief outward flourishes, 'mega-thermal' forests were on the back foot, retreating, in stages, in the face of dry, seasonally adapted grasslands and woodlands towards the equators.

Horses, giraffes and bats are just three of the many more mammal groups to have been influenced by these global environmental changes between the Eocene and Miocene, showing how the dynamism of warm wet forests, and their retreat towards their modern tropical distribution, influenced some of the more characteristic mammalian lineages we know and love today – including our own lineage, the primates.

The earliest true primates to appear in the fossil record were those of the *Strepsirrhini* 'wet-nosed' group, which includes modern lemurs and lorises. These small, arboreal mammals made the expanding lush, tropical angiosperm vegetation their own in the Paleocene and early Eocene. Diversification continued with the tarsiers (~58 to 55 million years ago) and then the simians (~40 million years ago), a group which includes all monkeys and apes.[70] As we have seen in the case of other mammals, primates found protection and opportunity in the flourishing tropical angiosperm vegetation of the Paleocene and Eocene. Like other mammals, it seems that the course of early primate evolution was characterized by teeth and body shapes that could better exploit the nutrient-rich fruits and seeds of these novel plants.[71] And like other mammals, primates were forced to contend with shrinking 'megathermal' forests from the late Eocene onwards. The simians emerged in Asia before moving to Africa by 35 million years ago, where they became smarter, bigger and more aggressive as they dealt with variable resources and threats in the fluctuating forest habitats of the Miocene. The primates are another clear example of how early mammalian evolution was tied up with the fates of tropical forests and the interlinking of land-bridges between separating continents. However, the Miocene was only the beginning. Ongoing, but shifting, interactions with the forests and woodlands in and around the equator were to set this group on an entirely new evolutionary course, one that would see them become arguably the most successful mammal that has ever existed. Us.

5. The leafy cradles of our ancestors

When highlighting the perilous state of tropical forests today, conservation agencies will often emphasize the dangers of deforestation for chimpanzees, bonobos, gorillas and orang-utans in the hope that people will sit up and take notice of the plight of our closest living relatives and their lush surroundings. Their complex social lives and ability to use tools, to show emotion, to care for their dead, and even perform sign language[1] make these non-human great apes so easy to sympathize with. From watching them play in captivity to immersing ourselves in documentaries that follow their feuding 'dynasties' we can rapidly relate to these animals with whom, in the case of chimpanzees, we share 99% of our DNA.[2] Through various iterations of the film *Planet of the Apes*, for example, we have grown used to seeing them as our probable planetary inheritors should we ever slip up. Somehow, however, there's a widespread feeling that these apes ultimately represent what we *were* rather than what we are. That they look and behave how we used to. And that they are somehow stuck in time, living examples of an age gone by. The same is most certainly true of their tropical forest homes, which we think of as so foreign to ourselves, environments that we abandoned as soon as we were able, early on in our evolutionary journey.

It is generally accepted that some time between 13 and 7 million years ago a new type of great ape began to appear in Africa – the hominins. These novel 'human-like' apes were our great great great ancestors and, ever since Charles Darwin wrote about our 'descent',[3] we have associated their evolution with the leaving behind of tropical forests and the other great ape species. According to this narrative, striding out across grasslands, our predecessors started to specialize in upright walking. Moving more efficiently over long distances, they could pursue and hunt increasingly large animals out on the African savannah. With their hands now free, they could make tools and use fire to cook meat and fuel their growing brains, as their canine teeth became smaller and started to resemble our own. Measurements of hominin fossil limbs, the geochemistry of hominin teeth, and discoveries of plants and animals found alongside hominins, have all been used to support this so-called 'savannah hypothesis'.[4] Between ~7 million years ago and the origins of our own genus *Homo* within the hominin line, ~3 to 2 million years ago, the narrative has commonly been one of decreasing climbing and use of forest plants, and increasing bipedal walking and hunting of grazing animals. This idea is so deeply embedded in academic thought that even the first hominins to move 'Out of Africa' ~2 million years ago are thought to have done so only when shifts in the Earth's climate encouraged the expansion of a grassland highway all the way from Africa to eastern Eurasia.[5]

Against the broad backdrop of Miocene – Pleistocene environmental change – namely the decline of tropical forests at the expense of dramatic C_4 grassland expansion – on the face of it, this all seems to make perfect sense. Yet the latest work in palaeoanthropology, archaeology and environmental science shows that tropical forests may have played a far more active and persistent role in the emergence and evolution of our early ancestors than has often been appreciated. The origins of the 'hominins' certainly seem to have occurred *within* tropical forests, albeit ones that were changing quickly. Our star-studded cast of ancestors were born into a world of climatic and environmental diversity, from the earliest hominins to the specialized walker 'Lucy' and, later, *Homo erectus* – the hominin with the greatest geographical range prior to our own species, reaching from the United Kingdom in the west to Indonesia in the east. Although the different members of our family tree undoubtedly became increasingly at home in savannahs, it is clear that they continued to maintain an attachment to their earlier forest origins, something that left a trace within their hands, feet, limbs, and even the chemical make-up of their bodies. In their negotiation of the diverse and fluctuating environments around them, we can see the first traces of a behavioural flexibility that was soon to reach whole new levels as we evolved and expanded across Africa and the entire world.

*

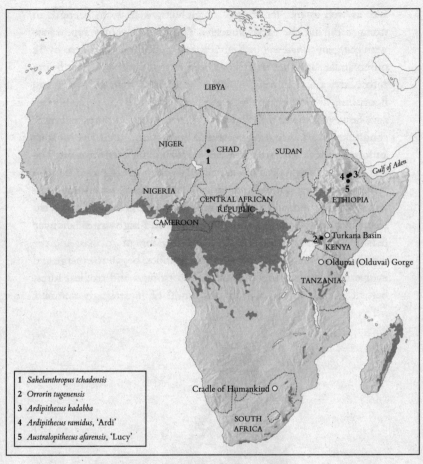

Map of major Miocene and Pliocene fossil hominins in Africa, featured in detail in Chapter 5, plotted against tropical forest distributions.[6]

During the early Miocene, 23.04 to ~15.99 million years ago, the Earth truly was the 'Planet of the Apes'. Warm global temperatures encouraged a relatively 'brief' period of forest expansion. Temperate forests inhabited Greenland while warm, wet forest environments spread across much of Africa and Eurasia.[7] Unsurprisingly, the primates took their chance, including the relatively new ape or 'Hominoid' branch of the Oligocene simians. These hominoids, including all the great apes as well as the 'lesser apes' or gibbons, went from strength to strength during the early Miocene. From a likely first appearance with the genus _Proconsul_ in Kenya ~31 to 23 million years ago, by 14 million years ago there may have been over 100 species of ape living across Africa and Eurasia, making the most of a warm planet and flourishing angiosperm fruits, leaves, and tubers.[8] By 10 million years ago, however, cooler middle to late Miocene temperatures and drier conditions had forced 'mega-thermal' forests, and their ape inhabitants, into retreat. The great apes (or 'hominids') suffered the most. The majority had disappeared or had started to disappear outside Africa by just 2 million years later. Iconic species, such as _Sivapithecus_ from Pakistan,[9] became extinct. Orang-utans were left as one of just two remaining representatives of the great apes in Asia, where they have persisted to the present day, while the ancestors of gorillas, and the common ancestor of chimpanzees and bonobos, began to emerge and evolve in the increasingly restricted sub-tropical and tropical forest belt of Africa. It was within this world of increasingly complex

Artist's reconstruction of _Sivapithecus indicus_, a species of hominoid (ape) identified in the Miocene fossil record of South Asia. This canopy-dwelling taxon appears to have gone extinct as 'mega-thermal' forests contracted during the Miocene.

'mosaic' patterns of tropical forests, drier woodlands and grasslands that the first 'hominins', and our ancestral lineage, split off from the rest and began to take on new, pioneering forms.

Palaeoanthropologists often identify fossil 'hominins' on the basis of the similarity of their teeth and skull shape to our own, and these skeletal elements are often preserved best through the ravages of time. Threatening canines used by other great apes in threat, bluff and social display reduced in size among hominins. Adaptations to 'bipedalism', or upright two-footed walking, are also considered a specific feature that might define the beginnings of our family tree, and reveal something about its changing environmental context. However, poor preservation means that scientists often have to make do with studying a mish-mash of these features when examining fossils contesting for the title of the 'earliest hominin', and where and how different hominin species moved around is frequently up for intense debate.[10] In 1994, a team of researchers trudging through the desolate, arid environment of the Awash River Basin in Ethiopia came across one of the most significant nominees for this award. Then-graduate-student Dr Yohannes Haile-Selassie, of the Cleveland Museum of Natural History, saw a partial hand bone poking out of the ancient silt. Subsequently, he and the rest of the team, including a seasoned veteran of hominin fossil discoveries in Africa, Professor Tim White of the University of California, Berkeley, went on to excavate the skull, teeth, pelvis, hands and feet of a female 'human-like' skeleton dated to 4.4 million years ago.[11] Named *Ardipithecus ramidus*, or 'Ardi' for short, this is one of the most complete pre-human hominin skeletons to have ever been found. In local Afar language, Ardi's name literally means 'ground floor' and 'root' and this is potentially exactly what she is.[12] In 2001, Yohannes also went on to discover and name an even earlier ancestor of 'Ardi', *Ardipithecus kadabba*, in the same region, dating to 5.8 to 5.5 million years ago.[13] This means that her lineage reaches back further, approaching most genetic estimates of the last division of our line and that of chimpanzees ~7 million years ago,[14] placing her right at the very beginning of hominin experimentation in Africa. However, as Yohannes tells me, 'just because "Ardi" belongs to the hominin group from which our own species eventually emerged, this did not mean that she had fully committed herself to the ground. She probably spent as much time in the trees as on the ground.'

Ardi had a small brain, one-fifth of the size of our own, and slightly smaller than that of a chimpanzee. However, the shape of her pelvis and feet mean that she was definitely much better suited for bipedalism than chimpanzees.[15] Meanwhile, Ardi's teeth and skull shape show the faintest beginnings of a trend towards those of later hominins and our own species. Yet, as Tim White argues, these features are not necessarily the most significant parts of Ardi's discovery. 'It is the environment that surrounded her that truly turns existing theories of hominin evolution on their head,' he says. Ardi's fossils were found near the skeletons of animals which show that when she was alive, tropical forest, and maybe even closed spring-fed forest, covered the now desertified region.[16] Thanks to the rare completeness of her skeleton, we can also see that Ardi had divergent big toes and long fingers that were well adapted for slow climbing among the trees. In other words, and contrary to beliefs widely held by both the public and academics, 'bipedalism' may have originated in tropical forests, not savannahs. More detailed analysis of the animals found around her and other fossil members of *Ardipithecus*, as well as geochemical analysis of Ardi's teeth,[17] have shown that they may have inhabited forest patches within mixed forest-woodland-grassland settings.[18] This would still be a far cry from popular assumptions of hominin origins out in the open.

Ardi is certainly not the only contender for the title of the 'earliest hominin' and, as is usual in palaeoanthropology, some have even questioned whether she is directly related to us at all. Another nomination comes in the form of *Sahelanthropus tchadensis*, a species largely represented by a series of jaw bones discovered in northern Chad by a French team led by Michel Brunet. Dating to ~7 million years ago, like Ardi, its teeth and skull-shape show a mixture of 'hominin-like' and other 'ape-like' characteristics, with some tentative suggestions of bipedal movement, in a mixed forest, woodland and grassy lake-edge environment.[19] A further competitor, *Orrorin tugenensis*, was found by the Franco-Anglo partnership of Brigitte Senut and Martin Pickford, in the Tugen Hills of Kenya, and dates to 6.1 to 5.7 million years ago. The femur, or upper leg, of *Orrorin* has a shape that is

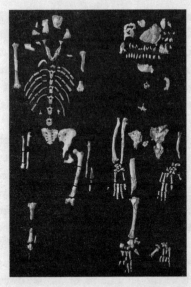

Fossil skeleton of *Ardipithecus ramidus*, showing long fingers and toes perfect for climbing slowly among trees. The fossil is curated in the National Museum of Ethiopia, Addis Ababa.

typical of later hominins, including our own genus, '*Homo*', suggesting that it was a common bipedal walker. Nonetheless, once again, it retains arm bones and fingers with shapes that are best linked to climbing among trees,[20] and evidence from the shape of its teeth, as well as other animals found nearby, suggests that it ate leaves, fruit, seeds, roots, nuts and insects within an area dominated by drying tropical forest. Amazingly, then, although all three challengers for the crown of first hominin were found in modern deserts or arid grasslands, they confirm the ongoing importance of tropical forests during the earliest stages of our ancestors' evolution, both as a safe haven away from predators and a likely source of food.

The fossil fragments left behind by some of the earliest proposed hominins in just a handful of locations ~7 to 5 million years ago mean that it will always be difficult to properly reconstruct our earliest evolution in Africa. What is clear, however, is that the early hominins of the late Miocene developed bipedalism at the same time as they were still climbing trees and using forest or woodland patches in the tropics. A forested origin for bipedalism is perhaps not so surprising when we look at long-term studies of modern non-human great ape movement. Astoundingly, the most bipedal of our living relatives are not the ground-dwelling, hunting chimpanzees. Rather, it is the orang-utans that most comfortably and most frequently use bipedalism to navigate delicate branches and retrieve fruit in forest contexts.[21] Not only that, but the fact that African non-human great apes only split from our own lineage around ~7 million years ago, combined with the fossil record approaching this split, allows us to see that instead of being 'left behind', chimpanzees have themselves evolved, changed, and survived,[22] ending up with a preference for living in tropical forests, a specialization that has only brought them to the brink of extinction because their close relative *Homo sapiens* is destroying this forest habitat.[23] This is a very different perspective indeed on the notions that chimpanzees are good proxies for models for early hominins, or that we diverged from chimpanzees simply by walking through grasslands.

Ardi was not the only, or even the first, female fossil star to rise out of Ethiopia or the Awash River Basin. In 1974, another team of palaeoanthropologists, directed by Maurice Taieb, was exploring a drier part of the same valley. Two of the expedition's members, Donald Johanson and his student Tom Gray, had spent two hours scrabbling around in the increasingly dangerous heat. Just as they were about to leave and seek shelter, Johanson noticed an arm bone poking out of the side of a dried-up river gully. As more bones emerged, nearly half a complete skeleton, the two men began to realize the significance of their find. In the subsequent celebrations that evening back at an exhausted but rejoicing camp, the Beatles' 'Lucy in the Sky with Diamonds' inspired a more personal name for the find that, until then, was known as AL 288-1.[24] The rest, as they say, is history. This skeleton, or 'Lucy', now known as *Australopithecus afarensis*, showed that by 3.2 million years ago our ancestors were, quite literally, strutting around eastern Africa. Not only was Lucy remarkably complete, but careful study of the shape and dimensions of her hip, knee and thigh bones show that, by this point in human history, hominins were now built to walk almost entirely upright along the ground of an increasingly open tropical landscape.[25]

In fact, Lucy is part of a clear, classic sequence of diverse, upright hominin bipeds from the Pliocene (5.33 to 2.58 million years ago) to the early Pleistocene (2.58 to 0.78 million years ago). She was preceded by *Australopithecus anamensis* in Ethiopia, and also Kenya, an ancestor whose limited preserved limb bones tentatively hint at frequent bipedalism. She was succeeded by the delicate-looking *Australopithecus bahrelghazali* in Chad, *Australopithecus africanus* and *Australopithecus robustus* in South Africa, and the thick-jawed *Paranthropus boisei*, or 'Nutcracker Man' in eastern Africa, which all document the spread of specialized upright hominins across the African continent.[26] At the end of this sequence came our own genus, *Homo*, perhaps as early as ~2.8 million years ago, represented later by the energy-efficient walking of, first, *Homo habilis*, and then *Homo erectus* – whose name directly hints at its upstanding nature. This trajectory was also accompanied by an increasing brain size, from around 300–500 cubic centimetres in Lucy to 1,000 cubic centimetres in *Homo erectus*.[27] Now comfortably walking, and needing to feed larger bodies and brains, hominins also began to fashion stones into tools to procure meat and marrow around 3.3 million years ago.[28] Traditionally, this gradual but relentless movement towards upright bipedalism, growing brains, the need for technology, and even larger social groups, have all been postulated as responses to the heat, openness and larger animals, including prey but also predators, that can today be found in the open, arid savannah-like settings where almost all these fossils had been discovered.

As we have already seen, however, modern environments are not always a good example of what an area looked like in the past. Fossil animals found in the same geological layers as hominin fossils provided some more direct clues based on feeding preferences, and generally supported a shift to more open habitats towards the Pleistocene, though it remained hard to work out how much of a type of food a particular hominin species ate. Up stepped my own methodological field, stable isotope analysis. Stable isotope analysis relies upon the fact that the abundant elements like carbon can have different atomic forms. These 'isotopes' (in this case ^{13}C and ^{12}C) have different masses, causing them to react differently during important biological processes such as photosynthesis. Crucially for us, when carbon dioxide is processed by the different types of photosynthesis practised by the C4 grasses that dominate sub-tropical and tropical savannahs, and the C3 herbs, shrubs, and trees that dominate sub-tropical and tropical forests, the heavier isotope, ^{13}C, is more weakly or more strongly discriminated against, respectively. This leads to different isotopic ratios in the plant tissues formed that are so clear that they can be measured, not just in the plants themselves, but also in the well-preserved teeth of the animals that eat them, including hominins.[29] First pioneered using hominin fossils in South Africa in the 1980s and 1990s by my former supervisor, Professor Julia Lee-Thorp, the now significant dataset resulting from this method, covering Ardi through to *Homo*, shows a clear shift from reliance on forest or wooded habitats to savannahs through time.[30] While this revolutionary work could not tell the difference between whether hominins ate animals feeding on C4 plants or the C4 plants themselves, alongside the shapes of hominin bones, fossil animals, and stone tool evidence, it apparently directly confirmed an increasing role for savannah in our evolution between 4 and 2 million years ago.

But before we dismiss tropical forests from this part of the human journey entirely, there are more than a few hints that hominins kept connections to their ancestral homes. Where limb bones are preserved, they show that a number of hominin fossils, including Lucy, still had arms and hands that suggest that they performed a substantial amount of climbing.[31] Even the first well-studied member of our own genus *Homo*, *Homo habilis*, had notably strong wrist and ankle bones that indicate some lasting connection to trees.[32] Moreover, within a gradual trend towards more C4 foods, the hominin stable isotope data shows significant variation. Data from *Australopithecus anamensis* indicates that it used mixed environments of forest-woodland-grassland, while study of wear on its teeth shows similarities to modern gorillas. Isotopic data from individuals of Lucy's species show that although some individuals were consuming mainly savannah foods, others lived almost entirely in woodland, or forest, habitats. Some hominins, such as *Australopithecus africanus* and *Paranthropus boisei*, even show massive variation in dietary isotopes associated with woodland/shrubland versus grassland habitats within a given year![33] In fact, the linear trend towards savannah is also broken by *Homo*, which shows a much more mixed diet than its contemporary savannah specialist *Paranthropus*. It is therefore clear that even though the world around them was broadly changing, from one dominated by forest to one dominated by more open grassland, many hominin species and individuals were more than capable of selectively using very different environments when they were available. As we will see in Chapter 6, it is possible that the hominin clade never actually truly completely left tropical forest habitats, right up until the origins of our species.

Palaeoclimatologists have also revealed complexity in the environmental context of hominin evolution when they have looked at more broad-scale changes in the presence of grassland on the African continent from 12 million years ago. Professor Sarah Feakins at the University of Southern California has developed an innovative approach to the study of highly resistant biomolecules of plant 'waxes'.[34] Studying the carbon isotope ratios of these waxes, alongside fossil pollen, that have been blown from across eastern Africa into the ocean and preserved in well-stratified off-shore marine cores from the Gulf of Aden, Sarah and her colleagues have been able to look at the trajectory of vegetation from across this portion of the continent, one unaffected by biases in discovered fossil location and damaging processes of erosion on land. What they have found has dealt a further blow to the 'savannah hypothesis'. First, by combining pollen and plant wax isotopes, they found that at 12 million years ago, C_3 *grasses* were already extensive across the region. As a result, the later expansion of C_4 grassland, from around 10 million years ago, actually replaced pre-existing grassland ecosystems as much as forest habitats. Furthermore, 'both of these processes took place *before* we see the emergence of bipedalism in the hominin fossil record', says Sarah. Secondly, the expansion of C_4 grasslands, when they arrived, was not a uni-linear, directional one.[35] It is clear from the ocean records that there are wetter and drier periods over the last 5 million years, with varying proportions of grassland ecosystems.[36]

'Ultimately, even when we "zoom out" to this continental picture provided by the marine cores, it is clear that hominin evolution in Africa was accompanied by a complex environmental patchwork, from the first hominins ~7 million years ago, through to the emergence of our genus ~3 million years ago,' concludes Sarah.

Returning to land, using methods such as Sarah's, as well as stable isotope analysis of carbonate nodules preserved in soils and the fossil teeth of herbivores, scientists have revealed similar patchworks on a diversity of regional and local scales.[37] It is clear that African geography and tectonics, such as steep rift valleys, resulted in wetter conditions and the formation of lakes in certain valleys and basins, even as the eastern portion of the continent was, overall, becoming more arid.[38] There would also have always been tree corridors along rivers and lakes among the grassland or dryland ecosystems, and forested mountains. This regional climatic variability also meant that C_4 ecosystems would have appeared and dominated at different points in time depending on where you were in eastern Africa.[39] On a more site-specific basis, a reconstruction of past environments using animal fossils and isotopic analysis of ancient soils associated with *Australopithecus afarensis* fossils at the Asa Issie locality in Ethiopia suggests this hominin was, at least in some locales, still living and dying in the same kinds of forested habitats that Ardi preferred.[40] Mammalian fossils and ancient remains of plants also show a mixture of forest, wetlands, grasslands and dry shrubland in association with a Pliocene hominin skull at Woranso-Mille in Ethiopia ~3.8 million years ago.[41] Similarly, scientists have found that some of the most important Pliocene and Pleistocene fossil and stone tool sites were actually surrounded by tropical woodland rather than grassland at the time of hominin occupation.[42] Finally, as Sarah reminds us, 'it is important to bear in mind that so-called "savannahs" refer to a range of plant types'. There are trees, and even forest patches, in savannahs today depending on local rainfall and geology, and this would certainly have been the case in the past.[43] Although the evolution of our genus was perhaps driven by more open, drier environments, it was also influenced by variability and a persistent use of tropical forest and woodland settings. Adaptations to this variability definitely stood early members of *Homo* in good stead as they went on to seek new Pleistocene horizons.

From 1.8 million years ago, discoveries of fossils in the deep, dark Atapuerca Caves of Spain[44] and the cool temperate climates of Dmanisi, Georgia,[45] as well as stone tools in rainy Norfolk in the United Kingdom,[46] show that some members of the striding, tool-making, large-brained genus *Homo* had moved 'out of Africa' to seek pastures new in northern and north-western Eurasia. Beautiful pear-shaped 'Acheulean' hand-axes, named after the site where they were first found in France, are found across Africa from 1.7 million years ago, South Asia and the Levant by 1.5 million years ago, and Europe just after 1 million years ago.[47] These remarkable pieces of stone are the breadcrumbs left behind by one of the most prolific of these early hominin adventurers, *Homo erectus*. These early hominin expansions beyond the African heartland have often been linked to grassland expansions across the Middle East, Europe and South Asia, providing new, widespread environments for the medium and large mammals that had become so important to hominin diets.[48] Nevertheless, one major outlier in this argument has always been the remarkable fossil finds from Southeast Asia which show that *Homo erectus* had made it to the Asian tropics by 1.5 million years ago in Java, Indonesia,[49] implying that perhaps these new pioneers also had more forested tastes than has commonly been assumed.

Dr Kira Westaway, of Macquarie University, Australia, is one of the leading authorities on hominin migrations into the Southeast Asian tropics. A former diving instructor, she is no stranger to working in suffocating humidity, bat-filled crevices, and long, dark, seemingly endless caves in her time investigating the environments and chronologies of the remarkable variety of different hominoids now known to have roamed this part of the world during the Pleistocene. While much attention has rightly been placed on Africa as the centre of hominin evolution, after two decades of remarkable discoveries Southeast Asia is now not far behind in terms of yielding frequent, dramatic advances in our field. Crucially, Southeast Asia provides a test case for us to determine how early members of *Homo*

Acheulean hand-axe, 500,000 to 200,000 years old, collected from the Dordogne region of south-western France. These tools have often been associated with *Homo erectus*, the first hominin to move beyond Africa.

might have adapted to, or avoided, tropical forest settings. Understandable caution when applying stable isotope methodologies to the few precious hominin fossils, an absence of relevant long-term land-based climate records when compared to Africa, and a limited understanding as to how preserved animal fossils relate to often-isolated hominin finds have all made this problem especially difficult to explore in the region, however. One approach has simply been to look at how hominins fit into wider patterns of changing plants and animals in Southeast Asia over the course of the Pleistocene. In particular, Kira and her team have focused their attentions on the fortunes of one remarkable Pleistocene giant: *Gigantopithecus*, the largest ape ever to have lived.

'Giganto', as Kira calls it, is found across China, and perhaps also parts of Southeast Asia, from 2 million years ago. Eating tropical forest seeds, fruits and bamboo, it could weigh in at a maximum of nearly 300 kilograms (around twice the size of a modern gorilla[50]), and

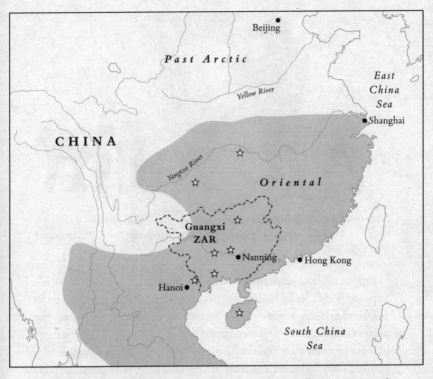

Map of known *Gigantopithecus* fossils dated to between ~2 million years ago and 300,000 years ago in southern China and Southeast Asia (stars show the site localities), with the reconstructed extent of tropical forest shaded in grey.

proved to be remarkably resilient, with its last appearance in eastern Asia occurring ~400 to 300,000 years ago.[51] 'However, something happened to this gentle colossus from around 700,000 years ago, as they were pushed back into a small part of southern China,' says Kira. The range, and presumably also the population size, of *Gigantopithecus* began to dwindle, alongside other early Pleistocene animals that favoured dense tropical forest habitats, including an early relative of the panda (*Ailuropoda wulingshanensis*).[52] This seems to have been caused by changes in sea-level and atmospheric carbon dioxide that resulted in the expansion of drier grasslands and woodlands across mainland and island Southeast Asia during the late early to middle Pleistocene

Size comparison of different hominids (great apes), from left to right: *Homo sapiens*, *Gigantopithecus*, orang-utan (genus *Pongo*), gorilla (genus *Gorilla*), chimpanzee (*Pan troglodytes*).

(~1 to 0.7 million years ago). Assisted by land-bridges across the now-exposed Sunda Shelf, which connected modern islands like Borneo to continental Asia, a whole new variety of mammals, including the ancient elephant-like *Stegodon*, spread across the region. These changes eventually took their toll on 'Giganto'. But one ape's misfortune was another's opportunity. Scientists believe that these new grassland conveyor belts, and their meaty herbivorous commuters, may have taken *Homo* right into the heart of the Asian tropics. Such a scenario would fit neatly within perceptions of an increasing hominin preference for resources of grassland habitats.

Yet, as we have already seen, Pliocene-Pleistocene hominin habitat preferences are a little more complex than such models allow. Where actual fossil plants and animals are found alongside *Homo erectus* in Southeast Asia, including pygmy hippos, turtles, deer, wild pigs and different types of *Stegodon*, they show that it was likely living within more mixed lake-edge marsh, grassland and woodland environments, with some tropical forest fragments.[53] Complex environmental mosaics hosted another star of Pleistocene Southeast Asia, the 'Hobbit'. Following discovery at Liang Bua Cave in 2003, by a team Kira was also part of, these diminutive hominins from the island of Flores, Indonesia, quickly became even more famous than their fictional namesakes from the Shire. Dating to 190–60,000 years old,[54] these tiny hominins, growing up to a maximum of about 1 metre in height, seem to descend from a small-bodied group of hominins that made it to the island 700,000 years ago and may, in turn, have been related to *Homo erectus*.[55] The survival of this unique hominin, right up to the arrival of our own species in the region, has been considered remarkable not only because of the potential for interaction, but also because it might tell us something about their island environments. Small stature is often linked to habitation of dense tropical rainforests, and Kira and her team were able to show that the occupation of Liang Bua by the 'Hobbits' was most intensive during wet, more forested periods.[56] However, it looks as though this intensification represented these hominins seeking

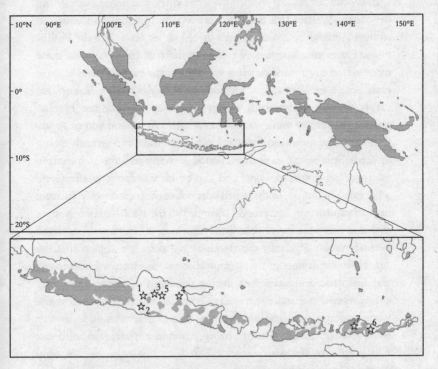

Map of major Pleistocene hominin sites in island Southeast Asia discussed in Chapter 5 against tropical forest distributions.[57] 1. Sangiran, 2. Ngandong, 3. Bapang and Trinil, 4. Mojokerto, 5. Kedung Brubus, 6. Mata Menge, 7. Liang Bua.

respite from the rain. In fact, once again, fossil plants and animals left behind by the foraging and hunting of our diminutive cousins, including miniature *Stegodon*, Komodo dragons and giant rats, indicate a preference for foraging in drier, mixed tropical woodland and grassland habitats.[58] The change in size therefore seems to be down to the more general well-documented process of 'island dwarfism',[59] linked to the range of the population being strictly limited on an isolated location, rather than to a more specific dense rainforest specialization.

Even though things were certainly more complex than sweeping 'savannahs', it is clear that an opening of environments, and expansion of drier woodland and grassland habitats at the expense of denser rainforest, aided some of the earliest hominin arrivals into mainland and island Southeast Asia. It enabled them to expand and diversify, reaching beyond the Wallace Line into island settings that remained unconnected to the mainland even during periods of low sea level, like the Philippines, Flores and Sulawesi. We can see the true importance of these more mixed environments when we consider the eventual fates of these early hominins. Kira and colleagues have now shown that *Homo erectus* survived in the region until ~100,000 years ago,[60] while the 'Hobbit' persisted on Flores until ~60,000 years ago.[61] At around 100 to 80,000 years ago, communities of tropical rainforest plants and animals sprang up across Southeast Asia during a period of wetter conditions. Including tapirs, orang-utans, gibbons, and sun-bears, this forest is effectively what exists in humid, undisturbed portions of the region today. In 2020, my colleague Julien Louys and I compiled all the available stable isotope data that exists for Pleistocene fossil mammals in Southeast Asia in the journal *Nature*. Studying the trends in the data, we argued that the aggressive expansion of these tropical forest ecosystems across island and mainland Southeast Asia during the late Pleistocene apparently overwhelmed our more 'generalist' hominin relatives, as well as the woodland and grassland animals that had walked alongside them.[62] Indeed, while the list of our Pleistocene hominin relatives in Southeast Asia keeps growing with remarkable finds on a near-yearly basis, only one hominin actually survived this major climatic and environmental swing back to extensive tropical canopies.

*

Since the Victorian era, savannah landscapes have been the crucial backdrop for the grand story of human evolution. All the while, tropical forests remained as out-dated backwaters, unattractive for increasingly upright, innovative, hunting hominins and better associated with non-human great apes. New research, however, has shown that tropical forests provided the vibrant cradle for the earliest great apes and, later, the earliest hominins, experimenting at a time of increasing environmental change in the late Miocene through to the Pleistocene. Hominin bipedalism itself may have evolved within and below the trees rather than out in exposed, sweltering grasslands. Certainly, drying and opening landscapes stimulated hominin diversity, specialization in bipedalism, tool-making, and, eventually, the expansion of our earliest ancestors beyond Africa and across much of Europe, the Middle East and Asia by the middle Pleistocene. But this was not a one-way street. Many hominins kept mementos of their forested beginnings, both in their skeletons and their chemical composition, often using nutritious forest or woodland patches within complex savannahs – not the uniform, sweeping plains we are used to seeing on TV today. Similarly, when some of the earlier members of our own genus ventured into the Asian tropics, it was mixtures of forest, dry woodland and moist lakeside grasslands that supported their populations, large and small. Monotonous grasslands were likely avoided by these populations just as much as dense, stifling tropical lowland rainforests, with trees continuing to provide a source of food and refuge for our ancestors.

In fact, it was climatic and environmental *variation* that framed the evolution and differentiation of our hominin ancestors.[63] Efficient bipedalism, increased brain size and social groups, and stone tool development were all key adaptations to a rapidly *changing* world, rather than the product of monotonous existence. The Miocene–Pleistocene saw not only grassland expansion but also increasingly dramatic swings in global climate states, driven on by the powerful Milankovitch cycles discussed in Chapter 2. Our ancestors adjusted to this unpredictability by adapting to the plentiful meat opportunities drying savannahs had to offer. But all the time they kept their eyes on profitable forest habitats and woodland edges – hedging their bets. This flexibility, however, had its limits. The late middle and late Pleistocene brought with them more 'extremes' in both climates and rapidly diversifying environments. In tropical Southeast Asia, as the late Pleistocene brought with it an expansion of dense tropical rainforests, its resident hominin generalists were faced with a thick, expanding green wall. One they could not ultimately overcome. Just prior to this a new hominin emerged into this increasingly shifting brave new world in Africa. It went on to march out across almost all the planet's continents, which, by now, were fully formed. It was, of course, our own species, *Homo sapiens*.

6. On the tropical origins of our species

The tendrils of social media, the widespread availability of mobile phones, the cult of 'celebrity' and reality TV, and our busy work and social lives, all mean that we grow up in a very human-focused world. Our widespread geographical range, from the ice of Greenland to the mountains of the Himalaya, from the bustling metropolis of Beijing to desert-fringed Dubai, leads us to simply assume that *Homo sapiens* is one of the most successful species to have ever walked the Earth. Not only that, but our huge global population, extensive cities and dominating impacts on the natural world seem to give us a certain planetary 'weight'. Yet we are relative newcomers to life on Earth, especially when compared to the ~300-million-year journey of tropical forests that we have been following. The first fossil skeletons with distinctive 'human' (in this book I reserve the word 'human' only for our own species, *Homo sapiens*) features, including a shorter, flatter face, round skull, and large brain (over three times bigger than 'Lucy's'), appear in Africa between 300,000 and 200,000 years ago.[1] This fits with estimates based on 'molecular clocks', akin to those used for plants and mammals in Chapters 1 and 4, respectively, calculated using modern and ancient human DNA.[2] Whichever way you look at it, this makes us practical new-borns in an evolutionary context. Indeed, two extinct hominins we commonly think of as having eventually 'failed', *Homo erectus* and *Homo neanderthalensis* (or the Neanderthals), both actually lasted longer on the Earth's surface before their demise than we have to date.

It is maybe not too surprising, then, that palaeoanthropologists and archaeologists have seen *Homo sapiens* as simply the most recent point in a longer-term evolutionary trajectory of hominins and the genus *Homo*, towards open, savannah contexts and away from supposedly protein-poor tropical forests. In fact, anthropologists and archaeologists alike have often thought of tropical forests as 'barriers' to Pleistocene human presence.[3] Small protein-poor and difficult-to-catch prey, poisonous plants, a lack of dense carbohydrate resources, tropical diseases and rampant humidity, let alone poor preservation of archaeological sites thanks to high water flow and acidic soils, have meant that these so-called 'green hells'[4] have often been neglected in the story of human dispersals. Between 300,000 and 60,000 years ago, the archaeological record begins to show evidence for new technologies (the bow and arrow[5]), items made for exchange to maintain social networks,[6] and personal ornaments and artwork including engravings and shell bead necklaces[7] – all of which have been considered uniquely human capacities. Yet these innovations have been linked to the need for more efficient projectile hunting and the maintenance of useful relationships between groups living in open grassland settings, prone to drying in the face of ever-harsher swings between glacial and interglacial conditions, as well as shorter-term swings in climate systems, in the late Pleistocene (Chapter 2).[8] Alternatively, they are seen as the product of growing human populations thriving on reliable marine resources in coastal settings.[9] Like *Homo erectus*, the eventual dispersal of *Homo sapiens* into the Middle East, Eurasia and throughout Southeast Asia has also been associated with climatic periods when more challenging environments like deserts or tropical forests retreated and rich 'savannah-like' hunting grounds expanded.[10] Either that, or it has been suggested that humans tracked rich high-protein resources from the oceans, all the way from eastern Africa, around the Indian Ocean rim,[11] to Australia by 65,000 to 45,000 years ago,[12] and the Americas by ~30,000 to 25,000 years ago,[13] as part of a series of 'coastal highways'.[14] While undoubtedly new, in many ways, *Homo sapiens* has thus traditionally been considered as equally dependent on particular – non-forested – environments as its hominin forebears and contemporaries that we met in Chapter 5.

These narratives rely on a single, homogeneous origin, and a single set of either grassland or oceanic desires for humans. But one of the things that has always drawn me to studying the origins of our species is that, while we are a relatively recent evolutionary arrival, we are also the only hominin that went on to occupy *all* the planet's continents and to use *all* the world's land-based environments. Much of this seems to have happened very quickly, during the late Pleistocene (125,000 to 11,700 years ago), and seems to call into doubt assumptions that humans stuck to a single, easy or familiar 'route'. In this chapter we will see how the latest research in archaeology, palaeoanthropology, genetics and environmental science is building a more complex picture of *Homo sapiens'* entrance on to the world stage, one that covers a number of different parts of Africa, including the tropical forests of West, Central and eastern Africa. Not only that, but we will meet daring archaeologists and palaeoanthropologists who are beginning to explore tropical parts of the world that have often been ignored as unprofitable 'blanks on the map'. Their work is demonstrating that, as we moved beyond Africa, *Homo sapiens* repeatedly and persistently occupied some of the most extreme sets of environments the planet had to offer. Certainly, our species was a continuation of a long period of hominin adaptation to variable Pleistocene environments that we met in Chapter 5. However, it is in settings such as tropical forests, rather than sweeping savannahs or uniform coastlines, that we can see how *Homo sapiens* used its new cultural, technological and social repertoire to become the most flexible hominin yet. One that could deal with all the climatic fluctuations, between cooling and warming and drying and wetting, that the late Pleistocene world could throw at it, and one that grew into the relentless, global force we remain today.

The traditional view of human origins is that we evolved at one time, in one place, and in one population in Africa. In the 1980s, genetic variation in modern human mitochondria, little organelles within our cells that can only be passed down through the maternal line, was used to develop a 'family tree' of human populations, linking all living people back to a single population or even just one female ancestor – 'Mitochondrial Eve' – that lived in Africa between 200,000 and 140,000 years ago.[15] Although more recent studies, using powerful modern statistics and genetic sequencers, now analyse genetic variation across the entire human genome to produce more sophisticated models, they retain the same basic principles. Namely, that the variability in the DNA of modern human populations can be inserted into a simple 'tree-like' scheme that points back towards a single point in time and space when our species emerged and began to diversify as it dispersed around Africa and around the world. In a number of recent cases, these studies tend to zero in on KhoeSan populations who have the most diverse human genomes seen anywhere on the planet today, suggesting that populations of *Homo sapiens* first emerged and started to diversify in the southern portion of the continent 300,000 to 150,000 years ago.[16] In recent years, however, it has become clear that genetic models of our species' origins need to become exponentially more complex. We now know that *Homo sapiens* interbred with Neanderthals,[17] a second hominin group, called Denisovans (named after the cave where they were found in Siberia, which itself was named after a hermit called 'Denis' who once lived there),[18] and perhaps even a third hominin. Not only that, but they produced fertile offspring, complicating traditional determinations of a 'species' and leaving their mark on the genetic make-up of surviving members of our own species.[19] Furthermore, there is only one Pleistocene fossil individual in Africa that has produced ancient DNA,[20] since generally hot conditions tend to break down ancient organic molecules. Justifiably, museum curators, archaeologists and palaeoanthropologists are also inclined to protect limited fossils from destructive analyses,[21] so they have approached this application cautiously. As modern DNA samples cannot reliably inform us about the geography of ancient populations, given long histories of human migrations between regions and interactions between different human populations, without ancient DNA, it is almost impossible to use modern DNA to 'read back' exact past locations or timings of human origins with any level of accuracy.

Palaeoanthropologists and archaeologists have therefore tried to get their hands on something a little more tangible than DNA to mark the date and place of human emergence: fossils. As with DNA analyses, there has been an obsession with finding the 'oldest' example of something in order to crown a new region as ruler in the story of human origins. The oldest accepted fossils with distinctly 'human' features have for a long time been those found, like 'Ardi' and 'Lucy', in the drylands of Ethiopia, this time at the sites of Omo Kibish and

Herto.[22] Dating to 195,000 and 160,000 years old, respectively, these skeletons, to all intents and purposes, look morphologically 'human'. Animal bones found in the same place with known ecological habits, climate records and geological studies suggest that these early humans were living in dry, grassland environments, apparently confirming our origins in the eastern African savannah.[23] However, relatively complete hominin fossils, including those of our species, are always rare, scattered across the continent in different localities where chance (e.g. transport of remains into a sheltered cave system) or specific conditions (e.g. layers produced by volcanic eruptions or flooding) have shielded them from the ravages of time. As a result, some archaeologists have suggested it may instead be better to try to find more commonly preserved material features like art, symbols and complex stone tools, to determine the 'behavioural' appearance of humans. As mentioned earlier in the chapter, such evidence has tended to come from coastal settings in northern and southern Africa dating to around 100,000 to 70,000 years ago,[24] suggesting a marine focus, although more recent work in Kenya hints at the appearance of

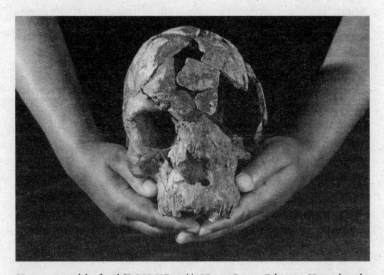

Homo sapiens idaltu fossil (BOU-VP-16/1), Herto, Bouri, Ethiopia. Housed in the National Museum of Ethiopia, Addis Ababa.

sophisticated human exchange patterns and technology 300,000 years ago, in a dry grassland setting.[25] In both cases, research has focused on trying to find the first skeletons or novel material remains (e.g. specific stone tool technologies, projectile technologies, symbolic beads and pigment use[26]) left behind by our ancestors. In doing so, most attention has thus centred on the long hominin-associated savannahs or on coastlines where caves that have eroded into cliff faces provide their own perfect micro-climates, sheltered from the elements, preserving traces of the past.

That is, until Dr Eleanor Scerri and her colleagues wrote a revolutionary paper in a leading scientific journal, *Trends in Ecology and Evolution*, in 2018. Although other scientists had already begun to dissent from the mainstream,[27] this paper challenged the simple traditional view that *Homo sapiens* appeared in one place at one time.[28] Eleanor heads the 'Pan African Evolution' Research Group at my own Institute in Jena and it does exactly what it says on the tin. 'We believe that between 500,000 and 200,000 years ago, the key timeframe for the emergence of our species, our ancestors lived in distinct, but interconnected, populations *across* Africa,' says Eleanor. In other words, we did not appear at a single point in time or in a single group. Instead, millennia of partial contact between populations led to a huge raft of human diversity. The dynamic mixing of these groups was to eventually lead to our species as we know it. 'To study this properly, we therefore need to look at *all* of Africa, not just a single mythical point of beginning,' continues Eleanor. In fact, casting the net a little wider we can start to see a number of hominin fossils which, like those from Omo and Herto in Ethiopia, have a diverse mixture of both clear '*Homo sapiens*' (e.g. flat face and rounded skull) and more 'archaic' (e.g. thicker brow ridges) traits, including those from Jebel Irhoud in Morocco, North Africa (~300,000 years ago) and from Florisbad in South Africa (~260,000 years ago). The 'full suite' of what we often consider to be distinctively *Homo sapiens* features actually only really appears to take shape between 100,000 to 40,000 years ago, though this may be a product of a lack of preserved fossils from this and the preceding time interval.[29] Significantly, in tropical West Africa, the earliest known human fossil dates to just

16,000 to 12,000 years ago at Iwo Eleru, Nigeria, and shows just how long considerable morphological diversity was maintained between different groups of *Homo sapiens* in different parts of the continent.[30] A re-examination of the existing genetic evidence, in light of the fact that genetic exchange probably occurred between different hominin lineages and that modern DNA gives very little insight into geographic patterning in the past,[31] seems to support this 'African multi-regionalism' proposed for the evolution of *Homo sapiens* by Eleanor and her colleagues.

Rather than a branching linear tree with a simple, single trunk, we therefore have fractured African roots. The same picture emerged when Eleanor and her colleagues looked at what they call a 'patchwork' of cultural materials left behind by *Homo sapiens* populations across Africa in the middle and late Pleistocene. Stone tools, thanks to the materials from which they are made, are reliably preserved over deep timescales. Not only that, but archaeologists can look at their shape and how they were made to infer things about behaviour and technology. Archaeologists have suggested that the appearance of our own species heralded a proliferation of new forms of stone tools that could be inserted into complex projectile set-ups and that required more sophisticated, thought-out manufacture. These so-called 'Middle Stone Age' toolkits have been found alongside human-like fossils at Jebel Irhoud and Florisbad, as well as at Olorgesailie in Kenya. Importantly, the Middle Stone Age toolkit varied by region.[32] In North Africa, the rapid expansion of grasslands under wetter climates led to growing populations of *Homo sapiens* and the widespread appearance of unique, complex, arrow-shaped (or 'tanged') stone tools ~120,000 to 80,000 years ago, alongside bone tools and shell beads, suggestive of complex cultural behaviours. In eastern Africa, by contrast, there is huge variation and continuity in Middle Stone Age stone tools from 300,000 to 60,000 years ago. The eventual transitions from these Middle Stone Age forms to diminutive 'microlith' stone tools, often linked to well-developed bow and arrow technology and hunting efficiency, between 80,000 and 40,000 years ago also occurs in highly diverse ways in different parts of Africa.[33] Sometimes there is complete, rapid replacement, other times there is a gradual change.

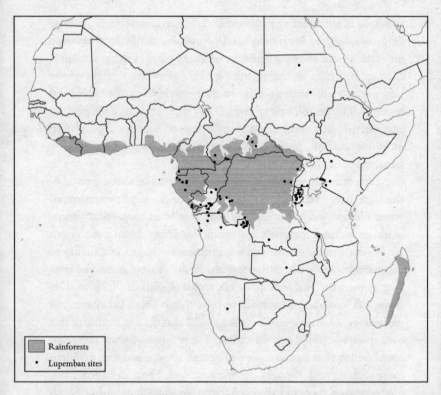

The distribution of all published 'Lupemban' sites, based on Taylor[34] (2016), plotted against the contemporary extent of Central African rainforest ecosystems.

The appearance and disappearance of 'human' productions like art and social display are equally regionally varied, and often come and go in the Pleistocene archaeological record right up until the Holocene (11,700 years ago to the present). It appears then, in contrast to the traditional story of our arrival on Earth, that there was no dramatic, consistent fossil or 'material' arrival of *Homo sapiens* at a particular place or time, nor was there a uniform, linear progression towards 'behavioural complexity'.

Things get even more interesting when we look at what Eleanor and her team think drove the dynamic separation and integration of these different, evolving human populations that gave rise to all of us, wherever we live around the world. The character and chronology of the fossils, stone tools and different material artefacts left behind by early *Homo sapiens* suggest that our species eventually appeared thanks to interaction between these diverse groups living in different parts of Africa. Eleanor and colleagues believe that it was undoubtedly the varied environmental settings of Pleistocene Africa, not just savannahs or coasts, that, at different points in time, could either keep these populations separate or bring them together. In fact, once we move away from a presumption of a uniform, single origin and environmental association, tropical forests begin to take a much more prominent role in this early, more complex evolutionary model. Indeed, as well as the fascinating Iwo Eleru fossil evidence for morphological diversity mentioned above, some of the most unique of the Middle Stone Age regional diversity comes from the modern-day tropical forests of West and Central Africa: large, heavy-duty pick-like forms, huge scrapers, and some pointed 'lance-like' points have been variously associated with digging for carbohydrate-rich tropical tuber plants, like yams, as well as hunting tropical forest animals.[35] Many of the sites where these tools have been found dating from between ~300,000 and 40,000 years ago are today surrounded by dense tropical evergreen rainforest or drier tropical forests and woodlands. There are hints from records and ancient plants and sediments in marine cores off the coast of West Africa that tropical forests also covered these same regions in the past.[36] More work is, however, necessary to confirm the ecologies of the populations that made these tools and the plants and animals on which they relied for survival.[37]

But while the role of dense tropical rainforests in early *Homo sapiens'* evolution in Africa awaits definitive proof, there is growing evidence that expanding and contracting tropical forests may have shaped past human genetic and cultural diversity. Studying the DNA of different modern tropical forest hunter-gatherers in Central Africa, and comparing them to neighbouring populations, suggests that they diverged from each other at least 70,000 years ago.[38] Not only that, but genetic differences found between different forest hunter-gatherer groups implies that they, too, diversified from each other, as a result of the waxing and waning of forest habitats.[39] At the cave site of Panga ya Saidi on the tropical coast of Kenya, where I was lucky enough to work on recovered archaeological animal materials, we found evidence that a resistant mixture of moist tropical forests, woodlands and grasslands buffered human populations against major climate change seen in the dry, grassland interior, allowing them to gradually experiment with their stone tool technology and symbolic material culture.[40] Indeed, despite being just 15 kilometres away from the present coastline, there was no prominent evidence for the use of coastal resources for subsistence during the Pleistocene. Elsewhere, similar mixed environments promoted the movement of growing human populations in and around the African sub-tropics.[41] Tropical forests were certainly not the only environment at play. Nonetheless, they were undoubtedly a key part of the complex 'Pan-African' skeletal, genetic and behavioural appearance of our species.

'Hurry up, Pat, it is about to rain!' my former PhD student, Dr Oshan Wedage of the University of Jayewardenepura, yelled as I staggered behind him through the dense, lowland, evergreen rainforest of south-western Sri Lanka. Drenched in sweat and covered in especially cunning leeches, which drop from the treetops as they feel your vibrations, I tried to pick up my pace. Not fast enough. The heavens opened and we were now trudging uphill through the mud, undergrowth and fast-running streams. The rest of the animal life was too smart to be out in this weather, giving this challenging habitat an even more 'barren' appearance. At that moment it was very easy to see how many leading archaeologists and anthropologists, and indeed the general public, might still think that the first human populations moving out of Africa would have actively avoided tropical forests. In fact, until the beginning of the twenty-first century, the earliest evidence for definite human occupation of a tropical rainforest environment was widely considered to be limited to the very final Pleistocene (~18,000 to 11,700 years ago) or early Holocene (11,700 to 8,200 years ago), when further technological innovations may have made dealing with these challenging conditions and food sources more palatable.[42] But then from up ahead Oshan shouted, 'We're here!' Out of the mist and water came a view of one of a cluster of cave and rockshelter sites in this part of the world that has been drastically challenging this anti-tropical forest bias.

Sri Lanka is an island at the southern tip of South Asia, just below the Indian subcontinent and just above the equator. Sitting in the middle of the Indian Ocean, it is at a key location in models that assume a rapid 'southern route' for humans from Africa all the way to Australia in the late Pleistocene. This express-way has been variously thought of as being driven by rich coastal resources, meaning that humans did not have to come inland to face more challenging environments, or by the periodic expansion of more familiar grasslands. But since Sri Lankan archaeologists, like Oshan and his earlier mentor Dr Siran Deraniyagala, transformed Pleistocene archaeology on the island, it has posed something of a problem for these blanket models. The site we came across at the end of our trek was Batadomba-lena which, alongside the nearby sites of Fa-Hien Lena and Kitulgala Beli-lena, remarkably preserves the oldest human fossils anywhere in South Asia and some of the oldest material culture that can be confidently linked with our species in the region, dating to 45,000 years ago.[43] As Oshan says, 'Although people have often ignored our "Resplendent Isle", it plays a major role in perceptions of adaptations of *Homo sapiens* as it moved beyond Africa.' Instead of on the coast or in the drier, grassland areas of Sri Lanka, the island's earliest, clear human sites are found in inland, wet tropical rainforests.

Sure, you might be thinking, but these rainforests might not have been there in the past? It is also just an island, so maybe humans moved around a lot, only briefly visiting the forests? It was to answer these questions that I first travelled to Sri Lanka during my PhD. I instantly fell in love with the country and its people and have rarely spent more than a year away ever since. Back in 2015, working closely with archaeologists like Oshan, we applied, for the first time, the stable isotope method I introduced in Chapter 5 to the rich record of human fossil teeth that had emerged from Sri Lanka's rainforest caves and rockshelters. Comparing our results to those from animal teeth found alongside the humans, we were able to show that, unlike the African hominins, from 36,000 years ago to 3,000 years ago, almost all the individuals living at these sites had isotope values that over-lapped with values of animals known to be inhabiting dense canopy rainforests, suggesting that they were fully reliant on tropical rain-forest resources. This dependence continued through periods of major climatic change, like the Last Glacial Maximum, and even beyond the arrival of agriculture in this part of the island 3,000 years ago. Our work proved that early expanding human populations *could* live completely in tropical rainforests.[44] Not just that, but they chose to, even when grassland and coastal settings were available nearby.[45] This specialized adaptation was also remarkably resilient, lasting over 30,000 years and perhaps even longer. Tropical Sri Lanka was revolutionizing our understanding of how humans spread so quickly around the world. Unlike many other hominins they clearly did not always seek out more open, mixed settings, and they certainly did not do everything the 'easy way'.

The next question, then, was what did these humans eat? And how did they get it? Fortunately, unlike many tropical parts of the world, Sri Lanka had some more gifts to offer global archaeology, producing long records of rich organic remains thanks to the cooler conditions inside the caves. Oshan's recent work, alongside a number of inter-national partners, has studied the evidence for the technology, plants and animals found in these archaeological sites alongside our special-ized ancestors in this part of the world. From the first human arrival in these rainforests 45,000 years ago, these groups were hunting small, fast rainforest animals. Amazingly, monkeys, as well as giant scamp-ering squirrels, with characteristic 'cut' and burning marks indicative of human processing, make up around 70% of all mammals found at the sites from 45,000 to 3,000 years ago.[46] Humans were also getting plenty of carbohydrate from the starchy nuts of the canarium tree, and maybe even wild banana and breadfruit, while snails living in the nearby, well-fed streams supplemented their diets. Oshan believes that this evidence shows that 'if used the right way and with the right knowledge, tropical forests can be just as rich, if not richer, than dry grassland environments'. In fact, prior to the British colonial period, some groups of the Wanniya-laeto Indigenous population of Sri Lanka still hunted wild forest game, fished, climbed to get honey, and obtained rich sugary plant resources in both the wet and dry tropical forests of the island. These were 'Forests of Plenty',[47] not barren 'green deserts'. Particularly for the highly flexible human populations that made them their home.

The tools left behind by these early human groups in Sri Lanka have also provided a new environmental perspective on some key technologies and innovations associated with *Homo sapiens*. Bows and arrows, and more efficient stone tools, have also often been linked to African grasslands or European tundra. But in Sri Lanka, humans were recycling monkey bones to make arrow tips, while tiny 'microlith' stone tools may also have been attached to projectile shafts. Similarly, clothing has often been thought of as a clever human adaptation to cold northern European or Siberian temperatures. Again, however, there is evidence for humans working animal skins, perhaps for protection against mosquitoes and leeches (!), 45,000 years ago in humid Sri Lanka.[48] Ochre and shell beads, and evidence for long-distance exchange with some human populations on the coast, all show a very 'human' repertoire of social connection, deep in the jungles of this South Asian island. The amazing finds made in the tropical forests of Sri Lanka encourage a more diverse exploration of human dispersals during the late Pleistocene. They show that we can no longer stick to searches on coasts or temperate and sub-tropical grasslands. Instead, as in Africa, we should turn our attention to little-investigated, extreme regions to reveal what it truly means to be human.

Moving away from Sri Lanka, the last decade or so of archaeological and anthropological research has begun to show the sheer diversity of tropical forests that were exploited and occupied by trailblazing groups of humans in late Pleistocene East and Southeast Asia, Near Oceania, and South America. Southern China is emerging as a potential hotspot for the earliest human fossils outside of Africa. Human teeth found at two sites with ages clustering around 130,000 to 100,000 years ago are found with animals that indicate a mixture of rainforest, mixed woodland, bamboo forest and grassland habitats, although their true ages remain hotly debated.[49] Turning further south and east, some early human teeth on Java and Sumatra ~100,000 to 70,000

years ago have been linked to the expansion of late Pleistocene rainforests in the region, though dating these teeth and determining their
relationship to animals found in the same caves is challenging at best,
given that these are not archaeological sites but rather collections of
fossils that are the product of past flood events which washed them
together, *post mortem*, into the locality.[50] One of the true goldmines of
Pleistocene human tropical forest adaptations is the Niah Caves on
Borneo. Here, 50,000 to 45,000 years ago, human fossils are clearly
linked to the hunting of wild forest boar and primates and the processing of toxic plant foods all within a unique mixture of Southeast
Asian rainforest, swamp forest and grassland patches that may have
been maintained by human hands.[51]

Unlike Sri Lanka, many early human sites in mainland and island
Southeast Asia are linked to a mixture of growing evergreen rainforest
and drier woodland habitats. Moving out to sea, into the Pacific, and
beyond the famous 'Wallace Line', the situation becomes more complex. Here, ground-breaking archaeologist Professor Sue O'Connor of
Australian National University has been undertaking excavations with
Indonesian partners for over two decades now. The islands here are
generally considered impoverished in terms of animal resources. Never
connected to the Southeast Asian mainland, most of these islands were
home to no animals bigger than a terrier during the late Pleistocene,
presenting a completely new tropical setting for humans to deal with.
In fact, Sue's work on small, isolated islands like Timor and Alor,
among others, has suggested that the earliest human groups practised
sophisticated seafaring adaptations. 'At the site of Asitau Kuru (formerly Jerimalai) on Timor we have found one of the earliest fishhooks
in the world dated to about 20,000 years ago and in the same site we
have evidence for the presence of tuna fish that usually live in open-sea
environments dating back to ~45,000 years ago, suggesting that older
fish hooks remain to be found,' Sue states. In this case, then, it seems
like a reliance on the ocean may have rapidly propelled some of the
earliest human groups through this part of the Pacific,[52] arriving in
Australia some time between 65,000 and 40,000 years ago.

Together, Sue and I performed the same stable isotope analysis on
human teeth from these Wallacean sites, the first time this method

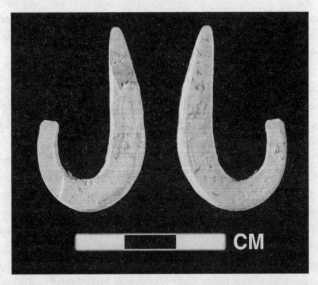

Fishhook made from marine shell preserved at the Pleistocene site of Lene Hara on Timor, Timor-Leste. Researchers believe these same technologies would have been used by humans to capture deep sea tuna from the earliest occupation of this part of the world.

had been used on Pleistocene hominin teeth anywhere in Southeast Asia. The results initially confirmed what she had expected. The earliest human tooth we had available, dating to ~45,000 to 39,000 years ago, showed an isotope value that we would expect from someone eating fish and molluscs from the sea.[53] However, from here things got more complex. During the dry conditions of the Last Glacial Maximum, explored in Chapter 2, tropical rainforests retreated in the face of more open, dry forest and grasslands. A return to wetter conditions in the terminal Pleistocene and Holocene saw them rapidly expand. It is at this point that growing human populations seem to have shifted their attention inland. From around 20,000 years ago, the stable isotope data shows that humans were increasingly reliant on these supposedly 'impoverished' forest environments.[54] What were they eating? Giant rats appear to be the answer! At around 5 kilograms, these rodents weighed about the same as an average house cat, and while to us they might not seem too tasty, they would have provided valuable protein. It is also highly likely that poor preservation conditions mean we are missing the large amounts of soft fleshy fruits and starch-rich tubers (the underground storage organs like we are used to devouring from sweet potatoes) from plants that humans must have used as they sought nutrition beyond the sea. Once again, *Homo sapiens*, albeit slightly belatedly, found a way to adapt to even the most challenging and isolated of tropical forest environments.

The pattern continues into the frosty mountains of New Guinea. Here, over 2,000 metres above sea level, humans were met by montane tropical forests that can receive snow in winter. Leading experts on the region like Chris Gosden have said that during periods like the Last Glacial Maximum they would have been particularly 'cold, difficult, and unpromising'[55] as temperatures fell by as much as 5°C. Nevertheless, from 50,000 to 45,000 years ago, a number of sites, particularly within the Ivane Valley of Papua New Guinea, show that early groups of humans occupied the region. Glenn Summerhayes and his team have faced death-defying flights in small planes to try to find the sites, deep in the heart of New Guinea, that allow them to discover how these Pleistocene populations survived under difficult conditions. Using a microscope to explore the soils left behind at these archaeological sites, they have found fossil starch grains left behind by carbohydrate-rich yams.[56] Charred, protein-rich pandanus nuts, apparently cooked by humans on a fire, have also been found at the sites, while the characteristic stone axes found in this part of the New Guinea highlands have even been associated with the deliberate modification of forests by humans. Alongside the likely pursuit of marsupials in more open settings, humans were able to refine their strategies to survive in these lofty environments.

Tentative linkage of increasing frequency of charcoal to human arrival and burning of the landscape, as well as archaeological evidence for the occupation of cave sites, suggest that humans had reached the dry and wet sclerophyll (a type of vegetation with hard, densely packed leaves along stems that famously includes eucalypt trees) subtropical and tropical forests of northeastern Australia by around 45,000 to 35,000 years ago,[57] though the wet rainforests of this region seem to have only been intensively occupied from around 8,000 years ago.[58] After the Antipodes, the tropics of Central and South America were the last to be met by dispersing humans during the Pleistocene. Arriving in the Americas ~30,000 to 20,000 years ago,[59] human populations seemingly rapidly dispersed across these regions by around 18,000 to 14,000 years ago, meeting the vast array of environments this part of the world has to offer. In Central America, terminal Pleistocene communities apparently focused on lowland, seasonal tropical forests and dry, cool highland forests. At the famous site of Monte Verde in Chile, humans made use of the diverse dense, temperate rainforest and open, dry forests of Patagonia.[60] By around 13,000 years ago, *Homo sapiens* had also made it to the rainy, humid evergreen rainforests of the Amazon Basin.[61] Finally, humans reached the supposedly inhospitable montane forests of the high-altitude Andes by 11,500 years ago, adjusting to the montane forests and low oxygen of heights over 4,500 metres above sea level in Peru that has left its mark on the genetic make-up of populations there today.[62] Tropical forests, then, in the Americas and around the Pleistocene world, provide a crystal-clear window into the early human taste for diversity, rather than monotonous, simple routes of environmental adaptations.

Tropical forests were, of course, not the only show in town. Archaeologists working in some of the most extreme environments around the world, including deserts, high-altitude habitats, and those beyond the Arctic circle, have all enriched our understanding of the immense threshold of human flexibility during the late Pleistocene. In the Sahara, the Arabian Peninsula, and in northern India, human entrances into these modern deserts seem to have been linked to changes towards wetter climates ~100,000 to 70,000 years ago. However, in the Kalahari and Namib Deserts of southern Africa, and the Central Deserts of Australia, Pleistocene humans seem to have also been able to prosper under dry conditions with limited water.[63] By 80,000 years ago humans had adapted to the cold, patchy environments of the African 'Mountain Kingdom' of Lesotho as well as the highlands of Ethiopia[64] and, by 40,000 to 30,000 years ago, had even reached the icy Tibetan plateau.[65] Bones with cutmarks found at 72°N

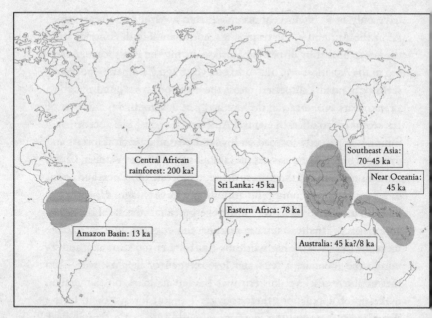

Central African
rainforest: 200 ka?

Sri Lanka: 45 ka

Eastern Africa: 78 ka

Amazon Basin: 13 ka

Southeast Asia:
70–45 ka

Near Oceania:
45 ka

Australia: 45 ka?/8 ka

Map showing the earliest known dates of *Homo sapiens*' tropical forest occupation in different parts of the tropics.

even tentatively show the advance of humans into the Arctic Circle by 45,000 years ago.[66] Of course, *Homo sapiens* used savannahs and coastal settings as well. They would have been foolish not to. But all of the above evidence, like the diverse tropical forest examples we have seen, shows that our species was not a one-trick pony and was, almost quite literally, *everywhere* by the close of the Pleistocene.

With this in mind, the new technologies, new social capacities, new ornaments, new cultural materials, and new exchange systems and ways of passing down knowledge that are often considered uniquely 'human' are best seen as a wide-ranging catalogue for dealing with diversity, rather than a one-size-fits-all tool for a single habitat. The advantage is obvious. Different, growing communities, all specializing in very different environments, meant that the fluctuating climates of the late Pleistocene would be highly unlikely to affect *Homo sapiens* at the species level.[67] Populations of our human ancestors almost certainly disappeared locally (or were 'extirpated'). In the Levant, for example, early evidence for human arrival from Africa between 200,000 and 100,000 years ago is followed by a marked absence until around 43,000 years ago.[68] However, seemingly unlike the Neanderthals, the Denisovans, the 'Hobbits' and *Homo erectus*, the *overall* human population generally kept growing, with communities in certain environments picking up the slack when other environments turned hostile. Using personal ornaments and social networks to communicate between populations inhabiting very different environments provided a crucial social fallback in times of stress. In this way we survived, but also continued to expand. With greater populations and greater numbers in different parts of the planet, we began to interact more intensively with the natural world, finding new, more manipulative ways of feeding ourselves and buffering the unpredictable Pleistocene world we had been born into. By the fall of the Pleistocene curtain we had adapted to all the land-based environments the world has to offer. What we did next would change them for ever . . .

7. Farmed forests

Following their arrival across a number of the world's environments during the late Pleistocene, humans began to make increasingly intensive interventions into the natural world in order to feed their growing populations and complex social networks. Using fire to shape the types of habitat in their surroundings, moving particularly nutritious plants, and carrying animals beyond their wild ranges as handy protein packages, human populations began to reconfigure the distribution of particular species and even entire ecosystems.[1] During the terminal Pleistocene and Holocene transition (~12,000 to 8,000 years ago), and perhaps also before, these interventions became even more intrusive. Human selection for docile, social animals or carbohydrate-rich, clustered plants eventually led to morphological and genetic changes in certain species which became formally 'domesticated' as part of what has been called the 'origins of agriculture'.[2] Western perceptions of what this 'agriculture' should look like – namely expansive fields of cereals like wheat or barley and dense, ranging herds of cattle, sheep and goats – have meant that much archaeological, archaeobotanical (the study of plants preserved in archaeological sites) and zooarchaeological (the study of animals preserved in archaeological sites) research into its origins has centred on the dry river valleys of the 'Fertile Crescent' in the Middle East, where our familiar domesticates emerged before spreading across Europe during the Neolithic period (~9,000 to 4,000 years ago).[3]

Tropical forests have, by contrast, frequently fallen by the wayside of such discussions. How could the dense green tangle and poor acidic soils of these environments provide our daily bread, our 'usual' at the pub, or our regular pints of milk or blocks of cheese? Let alone preserve traces of plants and animals we might use to retrace the footsteps of the first farmers? Surely these forests have to be avoided or removed for such activities to be successful? But here is the problem. We are so stuck in our Euro-American worldviews of what 'agriculture' should be – a settled economy, fully reliant on domesticated plants and animals, extensive land clearance and monotonous field systems – that we miss other, equally ingenious and equally impactful roads to 'food production'. If we instead focus on the farming *practices* that lead to domestication, we open ourselves up to a whole new series of possibilities.[4] Taking this approach, we can look at cultivation as human behaviours that promote the survival, reproduction and growth in certain plants, and at herding as human behaviours that shape the mobility, population and anatomy of certain animals, to provide a wider lens into how our species began to insert itself into the natural world.[5] Instead of relying on a narrow, confining definition of 'agriculture', by looking at the experience and process of farming or 'food production' we can gain a truly global perspective on one of the most pivotal periods in human history. One that, of course, can include some of the most biodiverse environments on the planet, tropical forests.

Far from being unproductive sideshows, it is in tropical forests that we see some of the earliest human manipulations of plants and animals in the name of food security as early as 45,000 years ago. These diverse habitats include montane or seasonally dry forests, with penetrating light, fertile soils, and where burning can be easily practised,[6] as well as wet rainforests. And they potentially hosted some of the earliest examples of deliberate cultivation across the terminal Pleistocene–Holocene boundary, as humans changed landscapes and planted trees with sugar-rich fruits and tubers in places they would not normally grow. We will see how archaeobotanists and zoo-archaeologists have shown that some of the most-used domesticates around the world today, produced by these cultivating and herding behaviours, actually have their origins in tropical forests. In fact, a significant portion of your weekly shopping lists comes from tropical forests without you even realizing. Eggs, cornflakes, tortillas, sugar lumps, chocolate bars, ham and pineapple pizza, pepper, cigarettes, and the food spreads or cosmetics that contain palm oil, are just some of the diverse products that owe their existence to the 'jungle'. In some cases, tropical farming practices did take the form of 'agriculture', with novel field systems leading to prehistoric deforestation and major ecological changes that provide a stark warning to us today. But, in others, domesticated resources were slotted into existing, sustainable hunting, gathering and fishing economies. From being environments where 'farming' has been considered doomed to failure, we now know that over the last 10,000 years human societies have modified tropical forest plants, animals and entire landscapes the world over to provide many of the supermarket items we all take for granted today.

Tropical forests have a reputation for being some of the most 'pristine' environments prior to industrialization. However, they actually have one of the longest, if not the longest, records of human modification. As early as 45,000 years ago, and the arrival of our species in the tropics and sub-tropics of Southeast Asia, Near Oceania and Australia, humans were burning tropical forest ecosystems to promote grassland patches and open up forest floors. This enabled them to pick and choose a cornucopia of resources from a variety of open areas, drier forests and denser rainforest.[7] Some groups were even managing plants by moving them. Starchy plants, like yams, were potentially carried across the Wallace Line as early as 40,000 years ago. Meanwhile the range of plants like the sago palm, various yams and taro across mainland and island Southeast Asia, and perhaps even Australia, by the start of the Holocene certainly has the ring of human handiwork about it.[8] Even tropical animals were not left in peace by our early ancestors. In Near Oceania, by 20,000 years ago, seafaring people carried furry bandicoots and cuscuses (not to be mistaken for couscous!) as snacks – moving them beyond their natural homes and on to more isolated islands such as those of the Bismarck Archipelago where they were to thrive.[9] Why then have archaeologists and the public alike so often ignored tropical forests as possible crucial locations where *Homo sapiens* performed early food production experiments? It certainly hasn't helped that many productive tropical plants are soft and fleshy, not prone to preservation over long time periods, while bones often face the harsh acidity of the soil upon burial.

Unlike more temperate or arid areas, where charred seeds or ample animal remains allow the observation of gradual changes from wild to domesticated features, similar detective work in the tropics has therefore certainly been at something of a disadvantage.

Nevertheless, archaeologists are stubborn and used to working with what they have. Archaeobotanists and geoarchaeologists, like Professor Tim Denham of Australian National University, have turned to their microscopes to try to identify uniquely shaped starch grains or characteristic, microscopic silica structures (phytoliths) left behind by particular plants in tropical soils. While difficult work, and prone to contamination from modern foods being eaten in and around the laboratory or excavation sites, these painstaking methods can revolutionize our understanding of food production in more challenging environmental contexts. In fact, using these approaches, Tim and his colleagues think that they have found a key, early heartland of independent plant cultivation and domestication in tropical forests. Not only does this heartland challenge the timeline of global food production, it also causes us to face our own biases and assumptions as to what 'agriculture' looks like – located, as it is, in the middle of a wet, sticky swamp. As Tim says, 'At Kuk Swamp, in the Wahgi Valley of Highland Papua New Guinea, archaeological excavations from the 1970s and 1990s have allowed us to reveal one of the earliest farming landscapes found anywhere in the tropics.' Marking out and excavating 200 trenches, archaeologists have uncovered a series of 'phases', between 10,000 and 4,000 years ago, that document the increasing lengths humans went to in order to modify the landscape to make it better for the cultivation of some of the world's modern tropical favourites – taro, banana, and even sugarcane. In fact, this evidence makes Kuk Swamp one of the earliest sites where farming practices by *Homo sapiens* have been observed. Not in temperate or dry river valleys or grasslands, but in highland, humid, tropical forest wetlands, around 1,500 metres above sea level.[10]

Phase 1 of Kuk Swamp, dated to ~10,000 years ago, has been famous since archaeologist Jack Golson identified a series of drainage ditches.[11] Back in the 1970s and 1980s, Jack put the cat among the archaeological pigeons by suggesting that these ditches were a deliberate attempt by humans to drain the local wetland and help the growth of carbohydrate-rich plants like taro and banana. This early evidence has proven controversial with a lack of direct verification of plant cultivation, suggestions that the ditches may, in fact, be natural, and an absence of similar landscape modification anywhere else in New Guinea raising some reservations. Nonetheless, banana phytoliths were found in this phase, in a more open environment than they would usually appear, hinting that some human gardening was in progress. Not just that, but a number of environmental records from the terminal Pleistocene and early Holocene have shown that 10,000 years ago saw the increase of human burning and opening of forest across montane New Guinea, including directly around Kuk Swamp,[12] conveniently providing the perfect environmental storm of forest patches for bananas, open forest floors for taro and yams, and grasslands for sugarcane. Some permanent settlements have also been identified in the region, suggesting growing populations and humans investing more in particular parts of the landscape, perhaps linked to the initial cultivation of these few key, managed crops,[13] though Tim remains sceptical of these.[14] The jury also remains out on this earliest phase of food production at Kuk Swamp. If proven, however, it would be some of the first deliberate food production and cultivation found anywhere in the world.

Modern genetic evidence has certainly shown that independent domestication of taro, the greater yam, and bananas all occurred in Near Oceania, before expanding out around the world. Today, the Eumusa banana found in the region is one of the most-used domesticated bananas globally. By Phase 2 (dated to ~7,000 to 6,400 years ago) and Phase 3 (~4,400 to 4,000 years ago), Tim and his team of 'microbotanists' (named because of the diminutive size of the plant parts they are working with) have provided clear evidence for the presence of starch grains and phytoliths of taro, bananas, yams and precursors of sugarcane,[15] planted in direct association with clear, planned and extensive digging of drains and ditches by the local human population, during a time when people also started transitioning to a more sedentary lifestyle. The well-thought-out maintenance of plants that liked forests and plants that liked more open grasslands would have provided a set

Present-day mound gardening of sweet potato in New Guinea.

of important food stores for growing human populations facing fluctuating climates in the cool highlands at the terminal Pleistocene–Holocene transition. It will always be challenging to try to identify exactly when these plants became 'domesticated' based on these tiny traces. But it is increasingly accepted, by archaeologists around the world, that the early Holocene plant management strategies at Kuk Swamp show a step change in human influence on plant life cycles and growing conditions – a new relationship with some of the plants many of us regularly have in our kitchens or put in our tea today. Kuk Swamp's importance to world archaeology was, in 2008, confirmed by the granting of UNESCO World Heritage status for the entirety of the site's remarkable 116-hectare span, making it a true giant of prehistoric farming.

Kuk Swamp is significant not just because it calls into question all our assumptions about the timing and location of 'agricultural' origins but because it also challenges our assumption about what the earliest farming societies looked like. The cultivation of banana, tubers and sugarcane at Kuk was not the rapid product of a single 'inventor' or 'inventors'. Instead, it was the end point of a long process begun by the first humans in Near Oceania that burned forests to maintain profitable mixtures of resources. Certainly this reached new levels at Kuk, but it did not happen overnight and had a long, tropical trajectory. Furthermore, the development of cultivation at Kuk did not lead to sweeping changes. Not everyone instantly wanted cultivation, and only later in the Holocene did it become a permanent fixture of New Guinea economies. With a team of researchers, I have worked at the site of Kiowa, in the same environments and same altitude of Kuk Swamp, to show that human populations just over 100 kilometres away continued to hunt possums, cuscus and tree kangaroos between 12,000 and 500 years ago, even as groups at Kuk were developing and intensifying 'food production'.[16] Evidently, then, not everyone saw the increasing tending of specific plants as the revolutionary, world-shattering idea we imagine it to be today. Looking on at the new experiments being undertaken at Kuk, some people, in one of the next valleys over,[17] were more than happy to continue to shape forest environments more broadly through burning, to gather nuts, and to hunt forest prey in a manner that, as we saw in Chapter 6, had enabled them to gain their first footholds in this tropical region. We often assume that farming was an inevitability, something humans grappled to attain once discovered. Here in tropical New Guinea, however, it was just one more way of managing landscapes that had reliably provided for humans for thousands of years.

From the early Holocene (11,700 to 8,200 years ago) onwards, a number of different societies began to perform farming practices in tropical forests. These interventions were to result in some of the planet's key dietary 'staples'. Today, maize, or 'corn', is the most produced grain in the world,[18] and has long provided the basis for many of the increasingly dense and complex societies that appeared across the Americas during the Holocene. Although it is often portrayed as an ideal arid-adapted crop, one that might sustain us to the brink of drought-filled doomsday scenarios, microbotanists, like Professor Dolores Piperno of the Smithsonian Tropical Research Institute and Smithsonian National Museum of Natural History, have shown that its origins actually lie in the warm, wet seasonal tropical forest habitats of Mexico. Traditionally, dates of maize domestication have been based on charred cobs found in ideal preservation conditions in the semi-arid highlands of Oaxaca dated to 6,000 years ago, alongside squash, gourds and beans. 'However, these cobs almost certainly do not represent the first cultivation of maize, and similar remains are simply not preserved in the wetter areas its wild ancestor is known to inhabit,' Dolores tells me. Instead, applying the same microbotanical approaches that were performed at Kuk Swamp, Dolores and her team discovered earlier evidence for maize cultivation 9,000 years ago in the tropical, humid Balsas River Valley,[19] in agreement with recent modern genetic evidence for the timing and geographical range of maize domestication.[20] Maize is not, however, the only important modern domesticate the tropical forests of the Americas have had to offer.

The stifling conditions of the wet tropical rainforests of the Amazon Basin were once thought to have made any type of farming activity absolutely impossible prior to colonial and industrial clearance.[21] Indeed, past human impacts to the Amazon rainforest have generally been thought to have been minimal. Nevertheless, starch, phytolith and genetic research has now shown that it was here that the significant root crop manioc, or cassava, was first cultivated and domesticated, in a process that began as early as 10,000 years ago.[22] Falling just shy of potatoes, a domesticate that itself emerged in the montane tropical environments of the Andes,[23] manioc is today the second most important root crop globally, feeding growing populations across Latin America, Africa and Asia. The importance of the Amazon for prehistoric food production does not stop there, however, and it was also the site of the domestication of pineapples, avocados and peanuts during the middle to late Holocene.[24] Chili peppers were also most likely domesticated just south of the Amazon Basin in drier tropical forest and woodland habitats.[25] Meanwhile, cacao was probably domesticated, or at least semi-domesticated, in the Amazon Basin[26] (although the Central American tropics have also been suggested as the geographical origins for this key ingredient of the chocolate that many of us cannot do without today).[27]

Sweet potato, which is becoming an increasingly popular food in trendy restaurants across Europe and North America, was also cultivated in the American tropics, somewhere between Colombia and Venezuela in the south and Mexico in the north. Although poor preservation of yet another squishy, but rich, tuber has made tracking down its origins challenging, starch grains of sweet potato, from residues of tools interpreted as having been used for plant processing, have now been dated to 7,700 years ago in the montane tropical forests of Colombia,[28] and firm data for domesticated sweet potato exists 4,500 years ago in Peru. By 1,000 to 500 years ago, this root crop was also being cultivated, not just across Latin America, but throughout the Pacific among the pre-colonial societies of the Cook Islands, Rapa Nui (Easter Island) and Aotearoa (New Zealand).[29] It is also worth noting that prehistoric societies across the tropics of the Americas also manipulated many wild, semi-domesticated or domesticated trees. The modern genetics and distribution of the hulking Amazon nut (Brazil nut) tree, for example, has been shown to pattern in close correlation with ancient archaeological settlements, suggesting that humans carried and planted Amazon nut for its protein-rich nuts, in a form of sophisticated agroforestry.[30] Peach palm was also domesticated somewhere in the Amazon Basin.[31] Finally, beyond foods, *per se*, the Neotropics have also produced some more 'recreational' crops. The Madeira River Valley in the south-western Amazon Basin was the origin of one of the major sources of modern crime across Latin America and, indeed, globally, coca (the constituent of cocaine). It also witnessed the domestication of the tobacco plant that fills many a cigarette or pipe around the world today.[32]

The complete shift in understandings of the tropical rainforests of the Amazon Basin as a centre of early plant cultivation is perhaps best encapsulated by the Llanos de Moxos area in the Bolivian Amazon. Today this sweeping region is characterized by a highly seasonal climate which sees significant portions of its 'savannah-like' landscape flooded for much of the year. However, all is not as it seems. When a team of scientists led by Umberto Lombardo cast their eyes over high-resolution satellite data for the area, as well as computer-based models as to how the terrain differed over space, they were able to identify some unusual alterations to the surface of the land. Together they identified over 6,600 sites that they call 'forest islands'. These are raised, forested patches that, when visited by archaeologists, also revealed clear traces of human presence and enrichment of soils with their waste.[33] These human-made mounds would have provided areas safe from the annual flooding, areas that were put to good use by past human populations. Publishing in the renowned journal *Nature*, Umberto and colleagues demonstrated that the soils of these 'forest islands' revealed a long record of human manipulation of plants. Heart-shaped manioc phytoliths were dated to 10,350 years ago, while spherical squash phytoliths dated to 10,250 years ago.[34] Whether these plants were fully domesticated by this time or whether this was just the beginning of a long process to producing crops that act as staples for populations of the Amazon Basin today remains unclear. However, as co-author Professor José Iriarte, of the University of Exeter, argues, 'this evidence, alongside the tropical forests of southwestern Mexico, undoubtedly means that this region should be considered, alongside the heavyweights of China and the Middle East, in global discussions of plant cultivation and domestication following the last glacial period'.

Turning our attention to the tropics of the 'Old World', South and Southeast Asia have also produced a number of domesticated plants, and also animals, that many of our kitchen shelves take for granted today. In the humid state of Odisha in India, a mixture of environments, from sweeping riverbanks to tropical forests, seems to have provided the setting for humans to locally domesticate Indian varieties of rice, millet, gourds and types of cucumber. Okra, a plant which will be familiar to anyone who has eaten a number of native Indian dishes, was also domesticated in this part of India.[35] Delicious, succulent mangoes originated as a tree crop in the tropics of Southeast and South Asia.[36] Meanwhile, black pepper, which flavours a vast variety of cuisines and dishes around the world, from morning eggs to spicy 'curry' dishes, owes its domestication to cultivators inhabiting the tropical coastlines of Kerala in southern India.[37] Cinnamon, another crucial spice for budding and professional chefs alike, appears to have been cultivated slightly further south in Sri Lanka. The first mention of this spice, 5,000 years ago, indicates that it was exported by local merchants to ancient China and, later, the Classical Mediterranean world. It even makes an appearance in the Bible.[38] Finally, although the exact origins of domesticated citrus precursors of the lemon remain hazy, and citrus fruits are notoriously difficult to keep track of archaeologically, it seems most likely that they were domesticated somewhere within Southeast Asia, within their wild range of growth. Although in all of the above cases, dating the exact arrival of a 'domesticated' form of these plants is immensely challenging, relying on rough, modern genetic 'clocks' and fragmentary remains left in tropical soils, much of our food and drink would taste very different without these tropical bounties that were shaped by prehistoric human hands.

Moving briefly away from plants, tropical forests may seem unlikely homes for domesticated animals, but they have also played a hand in the appearance of two of the animals most relied upon by farmers around the planet today. Water buffaloes are used by more humans than any other domesticated animal, with the large populations of Asia relying on these hulking beasts for milk, yoghurt and cheese. The two 'types' of these slow but strong beasts, the 'river' and 'swamp' buffaloes, today wallow in waterlogged settings flanked by tropical and sub-tropical forests, consuming freshwater plants as they do so. Although their origins remain obscure, and have received little attention compared to Euro-American cattle, the first signs of deliberate human herding of these bovids seem to occur around 5,000 years ago in India and 4,000 years ago in Southeast or East Asia, respectively.[39] The chicken represents perhaps a still more secure tropical forest domesticate. This bird has been morphed by humans into one of the most common animals on the planet today, with a population of over 25 billion. While you would not know to look at them, they are descended from the red 'jungle' fowls which, as their name suggests, have long stalked the tropical forests of Asia. A recent genetic study, based on the genomes of 863 chickens and sampling all species of wild jungle fowl, shows that the domestic chicken emerged from a subspecies of jungle fowl that lives in the tropics of southern China, Thailand and Myanmar.[40] After their domestication, these poultry were moved throughout Southeast and South Asia, further interbreeding with other jungle fowl species. While a complex history, the chicken, which forms the basis of many a fast food restaurant, sandwich or Sunday lunch today, undoubtedly has its origins in the tropics.

★

All the above examples refer to the domestication of plants and animals *within* tropical forests, as hunting and gathering communities living in these environments gradually increased their involvement in the natural world to become food producers. In many cases, as at Kuk Swamp, this meant that domesticated plants and animals were used to supplement existing consumption of wild plants and animals as part of diverse economies that operated on the basis of clear local knowledge of tropical forest life cycles and tolerances. In the Amazon Basin, for example, prehistoric farming practices represent a sophisticated early form of agroforestry. 'Gardens' of manioc and, when it arrived, maize, were cultivated at the same time as managed, but intact, forests were used for fruits, nuts and wild animals. It has even been suggested that wild Muscovy ducks and freshwater turtles were in some way 'herded' as part of these varied approaches to landscape management.[41] Although, as we will see later, this increasingly intensive food production shaped the soils, geography, forest structure and biodiversity of the Amazon Basin, it seems to have had a relatively limited impact on overall past tropical forest cover. Similarly, diverse and sustainable strategies of food production have also been documented in early and middle Holocene New Guinea and island Southeast Asia.[42] But what happened in prehistory when farming, particularly in the form of field-based agriculture, was introduced *into* tropical forests from the outside? Did it lead to massive deforestation, landscape degradation and cultural shifts, as it does today, as people tried to force their way of life on to these environments? Or did past societies manage to adapt external practices to local situations?

A classic example of a more 'agricultural' introduction into tropical forests is the spread of the type of rice most common around the world today. Genetic and archaeobotanical research suggests that rice emerged in the Yangtze River Basin in southern China around 9,000 years ago from a wild, wetland grass.[43] From here, the *Oryza sativa japonica* subspecies of rice diversified and moved both north, to the edge of its lowermost temperature-tolerance, and south, towards the tropical forests of South and Southeast Asia.[44] *Oryza sativa japonica* reached India, adding to the indigenous forms of rice (*Oryza sativa indica*) that had already been cultivated on the Ganges floodplain since 5,000 years ago, as well as island and mainland Southeast Asia, by about 4,000 years ago. Here, rice, although deriving from a wild sub-tropical grass which requires significant rainfall, also needs to be planted in the open, seemingly necessitating forest clearance if it is to be grown. This is particularly the case as more intensive 'wetland' farming techniques were applied to rice from around 6,000 years ago in China in order to harness its energy potential, with terracing and rice paddies requiring significant alterations to the landscape and year-round water-logging.[45] Foxtail millet was also domesticated in China and became rice's partner in crime as it moved southwards as part of a package. As this crop generally prefers drier, open cond-itions, it would have placed further demands on cultivated land, particularly in the tropics.

Unsurprisingly, the expansion of these crops into the tropical for-ests of Southeast Asia has generally been linked to large-scale forest clearance, resulting in dramatic landscapes, such as the UNESCO-protected Ifugao rice terraces of the Philippines, which tourists flock to see today. Nevertheless, pinpointing their arrival has been chal-lenging given that minuscule rice and millet grains are rarely preserved in archaeological sites and the diligent sieving required is only a fairly recent archaeological practice. The earliest archaeobot-anical evidence for rice in mainland Southeast Asia is dated to 4,000 to 3,500 years ago in the coastal portion of Thailand, with rice and millet reaching further inland by 3,000 years ago.[46] In island South-east Asia, the first evidence for rice is dated to roughly the same time in Borneo,[47] although some of this evidence is based on the impres-sions of pieces of rice left in pottery.[48] In areas with drier, seasonal tropical forests, there is evidence that burning occurred in order to clear the way for these new, productive crops.[49] That said, destruction was not guaranteed. In the lowland evergreen rainforests of main-land and island Southeast Asia, rice and millet farming were often combined with the ongoing cultivation of yam and taro tubers, as well as starchy sago palms, in smaller, more open, forest patches. Alternatively, it was ignored altogether by forager communities more interested in harnessing the local productive resources than investing energy to clear forests that had long provided them with shelter and food.[50]

The deep forests of Central Africa are another important setting where we can explore what happened when incoming agricultural strategies adapted to local tropical conditions. The arrival of farming in this part of the world has often been linked to the so-called 'Bantu' expansion, a migration of populations from West Africa that all speak languages within the same 'Bantu' family. These people are thought to have brought the crop pearl millet with them as well as iron technology, sweeping southwards through the African continent, all the way to South Africa, during the late Holocene (4,200 years ago to the present). Pearl millet is an arid-adapted grass that appears to have been domesticated in the river valleys on the southern edge of an expanding Sahara (just after its last 'green' phase) around 4,000 years ago. It then expanded south, moving with Bantu-speaking populations and reaching northern Ghana by 3,500 years ago, the rainforests of Cameroon by 2,500 years ago, and the Democratic Republic of the Congo in the last 2,000 years.[51] Assumed vulnerabilities of this crop to warm and wet conditions had meant that scientists have associated its appearance with the dramatic late Holocene 'rainforest crisis' of Africa. This was a time, from around 2,500 years ago, when lowland rainforest seems to have rapidly retreated in favour of more open settings. Some scientists believe this change was a climatic phenomenon. Others think it was actually caused by iron-wielding Bantu-speaking farmers who indiscriminately cleared the landscape to promote the production of their favourite food source.[52]

There is, however, growing evidence that disastrous forest clearance need not accompany the arrival of pearl millet in the Central African tropics. Scientists undertaking experimental farming today have shown that this crop can actually be effectively grown in wet humid plots next to rainforest.[53] Bantu-speaking communities, and pearl millet, also continued to move and occupy different tributaries of the Congo River long after the end of the supposed crisis. Interestingly, the first pearl millet in Ghana is joined by plants like oil palm and yams which had their origins in African tropical forests, suggesting a more mixed strategy. As we will see in Chapters 13 and 14, oil palm is now part of a global economy that fuels products as diverse as nut-based chocolate spreads and face creams. Local contributions also seem evident in the use of groundnut, castor bean and gourds, which were also domesticated in the tropics of Africa and continued to provide important food sources from these environments.[54] In the absence of evidence for pre-'Bantu' communities, at least in the western Democratic Republic of the Congo, these combinations of resources seem to show rapid farming ingenuity and local adaptation on the part of new human arrivals. In a recent study I was involved in, working closely with a prolific team at the University of Cologne who have undertaken decades of fieldwork in the Congo Basin, we were lucky enough to be able to study Bantu-speaking populations moving into the far, forested reaches of the region. Here, the earliest arrival of pearl millet occurred as part of a diverse strategy that included heavy reliance on freshwater fish and the use of wild and domesticated forest plants.[55] Similar 'mosaic' strategies of food production continue to be sustainable in this part of the Congo Basin, showing that there is no necessary single path to farming in tropical forests, even for incoming populations.

The diligent work of archaeobotanists (especially microbotanists), zoo-archaeologists, plant geneticists and archaeologists has demonstrated that tropical forests should now be seen as crucial sites for scientists studying the earliest prehistoric farming practices on a global scale. It is in tropical forests that humans apparently first began to interfere in habitat structures, plant growth and animal distributions. It is also in tropical forests that some of the earliest examples of farming practices occurred. Next time, before you put sugar in your coffee, before you peel a banana, before you apply moisturizer to your skin, or before you tuck into some sweet potato fries, remember just how many of our modern tastes and styles are fed by human activities in these environments that occurred throughout the Holocene. The 'Fertile Crescent', the dry highlands of Mexico, the rivers of North America, the great valleys of China, and the Saharan border of Africa have all proven to be key centres of domestication and agricultural origins, particularly in terms of animals and crops that are farmed in the way we think farming should happen from the view out of our Euro-American bubbles.

However, tropical forests also witnessed some of the most significant and effective examples of food production as prehistoric societies harnessed the huge diversity of species these environments had to offer. This is particularly the case when we move away from the popular myth that tropical forests are solely characterized by dense, light- and soil-poor rainforests, and look at the diversity of more seasonal forest environments that hosted some of the earliest examples of tropical food production discussed above. The changes to the landscape might not have been obvious to us but they were certainly real.

In fact, considering the issues of land-use sustainability we face around the world in the twenty-first century, tropical forests give us a remarkable lens into some of the most ecologically savvy of farmers there ever were. Locally domesticating plants and animals, early tropical farmers were able to manage these forms of food production within forests that were by-and-large left standing, and their wildlife left abundant, although their structure and species distributions may have been altered. Even some past farmers who moved into tropical forests with domesticated plants and animals from abroad, that often required more open, drier conditions, developed sustainable, mixed approaches that combined a wealth of local, wild tropical resources with these useful new species. These examples provide food for thought when we think of some of the disasters that governments, companies, and even our own choices wreak on these environments today in the name of 'development', 'infrastructure' or 'productivity'. The past shows that there are ways that tropical forests can be used and adapted, even within frameworks of farming, without being removed for ever. However, some prehistoric societies certainly did remove the forests, literally clearing the way for types of 'agriculture' that would have felt more familiar to us. This could have major consequences in the form of the destabilization of landscapes and soil erosion, collapses in biodiversity, and increased vulnerability to extreme weather events. This was particularly the case when new desires for 'food production' washed up on the shores of more isolated, tropical islands.

8.   Island paradises lost?

Like the forests they are home to, we are used to seeing tropical islands as untouched 'deserts' or blissful paradises. Whether you are a rum-sodden Jack Sparrow, an inventive Robinson Crusoe, a delivery man with a volleyball, or a troubled Jack Shephard trying to lead a ragtag group of 'Lost' lottery winners, conmen, artists and rock stars, tropical islands represent an inescapable nightmare of being stranded in maddening isolation. Reality shows like those hosted by Bear Grylls often use tropical islands as the setting for testing the survival skills, social interactions, and morals of twenty-first-century societies. For many of us, they are the dream, but perhaps impractical, holiday: clean blue oceans, fine white sand and quiet palm trees. Places where we could finally 'get away from it all'. For some sickeningly wealthy people who can afford to actually buy these landmasses, this is already a heavenly reality. Either way, it is probably a good thing that we widely view these tropical islands as arenas that are generally inaccessible to humans. Especially as we tend to also stereotype them as being biologically and ecologically fragile. They are often small, so their soils and forests can change rapidly across much of the island surface in the face of external threats.[1] While home to high levels of plant and animal biodiversity, unique species are often present in low numbers. Plants often have limited capacities to disperse their seeds widely. Meanwhile, isolated island animals frequently have few defences against predators and competitors. Close bonds between species also mean that a cascade of extinctions can occur should one begin to dwindle.[2] Together, this all means that island ecologies have been seen as particularly vulnerable to the invasion of new, competing species or predators,[3] including *Homo sapiens*.

These perceptions of islands have also pervaded academic thought. In archaeology and anthropology, island ecosystems have often been contrasted to their mainland counterparts as uncompromisingly sensitive to past human actions. Think of, for example, the biggest celebrity of tropical island extinctions, the dodo of Mauritius, which disappeared in the seventeenth century AD shortly after the first permanent European settlements were constructed on this Indian Ocean island.[4] On the one hand, seafaring hunters have been seen as decimating any island animal life seeing our species for the first time. Without protective strategies, and with no hope of escape or reinforcements from distant shores, mammals, birds, amphibians and reptiles could very quickly be erased from history.[5] On the other hand, islands have been framed as inherently 'resource poor', with human societies only being able to succeed if they brought new, domesticated plants and animals along for the ride. These same food producers faced further, if different, issues of sustainability, however. Limited island soils were prone to erosion and the loss of nutrients. Smaller forests, if not given time to regrow, could also rapidly disappear, leaving a barren outlook for settlers reliant on construction materials and firewood. Sedentary farming societies also often, intentionally or unintentionally, brought along hitch-hikers, like dogs, pigs and rodents, that wreaked their own havoc on local plants and animals.[6] Ultimately, in both cases, tropical islands have been seen as particularly prone to forces of human 'destruction'. The collapse of island environments as well as their prehistoric human societies must have been inevitable.[7] After all, the increasingly expanding infrastructure, settlement and profit-driven agricultural strategies of our own societies are unleashing irreversible changes on these habitats today.[8]

Nevertheless, if we want to properly study how past humans negotiated the colonization of uninhabited tropical islands, and if we want to determine the degree to which their increasingly intensive interaction with the natural world had impacts on island ecosystems, compared to human-driven changes seen in the twenty-first century, we need hard data. We also need to strip away our own value judgements of which activities and changes are inherently 'good' and 'bad'. Particularly, given what we have seen in Chapter 7, we need to take a step back from assumptions that all forms of 'food production' look like the monotonous, tree-clearing agricultural fields and pastures we see today. Instead, we can investigate islands as places where people experimented with the introduction of new plant and animal species, alongside the foraging, fishing, hunting and managing of the available natural resources.

Focusing in on some of the most classic of our island paradise fantasies, across the Caribbean, the Pacific and the coastlines of Africa, we will now delve into the blossoming sub-field of 'island archaeology'. We will see how archaeobotany, zooarchaeology, and modern and ancient genetic analyses are providing insights into how humans constructed new island 'niches' for themselves,[9] 'transporting landscapes',[10] as well as domesticates, from mainland settings. We will accompany scientists working in some of the most remote regions on Earth to look at how environmental records of landscape change, from lakes, swamps and archaeological sites, can show the changing scale and intensity of the subsequent impact of different practices on local or 'endemic' island plants and animals, as well as the soils and forests these populations came to depend upon for cultivation and raw materials. Certainly, difficult situations arose, sometimes in settings many hundreds or even thousands of kilometres away from the next human population. Yet, adaptable oversight, and innovative combinations of local resource use and domesticated plants and animals, could prove remarkably resilient in what have often been considered the most delicate of tropical environments. All it needed was some good old-fashioned on-the-ground understanding of local ecologies. Something that Euro-American societies are often disturbingly lacking in today.

We start our island circuit in the Caribbean, a series of 700 islands well-known for their tropical vegetation, spotless beaches, diverse Indigenous cultures and film-famous buccaneers. Surrounded by the Caribbean Sea and the North Atlantic Ocean, just east of North, Central and South

Map of the Caribbean showing the main waves of human colonization of different islands, with dates (the arrows show routes of dispersal that would have required seafaring). The first human inhabitants of the Caribbean ('Archaic') are thought to have arrived from South America, making it to Trinidad 8,000 years ago before reaching the Greater Antilles, the northern Lesser Antilles, and Barbados ~6,000 to 3,000 years ago. A later 'Ceramic' period of human expansion from South America, which reached Puerto Rico and covered the entire Lesser Antilles, occurred around 2,500 years ago. Here, BP stands for 'calibrated years Before Present' (the standard format for radiocarbon timescales).

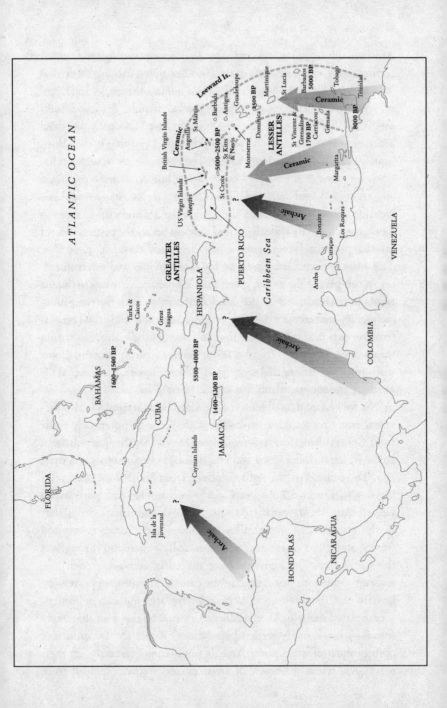

America, this region was occupied by a variety of hunting and farming communities over the course of the last 8,000 years, making them ideal 'laboratories' for observing how different human economies and forms of social organization can impact tropical islands. Professor Scott Fitzpatrick, of the University of Oregon, is well and truly an island archaeologist and has had the 'arduous' task of excavating in the Caribbean for much of his academic career. The envy of most of us, Scott has made his name exploring how ocean-going human societies colonized and adapted to, and impacted, islands across a region many would pay handsomely to visit just once. 'Beyond being a wonderful location to work, the Caribbean is ideally placed to test whether the combination of humans plus islands inescapably equals ecological catastrophe,' he says. This issue is particularly pressing for communities and governments across this part of the world, given that deforestation, threats to unique plants and animals, and the degradation of coral reefs all run rampant in the twenty-first century.[11] In fact, entire marine and land-based ecosystems are expected to disappear in many parts of the Caribbean within the next twenty years if action is not taken to curb pollution, global warming, over-fishing and tourism. Are these modern phenomena? Or are there ancient precedents that we can learn from?

The first human inhabitants of the Caribbean arrived via two different entry routes. One, so-called 'Lithic', group potentially came from Central America, arriving in the Greater Antilles (the northern chain which includes Cuba and Hispaniola) by ~6,000 to 5,000 years ago. The other, 'Archaic', group moved from South America, arriving first in Trinidad 8,000 years ago before moving into parts of the Greater Antilles, the northern Lesser Antilles, and at least two islands in the southern Lesser Antilles, Barbados and Curaçao, between ~6,000 and 3,000 years ago,[12] and eventually expanding throughout the Caribbean. These groups were mixed hunter-gatherer-fishers, making tools from stone, and, in the case of Archaic groups, relying heavily on coastal resources, as revealed by large mounds of human-accumulated shells (or 'shell middens') located on the coast that likely sustained increasingly settled populations. While we are unsure of their population sizes, these Archaic populations certainly left their mark. An extensive survey of environmental cores obtained from wetlands and lakes across the southern and eastern Caribbean has been used to suggest that these groups were active horticulturalists that cleared forests, perhaps using fire from their earliest arrival in order to manage environmental variation,[13] though this remains disputed. Certainly, in Trinidad, a present-day island that would have been connected to the South American mainland during the early Holocene (11,700 to 8,200 years ago), the opening up of patches of forest aided the growth of South American domesticates, such as maize, sweet potato and chili peppers, which have now been found at Archaic sites from as early as 7,700 to 6,000 years ago.[14]

These land-altering Archaic populations have been implicated in the extinction of native fauna in a region that has witnessed more Holocene (11,700 years ago to the present) mammal extinctions than any other. For example, the deliberate human burning of forest in Antigua may have stimulated the disappearance of three endemic small bat species from this island. Direct hunting of slow-reproducing large mammals, like native primates and the giant ground sloth, are also thought to have led to the extinction of species that had survived in the Caribbean well into the Holocene.[15] Their disappearance, following Archaic arrival, certainly piques the interest of archaeological investigators. Nevertheless, this remains debated, and there are few sites with clear evidence for hunting by these earliest settlers, with a combination of early Holocene climatic variability, human-induced habitat changes, and perhaps limited hunting more likely combining to bring these animals down in a more gradual manner.[16] In fact, these early Caribbean populations were remarkably in tune with their new island environments. As well as domesticates, they moved well-suited native wild species between the different islands, including wild avocado, wild fig and rodents. They cleared forests, not just to make way for domesticates, but also to promote nutritionally useful native taxa such as wild plantain, yams, the starchy roots of maranguey, arrowroot and the seeds of certain palm trees. On Puerto Rico they even seem to have modified soils to promote fertility by 3,000 years ago.[17]

The later 'Ceramic' period of human expansion across the Caribbean seems to have brought an intensification of certain types of activities along with it. Beginning 2,500 years ago in Puerto Rico and the Lesser Antilles, prior to spreading out across the rest of the Greater Antilles by 1,500 years ago, this wave of Early Ceramic populations, with an apparent genetic link to human populations in the Amazon Basin, involved the ancestors of the Taíno Indigenous group of the Caribbean.[18] As the name suggests, these populations introduced pottery to the region, although others have suggested it had already arrived with preceding Archaic groups.[19] They also introduced dogs and guinea pigs across its islands, and moved the agouti (a sort of long-legged guinea pig), the peccary (a pig-like mammal with hooves), as well as opossums and armadillos, out into the Lesser Antilles.[20] It was also during this wave of migration that archaeobotanists can show that South American crops truly took hold across the Caribbean, and maize, beans, manioc, peanuts, sweet potato, tobacco, chili peppers and guava can be found across the islands, fuelling the establishment of settlements that were increasingly distributed across island interiors as well as the shoreline, unlike their more coastal-focused Archaic predecessors.[21] The extensive farming,

Photograph of the coasts and corrals of Carriacou in the Caribbean today.

clearance and horticulture practised by these Ceramic groups had visible impacts on Caribbean ecosystems. By the time Europeans arrived, Indigenous populations were cultivating nearly 100 species of plants, had cleared forest in many island areas, and transformed hillsides with terraces and formed mounds for growing manioc.[22]

Unsurprisingly, this more intensive occupation and settlement could have significant impacts on local island ecosystems. In Puerto Rico and the Virgin Islands, exhaustion of local crab resources stimulated a switch to the consumption of molluscs. Molluscs were, in turn, over-harvested in Jamaica from 1,000 years ago. On the island of Puerto Rico, zooarchaeologists have shown a reduction of fish body size and a decline in the length of reef food chains at Ceramic sites over time that indicate overfishing. Evidence for the consumption of native amphibians, reptiles, seabirds and rodents, often in unsustainable fashion, is also clear at a number of sites across the Lesser Antilles.[23] These issues are something visible in many parts of the Caribbean today, particularly in relation to the over-use of reefs and forest clearance for agriculture. Nonetheless, the degree to which Ceramic populations led to sweeping and permanent change remains unclear, especially as it is difficult to estimate their population sizes. Many small animals being used by these groups, such as now-threatened tiny 'hutias' and other mice-like rodents, survived until after the arrival of European colonizers in the Caribbean.[24] On the islands of Anguilla and Nevis, it has also been shown that pre-colonial populations were sustainably, albeit intensively, harvesting marine molluscs.[25] We will never know how the Ceramic populations would have continued to fare on their island homes. Indeed, it is only with the spread of disease and atrocities perpetrated by European colonizers, which resulted in mass mortality among the Indigenous populations, that we begin to see a clear and complete disruption of long-term human–environment interactions and the onset of profit-driven extraction of resources and labour that we will turn to in Chapter 10.

In contrast to the multiple waves of human cultures with different economies moving into the Caribbean, the human colonization of the Pacific Islands by people speaking languages within the 'Austronesian' group has often been considered a more sweeping, classic example of a migrating human population with a uniform, introduced farming economy.[26] Ancient DNA of human skeletons from the isolated Near Oceanian and Polynesian islands of Vanuatu and Tonga has shown that their earliest human occupants, buried with so-called 'Lapita' pottery, descended from populations in East Asia.[27] Moving from Taiwan, human populations, with this characteristic 'dentate' style of pottery, reached the Philippines by ~4,000 years ago. True Lapita pottery, tempered from sand and shell, then appears on the Bismarck Archipelago by ~3,350 years ago before moving through the Solomon Islands, Vanuatu and New Caledonia, reaching Fiji, Samoa and Tonga by ~2,700 years ago.[28] From here there was something of a pause, with around 2,000 years passing before Polynesian descendants of these early populations went on to colonize Rapa Nui (Easter Island), Hawai'i and Aotearoa (New Zealand).[29] These expert seafarers navigated thousands of kilometres of open water, using sophisticated outrigger canoes like those made globally famous in recent times by Disney's *Moana*. In many cases they were the first human feet ashore, and what they brought with them could have major consequences for ecosystems, particularly on islands where there was no easy trip home and adventurous, exploring groups could become quickly cut off.

Despite common association with the expansion of the Austronesian language groups that spread across island Southeast Asia, there is no direct evidence for domesticated rice at any early Lapita site in the Pacific Islands. Instead, one of the main early domestic arrivals on these island outposts is an animal. Modern and ancient DNA evidence shows that domesticated pigs made the journey with Lapita-carrying populations out across the Pacific.[30] We do not often consider it when we see these animals rummaging around human settlements or farms, but they can actually have widespread environment-altering impacts. Today, feral pig populations on Pacific Islands reduce rare, native tropical plant biodiversity as a result of their indiscriminate foraging and encourage invasive species to take root by reducing local competition and dispersing seeds through their digestive tracts or by carrying them on their skin.[31] The foraging and trampling of pigs can also increase soil erosion, and scientists have invoked pigs, which are immensely important to many Pacific Island cultures and economies today, as a cause of increasing landscape instability on many islands following the arrival of humans.[32] Although without rice, these colonizing populations also carried horticultural 'transported landscapes',[33] using fire to open forests and promote the growth of other introduced plant domesticates such as taro, yam and banana.[34] Deliberate deforestation for houses, canoes, pig pens and other structures led to further landscape changes. On Tonga, Vanuatu and the famous Rapa Nui (Easter Island), with its haunting statues, tropical forest tree species declined in abundance following human colonization,[35] further exacerbating soil erosion as root systems were dragged away. Heavy human pressure on reef resources by these nautical maestros has also been observed by zooarchaeologists, particularly for later periods as settled populations expanded.[36]

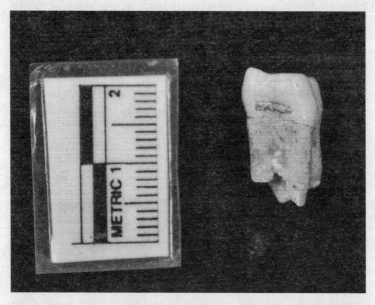

Human tooth with a deposit of dental calculus that was analysed by Monica Tromp in her study using a microscope. The particles trapped within this plaque showed that human individuals from Vanuatu were relying on local tropical forest foods.

The spread of Lapita pottery across the Pacific Islands also brought with it one of the most prominent examples of an island-altering ecological stowaway. The Pacific rat is commensal to humans. This means that it can rapidly become entwined with human settlements, relying on human food, and travelling closely with human groups where it is able. Archaeological and genetic evidence shows that during the expansion of Austronesian-language-speaking populations the Pacific rat was transported on to virtually every single island ecosystem in the Pacific. The results were often game-changing for island ecosystems. These rats were large, fast-breeding, omnivorous and numerous, used to feeding off human scraps. Alongside direct hunting by humans, rats posed a major threat to land and seabirds, given their particular fetish for the eggs of endemic birds. Unable to reproduce in safety, and unused to such menace, as many as 2,000 species of bird went extinct following human occupation on tropical Pacific Islands.[37] As well as birds, these feasting rodents may also have driven the extinction of native terrestrial snails and insects. Rats, gnawing on seeds and nuts, have also been implicated in the decline of native plant communities, including the reductions in forest cover witnessed on Rapa Nui (Easter Island).[38] As this latter example has often been used to show, together, intended and unintended human introductions to Pacific Island ecosystems could lead to the rapid transformation of island forests into landscapes with no trees, reduced diversity of land-based wildlife, slipping soils that posed obstacles to food production, and declining fish and shellfish. On islands that could be separated from other landfall by as much as 2,000 kilometres, human population collapse could become all but guaranteed.

Although this dramatic 'doomsday' scenario sells books, it also glosses over a number of instances where arriving farmers and horticulturalists tuned their strategies to their new island homes. Dr Monica Tromp, of the University of Otago, has applied the latest scientific methodologies to human remains from the island of Vanuatu to revolutionize our understanding of the diets of some of the earliest Lapita producers.[39] As we have seen, human colonization of much of the Pacific has often been seen as involving a complete overhaul of island ecosystems. 'However, by studying microscopic remains of plants trapped in the gunk between ancient peoples' teeth we gain another perspective on the region's earliest settlers,' says Monica. Instead of domesticated plants, Monica found that the vast number of preserved plant particles in this ancient tartar came from wild rainforest trees.[40] This is direct evidence that tropical forests not only supplemented, but often also dominated, the diets of the first settlers of Vanuatu. While cultivated crops were also introduced, Lapita-carrying seafarers knew the value of tending to the existing wild resources. This story is not unique to Vanuatu, and the microscopic remains of rainforest trees have also been documented at Lapita sites elsewhere in Near Oceania as well as Micronesia. Indeed, many signals of 'clearance' and burning may also be indicative of the promotion of local useful plants, as much as incoming domesticated crops and animals, as populations shaped local forests to their liking.[41] Elsewhere, the use of wild animals such as fruit bats, turtles, cuscus, rodents, lizards, frogs, snakes and crocodiles, some (albeit relatively few) of which survive to this day, has also been documented. Meanwhile, zooarchaeological evidence for the successful, diverse use of marine resources, from different ocean habitats, has provided a more sustainable, island-specific insight into fishing practices.[42]

In fact, it should not be too surprising given the sheer environmental diversity that spans the Pacific Ocean, but Lapita-carrying (and later Polynesian) settlers were clearly flexible as they felt out their new, yet also familiar, surroundings. They managed local and introduced tree crops, foraged for plant resources and hunted local animals, and introduced new domesticates into tended gardens. The fact that these populations inhabited this vast oceanic space, from the Bismarck Archipelago to western Polynesia, in just a few hundred years shows that they were able to ingeniously, and rapidly, adapt to a variety of island settings. Moreover, the paucity of remains of introduced domesticates like pigs, as well as chickens and dogs, in early Lapita sites may show the difficulties (or choices made[43]) when establishing stable domesticated communities across this diverse space.[44] In Near Oceania, in and around New Guinea, for example, Lapita settlements are located on the coast or on small islands, tuning in on marine resource use, perhaps as a way of avoiding the existing occupants of this region, who had been using inland resources since the Pleistocene (~45,000 years ago). Further east, on more remote islands, Lapita sites are located some distance inland, and it is increasingly evident that they managed and used forest environments at the same time as investing in opened spaces for cultivation and pig penning. The type of island could also make a difference. Volcanic islands, like those of Hawai'i, could provide productive waters for fishing and, where large enough, nutrient-rich soils for cultivation. Indeed, archaeological research on this tropical island chain has shown the ingenuity of past Hawaiian field systems that adapted to grow different cultivars and use different strategies on 'leeward' (facing away from prevailing winds and often drier) and 'windward' (facing towards prevailing winds and often wetter sides of ridges)[45] sides. Complex systems of ancient aquaculture, in the form of stone-constructed pools, were also developed in Hawai'i to manage marine resources.[46] Meanwhile, tiny coral 'atolls', like Pitcairn Island, had poor soils but vast reefs for fishing. Raised limestone, *makatea* islands, like those of the Solomon Islands, Fiji and Tonga, had high levels of inland biodiversity which, while potentially threatened by populations of humans and their animal companions, could also represent a significant resource if harnessed in the right way.

Though these expanding human populations were shaped by their environments, they could also make their own decisions when the going got tough. Tikopia is an island of just 5 square kilometres that was settled by people with Lapita pottery ~2,900 years ago. Over 2,000 years of forest clearing for the growth of yams and taro, the management of pigs and dogs, and the use of marine resources, led to a drastic reduction in local biodiversity and forest cover and the increasing intensity of soil erosion on what was previously a fertile island. Around 300 years ago, the local population decided enough was enough. They removed pigs and dogs from the island, managed forests instead of cutting them down, and eased pressure on vulnerable fish species. They also managed their population size. The result was relatively sustainable existence up to European colonization.[47] Similarly, as noted above, the classic tale of Rapa Nui (Easter Island) is usually one of human population explosion, complete deforestation, warfare, famine and collapse by 1680 AD, with stone faces of *moai* statues left behind on a now largely depopulated island as monuments to the over-exploitation of resources. However, detailed archaeobotanical analysis and renewed dating efforts have shown that, although humans and rats certainly did impact the forest between 1200 and 1650 AD, this was not the whole story. From this point, the local population, which likely never grew to more than a few thousand, began using grass and ferns for fuel, stopped the creation of their iconic monuments, shifted to joint management of wild plants and domesticates, and persisted, alongside some remnant tropical trees. The innovativeness of these populations can be seen in their use of mulch from volcanic rocks to supplement poor soils and formation of underground '*manava*' gardens to protect their crops from wind and sea spray.[48] In fact, it looks like it was actually the arrival of European colonizers on the island in 1722 AD, with guns, diseases and slavery, that led to the almost complete collapse of what had been undoubtedly adaptable Indigenous Polynesian populations.[49]

The expansion of Austronesian-language-speaking communities across the Pacific leads us into a surprising, but neat, segue into the final stop on our island tour, Africa's coastal islands. Madagascar, one of the last large landmasses to be occupied by humans, sits around 400 kilometres east of Mozambique. Remarkably, its modern Malagasy population has, among its ancestors, some of the same Austronesian-speakers travelling through Southeast Asia that gave rise to the first settlers of Polynesia. In what is perhaps the single greatest feat of seafaring in human history, beginning around 1,000 years ago, human populations from Indonesia, engaged in wider Indian Ocean trade networks, brought their language, crops and genes all the way to the African continent.[50] But we are getting ahead of ourselves. The timing of the first human settlement of Madagascar is, remarkably, not settled, and current suggestions range all the way from ~10,500 years ago[51] to ~1,200 years ago.[52] The most recent detailed study, based on radiocarbon dates from a number of archaeological sites, as well as clear evidence for butchered animal bones, suggests that humans were present by at least 2,000 years ago.[53] Whatever the case, these

Artist's reconstruction of the now-extinct elephant birds of Madagascar.

first settlers are famous for a very different reason to the aforementioned Austronesian-language-speaking arrivals. Carrying stone tools from the African mainland, and altering forests through burning, these groups have been linked to the disappearance of a variety of Madagascar's enigmatic 'megafauna', including giant lemurs, humungous elephant birds, and a native hippo.[54] By 2,000 years ago, long prior to any record of the cultivation of crops or clear dense settlements, it has often been suggested that human activities, either direct hunting or indirect landscape changes, had begun to have major implications for large animals across the spiny thickets, dry deciduous forest and montane forests of Madagascar.[55] By 500 years ago, all wild animals larger than 10 kilograms had disappeared from the island which, back in the 1980s, was used as a classic example of the 'blitzkrieg' impacts of human hunting on megafaunal populations encountering humans for the first time.[56]

As Professor Kristina Douglass of the Pennsylvania State University, a seasoned explorer of Madagascar's forests explains, however, 'the story is, once again, much more complicated than that.' Archaeological sites have reported finds of giant lemurs and hippo bones with

Teeth of now-extinct pygmy hippos from Madagascar housed at the Muséum national d'histoire naturelle, Paris.

cutmarks from between 2,300 and 1,100 years ago in different parts of the island.[57] Meanwhile, Kristina and her team have shown that prehistoric elephant bird eggs have clear signs of human exploitation.[58] Ultimately, though, these direct traces are few and far between. Peaks in deliberate human burning have been suggested to occur at various times in the different biomes of the island between 1,700 and 1,100 years ago as populations expanded into different parts of the interior,[59] though these changes seem to suggest a patchwork of human advance rather than a relentless frontier.[60] Furthermore, even if the later dates of human occupation (~2,000 years ago) are taken, then there was a period of at least 1,000 years during which human populations and megafauna coexisted without any obvious signs of decline, a period that would only extend further if the earlier dates are accepted. Many of the large 'megafauna' likely actually persisted on the island until at least 1,000 years ago. Not only that, but when these larger animals did begin to disappear, they did so in different ways, and at different speeds, across the hugely contrasting forest environments of Madagascar. As a result, it seems highly unlikely that human hunting alone eliminated Madagascar's megafauna, with more complex, multi-factor explanations, including recurring climatic variability and periods of drought,[61] seemingly more probable.

More recently, a 'subsistence shift hypothesis' has been proposed, which argues that it was a change from a hunting and foraging lifestyle towards food production, and notably pastoralism, that imposed more sweeping, island-wide pressures on these Malagasy megafauna,[62] particularly as populations expanded and Madagascar became part of expanding Indian Ocean networks from around 1,000 years ago.[63] The first more permanent 'residential' settlements appear from 1,300 years ago on the north-eastern coast.[64] In the next half millennium, villages moved to the interior as well as the shorelines, spanning from the eastern evergreen rainforests to the arid spiny bush of the south-west. Over approximately the same time period, cattle and, later, goats arrived, and we see an increase in forest clearance and grassland presence across Madagascar as these key herd animals were left to over-graze and stop tree regrowth, as they do in many parts of Africa today.[65] From 700 years ago, rice, the greater yam, and coconuts brought from Southeast Asia were grown in wetlands. Later, sorghum and cowpea from eastern Africa were used in more arid regions.[66] These farming activities, alongside growing populations, which began to produce elaborate settlements and hilltop fortifications, would certainly have increased the strain on native animals, perhaps even acting as the final tipping point for the disappearance of many endemic species by 500 years ago against a backdrop of increased aridity. Nevertheless, as we have seen, different species declined at different rates and in different areas, with many on their way out prior to this point.

In fact, the earliest food producers of Madagascar were remarkably resourceful. They combined the cultivation of incoming Southeast Asian and eastern African domesticates to different extents, alongside the use of wild plant resources, depending on their environmental context and social situations. The same is seen across the other eastern African islands caught within a wider, flourishing Indian Ocean trading network. On near coastal islands, like Zanzibar and Pemba, Asian crops were typically rare relative to African millets, sorghum and cowpea, with evidence for persistent hunting and fishing.[67] In Madagascar itself, villages adapted to local contexts. Some practised hill-slope and wetland cultivation, while others preferred the herding of cattle and goats. Coastal communities focused on fishing, which still drives Malagasy economies and supplements farming diets today. Today, over-grazing and agricultural landscape transformations are increasing soil erosion and deforestation, and have helped make the remaining native lemurs some of the most threatened primates in the world.[68] Yet, prior to 500 years ago, human populations remained relatively small. So too did their herds of goats and cattle.[69] The large-scale agricultural landscape modifications challenging sustainability on the island today only truly began following European arrival as colonizers and growing Indigenous polities grappled to control the increasingly contested tropical landscape.

Turning to the opposite corner of the African continent, although the Canary Islands are just north of the Tropic of Cancer, the moist, marine climate supports extensive tropical and sub-tropical vegetation, and they provide another useful example for exploring prehistoric human impacts on island environments. The Canary Islands are a volcanic group of islands in the Atlantic Ocean just off the coast of north-western Africa and include a number of favourite European holiday destinations. The earliest human inhabitants of these islands, who became collectively known as the 'Guanches' by later Spanish colonizers, arrived ~2,500 years ago.[70] Although previously characterized as stone-tool-wielding foragers, we now know that these first settlers had a rich toolkit of pottery, varied lithics, bone, leather, shell, wood and other fibrous plants. Genetic, linguistic and material comparisons indicate that these first inhabitants had a clear affinity to North African Berber populations, and they brought goats, sheep, pigs, and the trinity of dogs, cats and mice with them to all the Canary Islands. They also cultivated novel crops, like barley, wheat, lentils, beans, figs and peas, carried from the mainland.[71] As seen in Madagascar, and indeed in many of the islands we have touched upon so far, these farmers used fire to clear and open the landscapes for their introduced domesticates. This resulted in drastic reductions of pre-human forest cover and 'lost forests' in parts of islands or even across whole islands. Certain native tree species, like the laurels and the strawberry tree, disappeared on islands like Fuerteventura. Growing sheep and goat populations, on small islands, halted forest resurgence and led to soil erosion, while introduced mice and cats sent native rodents, large and small, into decline. Direct hunting also seems to have resulted in the demise of a large native lizard and a flightless quail.[72]

Scientists have suggested that, unlike some of the other examples we have touched upon, a general lack of useful native plant resources and limited numbers of native large animals meant that human arrivals were forced to intensively pursue food production on the Canary Islands, with its corresponding consequences. Yet, again, the story is more complicated. A distinction can clearly be seen between small, flat islands like Fuerteventura and Lanzarote, whose dry forest vegetation was more vulnerable to human clearance and all but disappeared, and the wetter of the small islands like La Gomera and La Palma and larger islands of Tenerife and Gran Canaria, where forests remained standing upon European arrival. In the case of La Gomera, a sediment core collected from a former lake bed demonstrated that there was no clear signal of deforestation with human arrival and that previous climate changes had led to greater forest disturbance.[73] On La Palma, prehistoric humans changed their choice of tree species for fuel, recognizing the growing problem near their sites and venturing higher up into the mountains to ease the pressure.[74] On Tenerife and Gran Canaria, the remaining forest regions might be seen as even more impressive, given that these two islands had developed larger populations and a clear social hierarchy by the time European travellers coasted by their shores. Overall, the arrival of farming practices on the Canary Islands had stark consequences for local animals, plants and landscape stability. The issues of deforestation and over-grazing still plague these islands today.[75] Nevertheless, prior to European arrival, these impacts were dispersed and varied by island and location. As we will see in Chapters 10 and 11, Spanish colonization, ranching and sugar manufacturing, driven by imperial demand, initiated truly sweeping changes across the archipelago, including perhaps local climatic shifts,[76] that stay with its inhabitants to this day.

Tropical islands have often been thought of as ideal, isolated 'laboratories' in which to study the impact of prehistoric human arrival on previously 'pristine' ecosystems. While by no means a complete tour of all tropical islands, our trip around the Caribbean, Pacific Islands and the shores of Africa has shown us that hunting and gathering, and particularly forms of farming practice, can have major, lasting impacts on the native plants, animals, and landscapes, of these often-remote settings. Over the course of the Holocene, tropical islands have experienced a massive loss of endemic birds, amphibians, reptiles and mammals, driven by human hunting, habitat change, and the invasive species humans brought along with them, either accidentally or deliberately. Rapid soil erosion and a loss of forest cover were also witnessed, particularly as populations grew and farmers sought openings to plant more productive crops or societies sought new landmarks to differentiate status and act as cultural symbols. Entire ecosystems and environments could also be changed, or even lost, as new, economically useful plants were carried or promoted, and others fell away. Beyond the land, over-fishing and over-use of shellfish led to a dramatic reduction in size and also the shortening of marine food chains. These laboratories show us that prehistoric human impact could leave a massive lasting impression on plants, animals and entire island landscapes. Consequential changes to soils and available wild resources could also leave the settlers themselves facing an ultimatum of population collapse or island abandonment.

Nonetheless, the growing field of 'island archaeology' has shown that these outcomes need not be inevitable. Not just that, but that they could be averted through flexible decision-making even in the most challenging of scenarios. Many of the societies we have met, including those that brought farming 'packages' with them, adapted their lifestyles to their new situations, or adapted the new situations to fit their lifestyles in what has become known as human 'niche construction'.[77] Some complemented their domesticated diet with the use of local tropical forest plants and animals, or fishing in the reefs or deep blue seas off their island coasts. Some managed the local landscapes to ensure a continuity of wild, native species alongside more tentative introductions. Where changes in ecosystem dynamics became obvious, groups could clearly also make decisions to rid themselves of domesticates and reduce deforestation, setting themselves on a new course towards sustainability. As a matter of fact, the starkest warning from these island examples is not that island environments were inherently vulnerable in the past and remain inherently vulnerable now. It is rather that in the past, populations were able to change their land use and their economies as they saw the landscape change around them. They noticed soils sliding, prey dwindling and forests receding and could make decisions. Today, the forces of poverty, climate change, political priorities and large industries wash up on island shores, limiting the hands of the local inhabitants. As we have seen hinted at the end of each example discussed above, it was invariably the disruptive arrival of European empires and mercantile interests that brought these pressures of global consumer desires along with them, something we will explore in more detail in Chapters 10 and 11. However, before turning to the beginnings of a 'globalization' that defines our current relationship with the tropics, it is worth first exploring the massive Indigenous urban societies that developed *within* tropical forests.

9.   Cities in the 'jungle'

Visions of 'lost cities' in the jungle have plagued Western imaginations since Europeans first visited the tropics of Asia, Africa and the Americas. From the *Lost City of Z* to *El Dorado*, a thirst for finding ancient civilizations and their treasures in perilous tropical forest settings has driven innumerable ill-fated expeditions. This obsession has diffused into Western societies' popular ideas of tropical forest cities, with overgrown ruins acting as the backdrop for fear, discovery and life-threatening challenges across a number of popular video games (the *Uncharted* series), horror films (*The Ruins*) and novels (*The Jungle Book*). Throughout all these depictions runs the pervasive idea that all ancient cities and states in tropical forests were doomed to fail. That the most resilient occupants of tropical forests are small villages of poison-dart-blowing hunter-gatherers. And that vicious vines and towering trees or, in the case of *The Jungle Book*, a boisterous army of monkeys will, inevitably, claw any significant form of monumental human achievement back into the suffocating green from whence it came. This idea has not been helped by best-selling books or apocalyptic films which focus on the blockbuster *Collapse*[1] of particularly enigmatic societies such as the Classic Maya. Ultimately, the decaying stone walls, the empty grand structures and the deserted streets of these tropical urban leftovers act as a tragic warning that our own way of life, communities and economies are not as infinitely secure as we would often like to assume.

The situation has not been all that dissimilar in academic considerations of the potential of tropical forests to sustain ancient urbanism. On the one hand, intensive agriculture, seen as necessary to fuel the growth of cities and powerful social elites, has simply been considered impossible on the wet, acidic, nutrient poor soils of tropical forests.[2] On the other hand, where the obvious rubble of cities can simply not be denied, in the drier tropics of Central America, South

Asia and Southeast Asia, catastrophic ecological insolvency has been thought to have been inevitable. Deforestation to build massive buildings and make way for growing populations,[3] an expansion of agriculture across marginal soils, as well as natural disasters such as mudslides, flooding and drought, like today, must have made experimentation with tropical cities a big challenge at best and a fool's gambit at worst. Overhauling these stereotypes has been difficult, as the large, multi-year field explorations usually undertaken on the sites of ancient cities, for example in Mesopotamia or Egypt, face unique tropical trials. Dense vegetation, mosquito-borne disease, poisonous plants and animals, and torrential weather have made it arduous to find, reach and excavate past urban centres in tropical forests.[4] Where organic materials, instead of stone, might have been used as a construction material, the task becomes even more taxing. As a result, research into past tropical urbanism, its forms, economies and extents, has somewhat trudged along behind similar research in the semi-arid and arid zones of Mesopotamia[5] and Egypt[6] and the sweeping river valleys of East Asia.[7]

Yet as we have seen in Chapter 7, and also Chapter 8, many tropical forest societies found immensely successful avenues to food production, in even the most challenging of circumstances. While these might not look like the 'agricultural' fields we think of as supporting cities and states today, they could sustain impressively large populations and social structures. We will now turn to some of the most famous examples of supposed tropical disaster, the Classic and Postclassic Maya, the Khmer Empire of Cambodia, and the cities of northern Sri Lanka to see how the last two decades of intrepid archaeological exploration, applying the latest science from both the land and the air, have stripped away canopies to provide new, more favourable assessments. We will find that not only did they flourish, but pre-colonial tropical cities were actually some of the most extensive urban landscapes anywhere in the pre-industrial world – far outstripping ancient Rome, Constantinople/Istanbul, and the ancient cities of China. We will find that even the supposedly most 'pristine' forest region, the Amazon Basin, was actually home to millions of people extended across urban-like settlement networks. Ancient tropical cities could be remarkably resilient, sometimes surviving many centuries longer than colonial- and industrial-period urban networks in similar environments. Although they could face immense obstacles, and often had to overhaul their social structure and settlement locations to beat changing climates and their own eventual over-use of the surrounding landscape, they also developed completely new forms of what a city could, and perhaps should, be. Extensive, interspersed with nature, and combining food production with social and political function, these ancient layouts are now catching the eyes of twenty-first-century urban planners trying to come to grips with tropical forests as sites of some of the fastest-growing human populations around the world today.

As with 'agriculture', we tend to view the concept of a 'city' through a blinkered Western lens. Based on our own experiences, we think of them as compact, densely populated areas, the home of the administrative and political elites, full of bustling trade and manufacturing, and fed by vast agricultural fields and animal herds that are often located at some distance from the city boundaries. For over fifty years, since the archaeologist Gordon Childe coined the agricultural and urban 'revolutions' as two major stepping stones in human history,[8] the same has been broadly true of archaeology. Excavation of vast 'tell' mound sites, formed from humungous accumulations of mudbricks and refuse, in the Near East, which represent the earliest cities dating to around 6,000 years ago,[9] as well as the cities of the Classical World and ancient China, have acted as the standard for tracking the origins and expansion of a settlement pattern that, to a large extent, still shapes the lives of many of us today. This 'model' of social complexity unsurprisingly seems somewhat out of place in tropical forests, where sweeping fields of uniform crops, grazing animals and dense, built-up settlements can lead to drastic deforestation, subsequent soil erosion and, eventually, starvation and social disintegration. As a result, where such seemingly 'compact' cities have been identified in the tropics, for example in the case of the Classic Maya of south-eastern Mexico, Guatemala, Belize, and western Honduras and El Salvador, scientists have tended to assume the worst. That they and their farming focus on certain key crops were too much for their tropical forest landscapes, leading to degradation, an overthrow of rulers, and ultimate abandonment.[10]

Maya urban forms began to appear around 800 BC in the so-called 'Preclassic' period. Cities, monumental stone architecture and writing all gradually emerged at certain key political centres, led by kings and fed by the North and Central American staple crops of maize, beans and squash. It was during the 'Classic' period that things truly took off, especially in a region that has become known as the southern lowlands (including northern Guatemala, Belize and south-eastern Mexico). Between 250 and 900 AD, growing populations, more cities, more monuments and more inscriptions appeared. The leaders of the famous cities of Tikal and Calakmul, which could have populations of as many as 120,000 people, engaged in warfare, extended political influence, and harnessed long-distance, high-value prestige resources in ways that would make many of our more devious administrations proud today.[11] Although many of these urban nodes were located on soils particularly suited to productive maize agriculture, there was one problem. Annual rainfall variation in the region could be as high as 2,000mm (for comparison the average annual rainfall recorded at Greenwich Observatory London between 1981 and 2010 was 621mm[12]), and the geology often made the capture and storage of precious water through the dry months challenging.[13] As a consequence, major droughts witnessed in lake records from the region have been argued to have brought the system to its knees in the 'Terminal Classic' period, between 800 AD and 900 AD in the southern lowlands. No response was possible given that the large centres and their political classes had already over-reached, cutting down trees to fuel their monuments and planting their corn on poor soils as the more fertile regions 'filled up'. Precariously sustaining their populations across the heavily altered landscape, there was nothing they could do.[14] People lost faith in elites, construction stopped, famine ensued, and the 'Classic' population dispersed itself across the landscape to eke out a living in a drier world. Passing by these former centres a few hundred years later, Spanish explorers such as the infamous Hernán Cortés did not even deem what was left significant enough to comment upon[15] as they waited for eventual 'rediscovery' by intrepid American archaeologists of the nineteenth century.

So often goes the classic story of the Classic Maya. Amazingly, however, as early as seventy years ago, surveys of Classic Maya cities by the archaeologists William Coe and Gordon Wiley had already questioned such an interpretation of their urban organization and landscape placement. Far from being compact, we now know that even the most well-known of Maya centres, like Copán and Tikal, practised what has elsewhere been called 'agrarian-based, low-density urbanism'.[16] What this effectively means is that instead of being dense, the population was relatively dispersed (1–10 persons per hectare). Instead of having fields outside and politics inside, fields were located throughout the urban infrastructure and residences. And instead of a small focal point, cities sprawled for over 100 square kilometres. To take Tikal as an example, recent survey data has illustrated a network of moats, dwellings, reservoirs and pyramid clusters which extend out from a single hill for up to 200 square kilometres into the surrounding landscape.[17] Innovative aerial survey methodologies have now made similar findings elsewhere across the Maya world, from Copán to Caracol.[18] In almost all instances, instead of isolated urban buds, scientists have found vast landscapes of small and large centres connected by dispersed agrarian landscapes, residential areas, causeways, and a complex, interlinking system of dams, reservoirs, sinkholes, channels and swamps that supported growing populations through even the driest of seasons. As leading 'Mayanist' Professor Lisa Lucero, of the University of Illinois, puts it, 'having been there for over a millennium already, the Classic Maya knew the importance of water and of fertile agricultural soils, the latter dispersed in variously sized pockets, mirrored by a dispersed agricultural settlement. This low-density approach to cities was a logical, innovative solution.'

The Classic Maya also had far more diverse and sophisticated economies than has often been appreciated. Alongside the key crops, archaeobotanists have shown that the planting of avocados, pineapples, sunflowers, tomatoes and manioc added to a more dispersed settlement and lifestyle, than popular imaginings of endless rows of

maize. Analysis of modern forests growing around Classic Maya centres also shows that the occupants of these urban networks actively
managed tropical forest plants to promote economically useful species.[19] They did not stop at plants. The Classic Maya are also known
to have penned, fed and fattened wild turkeys and deer for use as key
protein sources. Overall, instead of rolling agricultural fields of single crops, scientists have now presented diverse 'forest gardens' as the
sustaining source of these cities.[20] Based on ethnographic study of,
and testaments from, Maya communities today, this type of cultivation, called *milpa* (or *kol* in the local Yukatek language), involves the
use of multiple crops, and the movement of fields, allowing different
parts of the forest to grow back and patches of soil to rest and restock
before planting begins in a locality again. We also know that instead
of indiscriminately planting in soils of all types, the Classic Maya
actually followed rich veins of particularly productive soils known as
'mollisols', giving their field systems a winding appearance that
snaked along rivers and up slopes. They even added special plants,
like water lilies, to reservoirs. These plants are incredibly sensitive to
water quality, only growing under clean conditions, and allowed
people to monitor the build-up of stagnant water and thus guard
against disease.[21] 'Ultimately, the wider Classic Maya landscape can
be imagined as a patchwork of seasonal wetlands and well-drained
forests used for palms, dyes, fruits, animals and construction materials, and open areas filled with diverse fields (*milpas*),' states Lisa.

The difficulty of maintaining large populations and complicated
social and political structures in a highly seasonal tropical landscape,
where water was often in demand, did eventually take its toll in many
parts of the so-called Maya 'heartlands'. Despite well-accepted catastrophic scenarios, it is immensely unlikely that any single, dramatic
drought brought about the end of a given city. Nonetheless, detailed
research by climate scientists, working on growing records of past
precipitation from lake sediments as well as the gradual build-up of
stalagmites in nearby caves, has shown an increasing seasonality in
rainfall amounts, an increasing number of droughts, and a gradual
downward trend in precipitation.[22] The extent of pre-Columbian
deforestation is not agreed upon.[23] However, pollen records from

sediment cores from throughout the southern lowlands suggest varying degrees of deforestation and forest management at each of the
hundreds of Maya centres, which,[24] in some cases, may have exacerbated climatic drying.[25] In the southern lowlands, where surface water
was hard to find at the best of times, many cities, including Tikal, saw
failing agricultural returns, growing malnourishment and stress.
Since kings claimed such close ties to the gods, in the face of droughts
and failed crops, their source of political power was called into question, especially as inter-city violence grew more frequent.[26] People
refused to work on monuments, seeing that they made no difference.
Ceremonial centres were abandoned in the southern Maya lowlands,
the ruins left for scientists arriving centuries later. Looking at the
sequences of certain sites, this may certainly seem like a rapid, uncom-
promising disaster. But was it really? Given such long-term Classic
Maya knowledge of their ecosystems, their well-tuned economies,
and sophisticated water management, things that twenty-first-century
urban systems may often lack, is a sweeping, rapid 'collapse' truly
likely?

In fact, Maya centres actually flourished into the Postclassic period
(900–1520 AD) (Chichén Itzá, Uxmal),[27] some right up to Spanish
arrival, simply re-focusing on sinkholes fed by groundwater or coasts
in the northern lowlands,[28] near the relatively few lakes and rivers in
the southern lowlands,[29] and, later, in the highlands of Guatemala
(e.g. Q'umarkaj (Utatlán)).[30] Even in the drying southern lowlands,
we can see that the wealthy and powerful suffered – but what about
everyone else? In many areas, it actually seems like the lives of the
independent farmers that were the key links in the chain of vast Classic Maya urban systems continued, albeit with much reduced
population sizes. In the region of El Pilar, surrounding the ceremonial centre of Tikal, 'forest gardens' were managed by farming
communities. Remarkably, this management, and the presence of
large numbers of farmers, continued through the rise and fall of Tikal
itself.[31] This diverse, *milpa* agriculture also persists among Indigenous
Maya communities who still occupy many parts of the region today.[32]
These groups still practise traditional manufacturing and landscape
management. Urban archaeologists, like our societies in general, tend

to focus on the 'rich' remains of elite structures which can make a big media splash and leave a big mark. But when your tropical city is built on sprawling networks of independent farmers and craftspeople, you can often miss the remarkable resilience inherent in the foundations of the system beyond the more obvious stars of the show.

The Greater Angkor region of Cambodia is another prominent, tourist-attracting set of ruins emerging out of a 'jungle'. Gap-yearing students, jetsetters, elderly tour groups and interested locals all flock towards the prominent, monkey-covered temple of Angkor Wat, which became the religious centre of the Khmer Empire in the twelfth century AD. Few of them, however, realize that this monumental shrine is actually just the tip of the Angkorian iceberg. Urbanism began to emerge in this part of the world from the first millennium BC. Initially this was based on walled and moated towns of 5 to 20 square kilometres with a focus on the management of water to feed their growing rice fields through the often severe regional dry season, which boomed into the first millennium AD.[33] From the eighth century AD, the region of Angkor took this to new levels. Beginning as two separate centres, by the ninth century AD the new capital of the Khmer Empire, the dominant state of mainland Southeast Asia at the time, Yasodharapura, had formed. The capital was characterized by massive earthen-built reservoirs known as *barays* and a series of walled administrative palaces and Buddhist and Hindu temples which thrived for over half a millennium, until the fourteenth century AD.[34] Previously, like tourists do today, many archaeologists had focused on the seemingly important, compact 'ceremonial' centres like Angkor Thom and Angkor Wat, built by the various rulers of the Empire.

But then two things happened. First, the French archaeologist Christophe Pottier and his local collaborators undertook decades of on-the-ground survey work, even in the face of security issues with holdouts from the Khmer Rouge rebellion in the 1990s. In doing so, they were able to plot the presence of vast numbers of construction features, large and small, across an entire 'Greater Angkor' region. Second, came a new method: Light Detection and Ranging (LiDAR).

One of the leading experts applying this methodology in tropical archaeology today is Dr Damian Evans of the École française d'Extrême-Orient, Paris. Dense vegetation has often hindered the identification of ancient buildings in tropical forests, particularly in Southeast Asia, where dwellings are known to have often been constructed from wood or bamboo, leaving behind just subtle changes in soil colour or elevation. However, as Damian describes, 'LiDAR allows us to virtually "strip away" vegetation. Using a laser scanner attached to an aircraft we carpet the terrain below us with pulses of lasers, collecting billions of points. Some bounce back from the trees but others slip through, allowing us to build a model of what lies beneath.' What they have found at Angkor is simply mind-boggling: an urban residential area of over 1,000 square kilometres has emerged

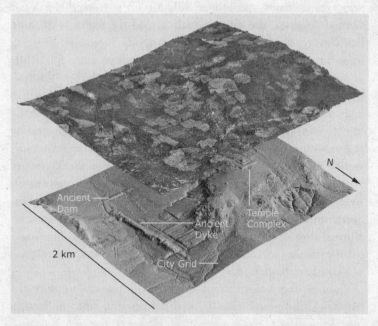

View of the central temple zone of Mahendraparvata, Phnom Kulen, Cambodia, from a 2015 LiDAR acquisition. Top layer: treetop surface elevation model textured with conventional aerial imagery. Bottom layer: bare-earth terrain elevation model rendered with a hillshade effect.

alongside a modified landscape of 3,000 square kilometres.[35] This makes Greater Angkor the most extensive pre-industrial settlement complex anywhere on the face of the Earth, and larger even than cities such as Paris today. It also revolutionizes our understanding of how this ancient megacity operated.

As in the case of the Classic Maya, where LiDAR has also exposed vast, settled urban landscapes, rather than being solely composed of compact ceremonial centres, Angkor was another sprawling example of 'agrarian-based, low-density urbanism'. At ceremonial centres like Angkor Thom, LiDAR filled in the apparently spacious courtyards to reveal a densely packed religious support team living in wooden pile houses within the stone walls.[36] Beyond focal points, numerous sprawling domestic mounds, smaller shrines and rice fields extend from right up against the boundaries of sites like Angkor Wat, out across the lowlands, and into the hills. Similar to tropical North and Central America, the urban populace used the forests, exploiting fruits in managed orchards, wild palms, and vegetables, ensuring some forest cover persisted.[37] Rivers and ponds provided fish, while zooarchaeologists have shown that pigs, cattle and chickens snuffled, brayed and clucked through the busy streets. This expansive human and animal population could only be supported among the seasonally dry forests that line the Siem Reap river delta by the human creation of a vast water drainage system. As well as buildings, the earlier surveying and later LiDAR work together show a dendritic network of water transport and storage, from giant reservoirs and temple moats to small, locally maintained channels. Ingenious design and application of sandy clay to parts of the probing tendrils of this vast construct allowed it to deal with both droughts and floods,[38] and the Angkorian leviathan persisted through many a dry season and political crisis until the fourteenth century AD.

Unsurprisingly, this vast urban metropolis placed a strain on the tropical landscape. Human-induced deforestation and soil erosion have left a clear signal in the region. However, the agrarian urban spread dispersed this across a catchment rather than intensively impacting any particular area. Angkor's time did, however, eventually come to an end. During the late fourteenth century AD, increasingly extreme climatic swings between drought and deluge ruptured parts of the water network and took their toll on farming yields.[39] Palaces and temples were abandoned and the urban landscape broke up, as farmers saw no reason to stay within a dysfunctional state-guided system of competing elites. But, again, instead of complete 'collapse', the result was more interesting. The rulers moved to new, more compact cities in the Phnom Penh area, where the current capital of Cambodia still stands to this day. Meanwhile, the diligent farmers populated more numerous small towns along the banks of the Mekong River and Tonlé Sap Lake, preferring more stable water courses for their future endeavours. As with the Classic Maya, then, the governing system certainly failed, particularly in the wake of climatic challenges facing a region prone to seasonal variation in rainfall. The eventual abandonment and upheaval seen at Greater Angkor is not something to be sniffed at. However, the elite thought up a new strategy and moved to new sites of power.[40] Meanwhile, the farmers that had fuelled the sprawl from its beginning continued on a landscape that had provided for them for so long, albeit focusing their efforts on more profitable areas for cultivation and herding. Once we cast our eyes away from the overgrown walls of temples, we can begin to see how Greater Angkor represented a vast, highly resilient approach to urban living in the tropics.

A comparable state of affairs can also be found in one of the greatest ancient urban tourist attractions in South Asia, Anuradhapura of the UNESCO 'Golden Triangle' in Sri Lanka. Capital of the Sinhalese King Pandukabhaya, construction seems to have begun around 500 BC.[41] Monumental Buddhist temples (*stupas*), including the largest structure made from bricks anywhere in the world, monasteries, palaces and state-directed reservoirs to support intensive rice agriculture characterize this centre from the third century AD. The water 'tanks' are particularly impressive, with the largest, Nuwarawewa, built in the first century AD, reaching as large as 9 square kilometres.[42] By the ninth and tenth centuries AD, Anuradhapura dominated the political landscape of the island. The three great monastic centres of the capital could have sustained 30,000 religious worshippers at any given time, documented by immense preserved stone vats that could have each fed rice to 1,000 people every mealtime. Comparably, only 1,000 worshippers were invited into the Hagia Sophia for prayers when it was converted into an active mosque in 2020.[43] However, once again it has taken dedicated survey work to see past Anuradhapura's stone and brick behemoths to the 500 square kilometre sprawl that extended out into the wider countryside, housing an estimated 250,000 people at its peak.[44] That's an area over four times larger than the modern city of Liverpool.[45] This spread was fed by timber and stone canals which linked the central tanks to water storage constructions in the hills nearly 100 kilometres away. While these larger layouts were state directed, as at Greater Angkor, small field systems, reservoirs and alterations to rivers were undertaken by local farming groups that occupied the surrounding landscape, independent of the larger state building works and political power shifts. These farmers also developed gardens and moved their plots of land on a frequent basis to maintain intact parts of dry tropical forest to diversify their diets and landscape use beyond rice.

Political and climatic fluctuations eventually spelled the end of ancient Anuradhapura as a ceremonial and political centre. The Sinhalese capital moved first to the nearby Polonnaruwa in the twelfth century AD, before rulers abandoned the drier north altogether by the thirteenth century AD and moved south to the wetter tropical forests of Dambadeniya and eventually Kandy.[46] This may have been stimulated by fluctuations in the Indian Ocean monsoon system, which brought intense drought to the reservoir-dependent cities of the north, a region that faces marked arid dry seasons in the summer months to this day. Nevertheless, although the state system faded locally, through relocation it persisted in the region as a whole. Furthermore, the more independent farmers once again ploughed their own course beyond the end of the political structure. In the later twelfth century AD, after the rulers had already left, there is evidence that some dogged farmers reoccupied different parts of Anuradhapura and dusted down the monuments. Moreover, although when British colonial bureaucrats arrived in the north in the nineteenth century they found ruin, they also found continued small-scale rice farming and water management across the Anuradhapura region,[47] which, incidentally, still persists as a city, albeit fuelled by tourism. In fact, having been occupied for around 1,700 years as an urban centre, ancient Anuradhapura stood far longer as a Sri Lankan capital than the current city of Colombo, which, while occupied by multi-cultural traders previously, was only truly established as a major urban centre following Portuguese, Dutch and British colonization from the sixteenth century AD onwards. Given that this period is also far longer than most of our European cities have stood the test of time, the Sri Lankan example should once again make us reflect on our common assumptions that tropical forests make bad homes for urbanites.

The lowland evergreen and semi-evergreen rainforests of the Amazon Basin are another challenge altogether for urban societies compared to the seasonally arid dry tropical forests of Central and North America, South Asia and Southeast Asia. In fact, the heat, sticky humidity and acidic, often flooded, soils led some archaeologists and anthropologists to assume that both cities and settled agriculture were environmentally impossible.[48] Most surviving Indigenous communities in these forests were, after all, living in small villages with no clear evidence for social hierarchy. Meanwhile, today, expanding infrastructure, development, plantation agriculture and ranching undoubtedly does untold harm to these environments. Nevertheless, Chapter 7 has already shown us that more innovative, ecologically sensitive paths to food production were under way in the Amazon for much of the Holocene. In fact, renewed survey by a number of dedicated teams of archaeologists, anthropologists, environmental scientists and Indigenous communities, as well as tragic deforestation stripping away protecting forest, has revealed substantial 'garden city' landscapes of earthworks, clustered structures, and 'road-like' paths across the Amazon.[49] In and around the Xingu River, these settlements reached their peak between 1250 and 1650 AD, just prior to European arrival. Intriguingly, they show a similar pattern to the cases of 'agrarian-based, low-density urbanism' we have already visited. In each case, clear, larger towns at central points were surrounded by monumental wooden walls and ditches, connected to a number of satellite villages by pathways cleared through the forest. Instead of mass deforestation, these settlements were separated by intact forest used for the management of fruit-bearing trees, ponds for farming freshwater turtles and fish, and more open fields for manioc and maize.[50]

Many years of research have also documented a similar series of urban-like settlements at Marajó Island, right at the mouth of the Amazon River. During the first millennium AD, mound sites were produced by the sustained occupation of large domestic residences, burial areas, elaborate pottery and waste disposal. Increasing in size and density until the start of the fourteenth century AD, population estimates for this region at its peak have been as large as 100,000.[51] Whatever the true size, as in the 'garden cities' of the Xingu, archaeobotany, zooarchaeology and stable isotope analysis of human remains have shown that the occupants of Marajó Island, as well as the nearby Maracá region, fed themselves through a mixed system of foraging for diverse plants, hunting of wild animals and fishing or penning freshwater resources in and around intact rainforest, with some limited cultivation of manioc and maize in more open patches perhaps occurring in addition.[52] A similar scenario is seen at the now famous settlements of Santarém, this time at the mouth of the Tapajos River. Here, a subsistence combination of manioc and other root crops, alongside tree fruits and fishing, and the occasional, perhaps ritual, use of maize once again drove the success of an interlinking network of denser, occupied settlements and more dispersed satellite villages. Interestingly, in this case, each satellite may have had its own specialism, with some showing a particular focus on the hunting of rainforest animals.[53] Ultimately, together these Amazonian forms of 'agrarian-based, low-density urbanism' may have enabled populations as high as between 8 and 20 million to have existed across Amazonia by the time of European arrival. Given that the population of all of Europe in 1492 AD is estimated to have been between 70 and 88 million,[54] this represents a considerable number of human lives being sustained within what must, undoubtedly, be seen as substantial urban types of settlement, albeit very different from our usual ideas of cities.

These Amazonian examples have perhaps been neglected in global discussions of urbanism because they lack the more obvious stone architecture that many more classic examples display. Yet they had their own 'monuments'. We met the early cultivators of the Llanos de Moxos and their 'forest islands' in Bolivia in Chapter 7. However, from the start of the eleventh century AD, their interventions in the landscape to protect their key crops from flooding began to become even more large-scale, as extensive mounds, earthworks and ditch systems started to emerge.[55] More ceremonial constructs by so-called 'earthwork builders' have now been found across north-eastern Bolivia, the uplands of the Ecuadorean Amazon, Acré State, Brazil in south-western Amazonia, and across the eastern Amazon.[56] Perhaps the most resilient 'monuments' of Amazonian urban life, however, are the vast changes made to the soils to support increasingly large populations. As Professor Eduardo Neves, a leading Amazonian archaeologist based at the University of São Paulo, puts it, 'the *terra pretas* (now known as Amazonian Dark Earth) soils are perhaps the late Holocene Amazon's greatest legacy'. These are clear layers of dark soils found full of charred bones and ceramics left over from occupation and seem to be deliberate attempts to maintain soil fertility through controlled forest burning and the mixing of fertile waste products.[57] They were likely created by food producers, within these wider urban systems, that kept moving their fields around in a so-called 'slash and burn' system, to try to maintain forest cover and also allow soils to locally rejuvenate every so often. Amazonian Dark Earth soils are almost always found in the growing settlement sites of the first and second millennia AD in the Amazon. 'Their enormous extent, now known to span across the eastern and western Amazon, stands as testament to the spread of populations, and their "garden cities", across this region,' continues Eduardo.

As well as being yet another example of more dispersed and integrated types of urbanism than we are used to, these Amazonian settlements are interesting for another reason. Despite showing many of the hallmarks of urban networks (craft production divided into different, dependent satellites, significant population size, and physical and ecological monuments), they almost completely lack evidence for a ruling class. At the sites of Marajó and along the Xingu River

there are no distinctive fancy 'palace' residences, no obvious division
of 'wealth' in terms of personal property, and no hoarding of elaborate grave goods by certain individuals. 'Nobles' and directed
'warbands' noted by European chroniclers imply some kind of centralized control, though these accounts could also have been impacted
by the biases of those observing. Indeed, overall, surviving images
and feasting contexts seem to be based more around shamanic rituals
and interaction with forest environments than a display of overt
political power.[58] A similar state of affairs has been documented in
sub-tropical Mali at the site of Jenné-jeno, where large numbers of
people, complex satellite settlement networks, and walled mounds
seem to have appeared by the ninth century AD without a clear
ruling class.[59] Likewise, archaeological records of dispersed, but connected, settlements can also be found in the Middle Senegal Valley
throughout the first millennium AD. Ethnography in West Africa, as
well as in Bali in Southeast Asia, has shown that it is possible that
complex works and large numbers of people can be coordinated by a
'heterarchy'.[60] In these systems, power is spread more horizontally,
rather than vertically, with decisions made by representatives of different crafts, ritualistic specialists or families. It is possible that such a
structure existed in the archaeological examples of Amazonia and
West Africa. It may even have helped the sustenance and maintenance of widely spread urban populations. Either way, the more
obvious divine rulers of the Classic Maya, the Khmer Empire, or
Anuradhapura are currently nowhere to be seen.

The degree to which the 'garden cities' and their other urban-like
cousins impacted Amazonian environments is currently unclear. Vast
earthworks, burning of forest to produce human-modified soils and
maintain open patches for crops, as well as the selection of certain
economically useful trees, must have had some wide-reaching effects,
particularly given the potential number of people undertaking these
activities. Nonetheless, there is, as of yet, no unified message from
environmental records taken across the Amazon Basin. Some scientists have argued that, in Bolivia, 'earthwork builders' simply used
already-open grassland landscapes produced by drier climates just
before 2,000 years ago.[61] Meanwhile, further south, environmental

archives seem to show no evidence for permanent clearance in connection to ceremonial landscape modifications. Others, however, have debated these findings and the case very much remains open.[62] In general, most records of Amazonian burning during the timeframe of these 'garden cities' seem to fit with a shifting form of farming (known as 'swidden'). The temporary clearance of certain patches for a few years of cultivation, prior to movement to a new area, supported dynamic, mosaic landscapes rather than complete clearance, and no evidence for severe soil erosion or environmentally induced collapse has been suggested. In fact, many of these urban forms had been in place for 1,000 years when European colonizers arrived. As we will see in Chapter 10, it was then that everything changed. Ultimately, all that was left were the small villages of Indigenous peoples met by anthropologists in the twentieth century. Assumed to have been timeless and pristine like their forest homes, all traces of their previous urban glories were erased from history until the last three decades of archaeological research and scientific methodologies, alongside better consultation of these same Indigenous communities and recognition of their stewardship and knowledge, have raised them from the earth, challenging widely held misconceptions.

'Agrarian-based, low-density urbanism' does not characterize all ancient cities in the tropics. Compared to more 'compact' city forms, which persisted from 6,000 years ago in the Middle East through to medieval Europe, and also emerged in the wake of the fall of some of these monstrous urban expanses in the tropics, it is certainly rarer and, on balance, more prone to disappear from a region. However, 'agrarian-based, low-density urbanism' provides an important avenue for looking at past cities in tropical forests as both obviously possible and incredibly inventive. The diverse use of wild plants and animals in managed forest patches, fishing in rich freshwater settings, and the mobile use of open areas for crops, provided sustainable pathways to food production sufficient to fuel what were some of the largest cities in the pre-industrial world. Dispersing the impacts of growing populations across satellites tempered the loss of unique tropical biodiversity and soil quality in the face of growing human pressures. They could certainly face challenges, particularly in seasonally dry forests where human-induced deforestation and climate change could tip the scale between precarious viability and failure. Nevertheless, even in these instances, urban centres and ruling classes emerged elsewhere on the landscape. Meanwhile, ecologically experienced food producers persisted, often in large numbers, on the still productive, though challenging, landscapes that had fed them for so long, even as monuments crumbled around them. Not only that, but while many of these types of cities might have fallen to ruin, ultimately, their hardiness can be seen in their often long spans of existence – longer than 500 years in the case of Greater Angkor and some Classic Maya urban centres, and nearly two millennia for Anuradhapura in Sri Lanka. This is far longer than the majority of industrialized tropical cities, as well as many modern European and northern North American cities, providing a very different perspective on our assumptions about their inevitable 'collapse'. Indeed, the timeless strength of 'agrarian-based, low-density urbanism' remains an attractive model for present-day urban planners looking for 'green cities'[63] that balance urgent conservation and environmental needs, political and cultural infrastructure, and growing urban populations across the twenty-first-century tropics,[64] something we will return to in Chapter 13.

There are many, many more examples of cities ruled over by precolonial states and empires that emerged across the tropics in North, Central and South America (e.g. those of the Triple Alliance of the Aztec Empire and the Inka), mainland and island Southeast Asia (e.g. Bagan of the Pagan Kingdom, Borobudur of the Sailendra Dynasty in Java), the Pacific (e.g. on Tongatapu at the centre of the Tu'i Tonga Empire), and West and Central Africa (e.g. the centres of the Oyo Empire and the Kingdom of Benin in Nigeria, the Ashanti Empire of Ghana, and the Kongo Kingdom of the Democratic Republic of the Congo) during the late Holocene, some of which we will meet in the next two chapters. A significant number of these, like the examples of the Amazon and the Postclassic Maya, were still flourishing at the time of European contact. They were often even actively admired by European visitors.[65] So why then do we now tend to think of tropical forests as so hostile to large, food-producing human populations? Why do all our popular assumptions fit ruins and small, isolated Indigenous bands of foragers to these environments rather than lively, humming streets, residential blocks and monumental constructions that could stand the test of time? The answer probably has something to do with what happened next. Arriving on ships across the horizon, Europeans did not just bring their notebooks with them to the tropics. They brought new diseases, new crops, new animals, new ways of using and seeing the natural world, and a political, religious and social agenda that sought to 'progress' from anything that had gone before. From the traumatic clash of worlds that followed can be found the origins of our globalized, but unequal, world. It is here that our modern economic, political and climatic reliance on the tropics, no matter where on the planet we live, began . . .

10. Europe and the tropics in the 'Age of Exploration'

'We were not discovered', read the statement released by a coalition of Indigenous peoples in Mexico's western state of Michoacan on 12 October 2020 – the anniversary of the day when Christopher Columbus (or Cristóbal Colón as he was actually known) first landed in the Americas.[1] 'Columbus Day' has been a Federal Holiday in the United States of America since 1971,[2] and forms part of a wider public consciousness in Europe and northern North America that has tended to hold up Columbus, as well as other individuals such as Ferdinand Magellan, Vasco de Gama and Walter Raleigh, as fearless pioneers of an 'Age of Exploration' that spanned the fifteenth and seventeenth centuries. Particularly within the tropics, they are often credited with 'finding' mineral wealth, exotic plants and animals, and new, productive lands that were ripe for the taking – boosting the global economic and political standing of expanding European empires as they did so. Away from individuals, particular royal lineages are similarly frequently focused on as the active shapers of 'New Worlds', inventing new ways to travel the oceans, and wielding unparalleled weapons, as they sought to extract new tropical resources and labour in a bid to gain religious and political prestige back home on the competitive stage of Europe.[3] Certainly, when Pope Alexander VI divided up the lands of the Americas and Asia between the 'Catholic monarchs' of Castile and Aragon and the Portuguese in 1493 as part of the Treaty of Tordesillas,[4] he had little thought for how the people already occupying these lands might react. This European focus can lead us to assume that existing Indigenous inhabitants were passive or insignificant throughout this process. Something that, as we have already seen in Chapters 8 and 9, could not have been further from the truth.

In the fifteenth century, Europe was actually something of a backwater. In the thirteenth and fourteenth centuries it had benefited from a renewed flow of spices and commerce along the Silk Roads that united Europe, the Middle East, eastern Africa and Asia, following the expansion of the Mongol Empire and its successors and the relative peace and stability that ensued (the so-called '*Pax Mongolica*').[5] However, by the fifteenth century, the expansion of the Ottoman Empire centred on Constantinople had blocked direct European contact with Central and East Asia and these crucial lines of commerce.[6] Meanwhile, in the tropics, powerful empires had emerged. In the Americas, Tenochtitlan (now Mexico City) was the centre of the Triple Alliance of the Aztec Empire.[7] In Asia, Vijayanagara was the capital of the Hindu Vijayanagara Empire in southern India.[8] In Africa, the city of Gao was the leading city of the powerful Islamic Songhai Empire in Mali.[9] This is not to mention the significant numbers of other empires, kingdoms and polities that also occupied the tropical portions of the Americas, Africa and Asia, jostling with these larger powers for prestige, wealth and survival, as well as the innovative resilient island societies we met in Chapter 8. Even in the Amazon Basin, Chapter 9 has shown us that well-populated 'garden cities' sprawled across lowland evergreen rainforests, and the latest estimates put the human population of the pre-colonial Americas at 60.5 million,[10] just shy of Europe's estimated total at the time. As the last two decades of historical and archaeological research have shown us, tropical landscapes at the time of Columbus's voyages were certainly not empty. Instead, they were very much full of active rulers, merchants and food producers who took part in vast exchange systems that already spanned across the Americas,[11] across Africa[12] and across Southeast Asia[13] prior to European arrival. Perhaps unsurprisingly, many Latin American nations therefore question the validity of celebrating misguided, and often murderous, European individuals, instead of the Indigenous achievements that preceded them.

So what changed? How did powerful tropical states and individuals, in stark contrast to their European counterparts, end up as the 'people without history'?[14] Contact between Europeans and the diverse societies inhabiting the tropics generated shockwaves that have echoed through the landscape of global history ever since. We will now zoom in on tropical forests as the crucial theatre for the

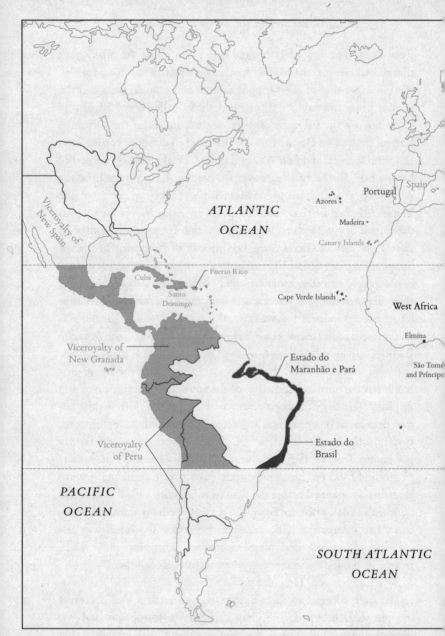

Map showing the pan-tropical extent of the Spanish Empire, prior to the nineteenth century and the Portuguese Empire during the Iberian Union (1581–1640) (note that shading of their extent is only provided for the tropics). Key regions and centres mentioned in the text are shown. The boundary of contemporary Brazil is also shown for reference.

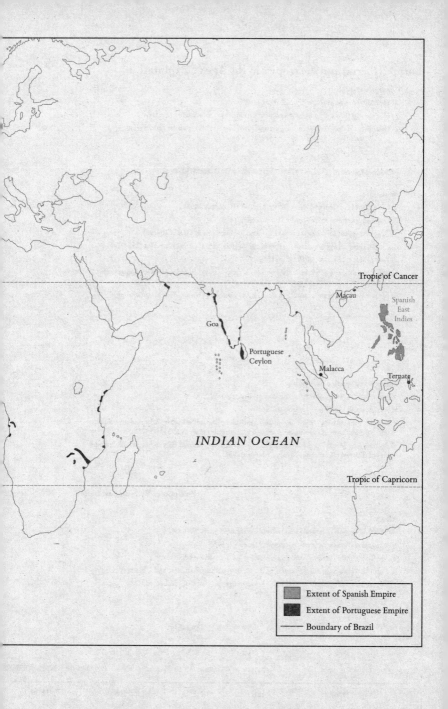

Tropic of Cancer

Macau

Spanish
East
Indies

Goa

Portuguese
Ceylon

Malacca

Ternate

INDIAN OCEAN

Tropic of Capricorn

Extent of Spanish Empire

Extent of Portuguese Empire

Boundary of Brazil

Europe (Iberia) and the tropics in the Age of Exploration

Canary Islands

a) Spanish arrive on the Canary Islands in 1402

b) Diseases rip through Indigenous populations in 50 years following first arrival

c) Gruelling campaigns against Indigenous Guanche populations across the different islands, see final subjugation of Tenerife in 1496

Canary Islands

Neotropics

a) Columbus lands and establishes settlement on Hispaniola 1492

b) Columbus returns to find settlement sacked in 1493

c) First permanent settlement on the mainland in Venezuela in 1501 (Cumaná)

d) Cattle, sheep and goats start to arrive in the Caribbean and Central America (1501–50)

e) Portuguese reach coast of Brazil in 1502

f) 150 years following Iberian arrival witnesses various waves of disease destroy Indigenous populations (a little later in Brazilian Amazon)

g) Initiation of sugarcane planting in fields in the Caribbean within 25 years of landing

Neotropics

Philippines/Spanish East Indies

a) Spanish attempt at colonizing the Philippines begins in 1565

b) First Spanish settlement at Cebu in 1565

c) Spanish conflict with 'Moro' populations including Borneo Sultanate, Maguindanao Sultanate and Sulu Sultanate. Ongoing resistance to Spanish rule between 1565 and 1898

d) Spanish take Manila (present capital (1571)) and establish Spanish East Indies, which includes bits of Taiwan, Guam, the Marianas islands and Palau

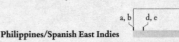

Philippines/Spanish East Indies

Africa

a) Mid-15th century, caravel invented and Portuguese reach West Africa

b) Expansion of sugarcane cultivation on Madeira in 1455

c) São Tomé and Príncipe found in 1469 and sugar plantations begin there

d) From late 15th century, expansion of Portuguese trading with West and Central African states

e) Portuguese traders settle at sites like Elmina under the watch of local elites

f) Maize reaches Africa around 1500

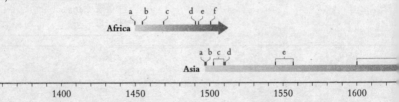

Africa

Asia

| 1350 | 1400 | 1450 | 1500 | 1550 | 1600 |

Year

h) Spanish with Indigenous allies defeat Triple Alliance and claim sovereignty over
   Mexico – 1519–1521 (Aztec)

i) Spanish and Indigenous allies capture Inka leader Atawallpa in 1532

j) Potosí mine (Bolivia) opens in 1546

k) Final defeat of the Inka Empire following gruelling war in 1572

l) Portuguese establish captaincies and sugarcane plantations are expanded rapidly in
   areas such as Pernambuco

m) Gold discovered in Minas Gerais, Brazil, in 1693, causing a rapid influx of population until 1739

n) Portuguese settle region that is now Amazonas state and formally incorporate into empire in 1750

e) Manila 'galleon trade' begins in 1571

f) Introduction of sweet potatoes, horses, etc. between 1630 and 1800 AD.
   Sweet potatoes taken up widely by Indigenous populations

g) Initiation of wet rice agriculture on the Ifugao terraces between 1630 and 1800 AD,
   with staunch resistance to Spanish rule in the Highlands

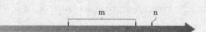

Asia

a) Vasco da Gama rounds the Cape of Good Hope between 1497 and 1498

b) Portuguese reach south-western India in 1498

c) Portuguese establish forts on Indian and Sri Lankan coastline and capture
   Goa amid a series of conflicts with local rulers 1503–10

d) Portuguese capture of Malacca in 1511 disrupts thriving spice trade between states in India,
   the Middle East and China

e) Portuguese conflict with China until China allows Portuguese traders to settle,
   enabling a new flow of trade between Europe, Goa and China 1545–57

f) 17th century, Portugal fades from prominence in the Indian Ocean due to defeats to
   local rulers and Dutch assaults

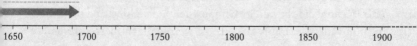

| 1650 | 1700 | 1750 | 1800 | 1850 | 1900 |

Year

emergence of a new, interconnected world. The latest archaeological, ecological and historical thinking reveals how the interaction between Europe and the tropics brought disease, warfare, forced relocation, and demands for labour and land that catastrophically impacted Indigenous populations and their traditional management activities. Not only that, but through both deliberate and unintentional 'Columbian exchanges',[15] plants, animals, people and beliefs began to move regularly between cities as far apart as Madrid, Mexico City and Manila, reconfiguring entire tropical ecosystems. The outcomes for landscapes and societies were shaped, not only by the political, economic and religious situation back in Europe,[16] but also by politics within and between Indigenous states, by local merchants seeking to tap into new global flows of goods, and by Indigenous food producers and hunter-gatherers incorporating or resisting new crops and animals. However, by the end, the varied, and often violent, processes of colonialism, and the overall European control of global flows of wealth across oceans,[17] led to a masking of the previous human achievements in the tropics. Ultimately, it was this collision of worlds that left us with the pervasive Euro-American assumption that tropical forests could only be successfully occupied by small-scale, mobile hunting and gathering communities – 'green hells'[18] needing to be profitably converted rather than productive, manageable 'forests of plenty'.[19]

In the fifteenth and sixteenth centuries AD it was the expansion of the Spanish (in the form of the combined 'Catholic monarchy' of Castile and Aragón) and Portuguese Empires that began the joining of two 'worlds' that had been separated since the division of the Pangaea supercontinent ~150 to 65 million years ago. Of these empires, the Spanish quickly became the largest. Following the gradual occupation of the Canary Islands (1402–96), and Columbus's landfall at Hispaniola in the Americas in 1492, Spain expanded its pan-tropical grip across the Caribbean, North and Central America, South America[20] and, towards the end of the sixteenth century, parts of Southeast Asia. So how did they do it? Gunpowder, horses, steel and shipbuilding innovations certainly helped.[21] So too did the ruthless enslavement of local populations[22] that began almost as soon as Columbus stepped ashore.[23] The manoeuvring of local political rivals, often using the Spanish to settle their own scores, was also key. Yet one of the most terrible products of the 'Columbian exchange', reconfiguring power dynamics both regionally and globally, was the fact that Europeans did not just bring themselves to the tropics. As we now know from extensive archival research of records left by colonizers and Indigenous writers, as well as the recent ground-breaking ability of archaeogeneticists to extract the DNA of not just past humans but also the diseases that debilitated them,[24] they also brought along the microbial causers of deadly measles, smallpox, typhoid, influenza and the bubonic plague. Where populations had never before been exposed to them, and had little to no immune response, the result could be catastrophic. Reconstructing exact population changes pre- and post-colonial arrival using census data, archaeological sites and other forms of historical records is notoriously contentious, and there have been fiery debates relating to estimates of pre-colonial populations in the Americas for the last fifty years.[25] Nonetheless, demographic and earth systems modeller Dr Alexander Koch, at Hong Kong University, tells me that he and his team's latest compilation of published estimates from colonial census figures, archaeological data (e.g. numbers of buildings, settlement size), and predictions based on assumptions as to how many people a given environment could hold[26] 'suggest a staggering ~90% of the Indigenous population of the Neotropics was wiped out within 150 years of Columbus' arrival by repeated waves of these diseases'. That is nearly 55 million people.

Two prominent, but very different, tropical examples highlight this reality. In 1492, Tenochtitlan (now, for the most part, buried under the present-day capital of Mexico, Mexico City) was one of the most densely populated cities on Earth. Tenochtitlan, the home of the Mexica ethnic group, lying just south of the Tropic of Cancer, sat at the centre of the so-called 'Aztec' Empire, based on the Triple Alliance of three city states, which extended out across the Valley of Mexico and into the sub-tropical and tropical lowlands and highlands beyond. The complex drainage system of this often-flooded area, specialized raised fields known as *chinampas* for growing a diversity of crops, including maize, chili peppers and beans, and canals for transport as well as fish and wild game, formed the basis of a vast 14 square kilometre anthropogenically constructed 'floating' urban landscape.[27] Royal courts, the gold, silver, jewels and exotic feathers of wealthy elites arriving through long-distance exchange networks, detailed inscriptions and texts, vast markets, and giant temples built for human sacrifices both impressed and appalled the Spanish arriving at the capital in 1519.[28] Numbering just 630, the Spanish, headed by Hernán Cortés, defeated this giant through political and biological opportunism. First, after initial encounters and skirmishes, the Tlaxcala, the traditional enemies of the Triple Alliance, entered into an alliance with the Spanish.[29] This was made easier by the fact that an enslaved Indigenous woman from the Gulf Coast of Mexico, *La Malinche*, gifted to the Spanish by a Maya state, acted as an interpreter, adviser and intermediary for Cortés.[30] Backed by military support from their local allies, the Spanish advanced on to Tenochtitlan, where they somehow managed to abduct King Motechuzoma I from right under the noses of his subjects. Following retaliation and ejection from the city, the Spanish returned with reinforcements, and another ally, this time from the nearby city of Texcoco, to lay siege on the capital. Nonetheless, given the ongoing staunch Mexica resistance that faced them, if it had not been for a smallpox pandemic tearing through Tenochtitlan prior to the crucial battle for the city, something documented in the colonial archival records of the time, they might never have succeeded.[31] Further pandemics, alongside the forcing of involuntary labour of enslaved Indigenous peoples in mines,[32] and brutal public executions,[33] reduced and pacified the population of the Valley of Mexico, making the subsequent maintenance of control more feasible than it might otherwise have been.

A similar end awaited the Inka Empire among the montane tropical forests and grasslands of the High Andes. By area, in 1492 the Inka had the largest empire on Earth, one that spanned lowland evergreen rainforest in the Amazon Basin to the deserts of Peru. A humungous 6,000 square kilometres (that's an area greater than 1,000,000 football pitches!) of agricultural terraces identified by archaeological landscape survey have been linked to pre-colonial Andean populations, perhaps even giving the mountain range its name (from '*Los Andenes*' for steps or ter-races) (though see note[34] for a discussion of alternative origins). Many of these likely supported an Inkan agricultural system based on the crops of potatoes, quinoa and maize, and domesticated llamas, alpacas and guinea pigs. A sophisticated archaeological and historical record of pic-tographs, winding roads, well-placed warehouses and knotted-string counting systems that acted as a means of conveying tax obligations, census records, calendars and military orders enabled the Inka to trans-port goods produced in different corners of the Empire to where they were needed.[35] All of this was ruled out of the mountainous city of Qosqo (now Cusco) with its glittering rulers, monumental plazas, giant stone blocks covered in plates of gold, shrines and tombs. In 1532 an even smaller Spanish party, 168 men led by Francisco Pizarro, moved on the Inka territory, having already seen its riches in 1526. While horses and cannon certainly helped, Pizarro, like Cortés, benefited from the local political situation. The Inka Empire had just emerged from a gruelling civil war. Pizarro made the most of the fact that the winning ruler, Atawallpa, left his mountainous citadel and met him at Kashamarka (Cajamarca). Here, he captured, blackmailed and then killed him, throwing his subjects into disarray. Pizarro advanced on to Qosqo. Local factions within the Empire saw an opportunity for advancement and supported the Spanish progression. Nevertheless, as Pizarro's own journals attest, he never felt that he was in a strong politi-cal position. Once again, this improbable conquest (it took until 1572 to complete) was spurred on by rampant smallpox, typhus and influenza epidemics, vividly recorded in text and images by Inka and Spanish chroniclers alike,[36] that had already killed Atawallpa's father as well as 200,000 subjects, and continued to catalyse civil unrest and eventual capitulation to the Spanish. Even then, though, the conquest of what was to become the 'Viceroyalty of Peru' took decades, with the moun-tainous and forested terrain frequently invalidating Spanish military advantage.[37]

The Canary Islands, in many ways, provided the sub-tropical step-ping stones for the Spanish on their path to the Americas. And here, once again, the spread of diseases was crucial to conquest. Control of the Canary Islands significantly reduced the length of the treacherous Atlantic crossing for the Spanish to reach the Americas (in much the same way the Azores islands facilitated long-distance Portuguese voyages). As Friar Espinosa, a sixteenth-century Spanish historian and chronicler, wrote, 'if it had not been for the pestilence it [the conquest of the Canary Islands] would have taken much longer'.[38] Even then, fierce resistance by well-organized societies, particularly on the islands of Fuerteventura and La Palma, meant that it was a long, hard and often fortuitous slog to victory for the Spanish. Similarly, the first island footholds in the Americas within the Caribbean were taken only following tides of disease which tore through resident populations. Only fifty years after Columbus's arrival on Hispaniola, the local Taíno population, estimated at between 100,000 and 1,000,000 people,[39] and who had initially put up forceful resistance to Spanish settlers, had all but completely vanished thanks to the spread of disease, warfare and enslavement.[40] Slightly later in time, the rampant documented advance of smallpox and measles throughout Brazil from the sixteenth century onwards, along with the enslavement of Indigenous groups, murder, and forced relocation to towns that exacerbated microbial transmission, reduced what had been a staunch resistance against Portuguese expansion into the evergreen rainforests of the Amazon Basin, which numerous European expeditions had shown themselves completely unequipped to deal with.[41] In the absence of these microbiological stowaways, it is hard to see how some of the most magnificent cities and largest, most powerful states that existed anywhere on Earth at that time could so rapidly succumb to invasion. Clearly, in many of the cases mentioned, Indigenous political rivals often exploited European arrival for their own advancement. Nonetheless, cruel European corralling of Indigenous populations as enslaved labour and malnutrition resulting from warfare and the breakdown of existing political and social structures,[42] as we know from more recent pandemic experiences, undoubtedly would have paved the way for what is perhaps the worst epidemiological disaster ever recorded.

These undoubtedly horrific impacts are now well attested, and these diseases have become a major part of discussions of the process

of early European colonization. However, this does not mean that
the exact same patterns of 'Columbian exchanges' played out everywhere in the tropics, and Europeans and Indigenous populations both
negotiated these new connections in a variety of different ways. For
example, in the case of the Spanish arrival in the Philippines in 1565,
Miguel López de Legazpi and his navigator Andrés Ochoa de Urda-
neta y Cerain settled Cebu and had conquered Maynila (now Manila)
by 1571. While European diseases may still have had an impact, the
existence of greater local immunity, dispersed Indigenous settlement
patterns, and different political structures led to distinctive patterns
of Spanish presence and colonization.[43] Slow, morbid warfare against
the existing Sultanates of Maguindanao, Lanao and Sulu failed to
yield Spanish economic or political control in these parts of the
Archipelago.[44] Indeed, these fights arguably never ended, given
ongoing battles today between the Philippine government and
Islamic extremists on the island of Mindanao. The Spanish committed many atrocities in the Philippines. Even during the first five years
of settlement on the island of Luzon, soldiers sacked villages, seized
food and enslaved inhabitants. Similarly, in the late seventeenth century, Indigenous converts to Christianity disappeared from founded
missions in the Cagayan Valley as they sought to avoid documented
ill-treatment from a garrison of soldiers at Itugud.[45] Once again,
Indigenous resistance is evident. Archaeological excavation and
scientific dating methods have shown that the majestic UNESCO-
protected Ifugao rice terraces we briefly met in Chapter 7, while used
for taro in pre-colonial times, were expanded and re-purposed by large
Ifugao populations that moved into the highlands to escape and resist
Spanish rule in the lowlands.[46] A similar pattern can be seen in early
Portuguese interactions with West and Central Africa. Here, local kingdoms such as those of Benin and Kongo actually seem to have grown in
response to new trading opportunities in the fifteenth century, materi-
ally displayed in the form of remarkable ivory objects deliberately
designed for export to Europe[47] and the finding of European pipes in
dwellings at the site of Savi in Benin.[48] The Portuguese participated in
intermarriage and local politics, and constructed their buildings adja-
cent to existing royal enclosures in traditional architectural forms, only

ever really maintaining a firm, settled foothold on the coast and largely relying on African associates to extract resources and enslaved labour from locations further inland.[49] In this case, as we will see in Chapter 11, it was the coordinated expansion of the transatlantic slave trade, particularly between the seventeenth and nineteenth centuries, that was to take the truest toll on local populations across the African tropics and encourage, alongside the discovery of quinine as a potent anti-malarial drug from the nineteenth century,[50] later European conquests into the interior.

The 'Great Dying' of the Americas, as it has become known, turned the global demographic and political status quo on its head. Once thriving populations collapsed, with between approximately 80 and 95% of Indigenous populations being lost in different parts of the Neotropics, based on a number of different estimates.[51] Today, the sheer scale of this destruction is hard for us to comprehend. The decline of Indigenous populations, from the Caribbean to Central America, from the Andes to the Amazon, led to European writers later documenting failing or small populations in tropical forests that still dictate our stereotypes of these environments today. The initial arrival of the diseases may have been unintentional. However, the historically documented abuses that occurred at the hands of European invaders which exacerbated and followed this disaster were not. For example, the Quechua nobleman Felipe Guaman Poma de Ayala (or Huamán Poma) wrote a famous treatise in Spanish entitled *El primer nueva corónica y buen gobierno* (or *The First New Chronicle and Good Government*), which chronicled and denounced the ill-treatment of Indigenous populations within the former realm of the Inka in Peru.[52] The ongoing marginalization of Indigenous groups up and down the American tropics stems from these initial, disastrous, genocidal encounters. Archaeological and palaeoecological evidence for the active introduction of new plants, animals and land tenure had similar impacts on traditional use of tropical environments. Much of the Americas, as well as the Canary Islands, saw widespread abandonment of well-honed strategies for food production in sub-tropical and tropical forests, as European agricultural species and ideas were

imposed on landscapes, and Indigenous populations dwindled. Nonetheless, groups that did survive across the American tropics, as well as
populations in Asia and Africa, often actively resisted relocation, dictated trade, and gave or did not give the local ecological knowledge
that would enable European colonizers to settle different regions.
The arrival of Europeans, as well as the diseases, plants, animals and
worldviews they brought with them, had major, sometimes irrepar-
able, impacts on societies and landscapes in the tropics. Yet, given the
often slow and limited arrival of Spanish or Portuguese settlers,
Indigenous populations, as well as new cultures and groups formed in
the melting pot of an interacting world, played a major role in how
these things were translated on to varied tropical canvases.

The creator of the term 'Columbian exchange' was the historian
Alfred Crosby, who, in 1972, revolutionized historical research into
European colonialism by focusing on both the cultural *and* biological
consequences of the meeting of two 'worlds'.[53] Crosby was not only
one of the first to highlight the potential role of transported European diseases in shaping a 'New World', but also discussed how plants
and animals from both sides of the Atlantic impacted landscapes
and cuisines the world over. In the context of the tropics, the arrival
of the Spanish, who took a particular interest in implanting their
own forms of 'agriculture' and 'pastoralism', could have major cultural and environmental results. Wheat, and grapevines, were
particularly important to them, given their association with Christian sacraments. Historians have discussed how many Spanish settlers,
including Hernán Cortés, attempted to plant these Middle Eastern
and Mediterranean crops which ultimately failed, particularly in
humid, lowland forest areas. Several faced severe malnutrition as a
result, and historical records document how, despite resistance, they
were forced to eat the local staple crops of maize, manioc, peanuts
and pineapples like their Indigenous neighbours.[54] Indeed, a number
of these Neotropical crops still form the basis of cuisines in many
North, Central and South American nations today. Other crops
introduced by Europeans were more successful. By the sixteenth

century, observers in Peru recorded how lettuce, cabbage, radishes, peas, onions and turnips were all being cultivated in the Andean highlands.[55] Meanwhile, the Spanish and Portuguese also brought ecologically well-suited crops from their other tropical outposts in Africa and Asia. As a consequence, bananas, mangoes, coconuts and oranges are all successfully cultivated up and down the American tropics today and, particularly in the case of oranges in Florida, have even often become popularly associated with this part of the world.[56] In this period of new oceanic connections, the tropics witnessed not only arrivals from outside their latitudinal boundaries, but also a re-shuffling of the plants and animals being grown in different corners of their equatorial realm.

More dramatic environmental consequences occurred as a result of the introduction of domesticated animals, including sheep, goats, cattle, donkeys and horses. Ranching of cattle and sheep, as well as the keeping of pigs, became perhaps the single greatest economic practice across the Neotropics, from Mexico, across the Caribbean, and south into the Amazon. In Mexico, the introduction and dramatic expansion of sheep, consuming all vegetation in their path,[57] has been shown, through palaeoecological analysis of lake records and historical reconstructions of herding, to have led to a drying out of the landscape, reduction of forest cover, increased soil erosion, and the introduction of invasive species in their dung.[58] Similarly, a pollen record from a dried-up river meander bed in the Dominican Republic shows how the introduction of cattle by the Spanish resulted in a decline in tropical forest density on Hispaniola.[59] Cattle were, and still are, a main driving force behind the clearance of the Amazon from the sixteenth century onwards.[60] Even closer to home on the Canary Islands, where sheep and goats were already present, Spanish introduction of new animals, like donkeys, reduced the landscape to such an extent that, in 1491, it is recorded that the Spanish decided to ride out on donkey hunts on Fuerteventura to keep the land profitable![61] On Porto Santo, the introduction of rabbits likewise consumed the Spanish crops, causing them to abandon the island. Nevertheless, in Mexico, zooarchaeological analysis of animal remains from colonial era farming sites has shown how Indigenous communities could effectively combine these new agropastoral species within traditional *milpa* forms of food production.[62] Similarly, in the Caribbean, analysis of linguistics and historically documented Indigenous cosmologies show how domesticated animals brought by Europeans were incorporated into long-held traditional schemes of adopting and taming a variety of animals, including parrots and manatees.[63] Sadly, all too often, the influence of these more ecologically canny approaches was dampened by Indigenous population decline and maltreatment.

The 'Columbian exchange' did not just operate in one direction, and many crops domesticated in the Americas also made it back into the Old World. We might often associate capsicum or chili peppers with spicy Indian food, Sichuan cuisine from China, or kimchi cabbage from Korea. However, as we have seen in Chapter 7, these were first domesticated in the Neotropics. Historical records show that chili peppers were brought across by the Spanish to Europe and Africa, becoming the basis of Hungary's national spice, paprika, before being carried eastward by the Portuguese in the early sixteenth century to be deeply embedded in the cuisine of South and Southeast Asia.[64] Meanwhile, tomatoes, shown by botanists to have been domesticated in South and Central America in the first millennium AD, are first mentioned in European texts in 1544. Cultivation of this fruit in Italy in the sixteenth century paved the way for the creation of some of the most globally distinctive sets of cuisines on the European continent.[65] The Portuguese also likely introduced maize into Africa in the sixteenth century, while manioc was similarly brought to the African continent as part of this process. Cacao, cultivated by farmers in Central America at European arrival, became a taste of kings and queens across much of Europe, and a widely distributed drink in the new globalized capital of Madrid, where it remains a famous tourist attraction as a drink in combination with *churros* to this day. You might associate vanilla with Madagascar.[66] However, historical archives and botanical research show that it too comes from the Neotropics, arriving in Africa via Europe through first Spanish- and then French-mediated hands. The onset of global addictions to cocaine and tobacco, as we saw in Chapter 7, can also be ultimately linked to the 'Columbian exchange' from the Neotropics, albeit the latter had already been passed up to North America prior to the arrival of the French and British, who developed a liking for it.[67] Zooarchaeological and genetic analyses also demonstrate that guinea pigs and turkeys, common pets or Christmas dinners across western Europe and northern North America today, arrived from the American tropics,[68] as a consequence of the new global networks that emerged between the fifteenth and seventeenth centuries.

Much discussion of the 'Columbian exchange' is focused on the Atlantic. However, by colonizing the Philippine Archipelago, the Spanish formed a regular global exchange system that united the Americas, Europe and Asia. A 'galleon trade' was created between Manila and Acapulco in Mexico (what was known as New Spain), which, from 1573, saw two fully-loaded ships take products from Asia to the Americas and back again in most years. While wheat was eventually successfully cultivated in the Philippines at Spanish direction later in the seventeenth century, less direct European control over agrarian activities than in much of the Americas led to prominent locally directed incorporation of novel plants and animals, both by Indigenous populations and the more independent religious orders, into subsistence activities. The sweet potato arrived from the Americas and quickly became widely used in local swidden fields, having similar properties to natively cultivated yams and taro. Their journey from the Neotropics is highlighted by the fact that sixteenth-century accounts show that Indigenous communities in the Philippines called them *camotes*, the Nahuatl (and Uto-Aztecan language) word for the crop.[69] Other arrivals included tomatoes, papayas, pineapples, agave, squash, tobacco, as well as potatoes, whose global journey we will read more about in the next chapter. Zooarchaeological analysis of Philippine sites shows that goats were already present in the Philippines in the pre-Spanish period. However, sheep and horses were likely introduced by the Spanish.[70] Meanwhile cattle keeping was more limited here, at least prior to the eighteenth and particularly the nineteenth centuries. As Filipino zooarchaeologist Dr Noel Amano of the Max Planck Institute for the Science of Human History puts it, 'long-term

Indigenous use of water buffalos for traction and manure, that still continues in the rural areas of the Archipelago today, meant that the new arrivals from Europe provided far fewer local benefits away from a few wealthy, Spanish landowners who could keep thousands of them in profitable herds'.

As the example of the Philippines demonstrates, the spread of crops and animals from their regions of domestication into novel ecologies was neither sweeping, uniformly European-dictated, nor necessarily desirable. Archaeobotanist Professor Amanda Logan, at Northwestern University, has further highlighted this point in the context of the spread of maize into tropical West Africa. The arrival of maize has often been seen as a 'grace' for a continent that has frequently been framed as doomed to scarcity.[71] Its rapid growth and high-yielding properties provided the opportunity to produce a large surplus and support local populations,[72] and monoculture fields of maize continue to be promoted as a means of alleviating food insecurity across Africa in the twenty-first century.[73] Nevertheless, working with the Banda community of West Ghana, Amanda has elegantly compiled ethnographic interviews, historical information, and archaeological and archaeobotanical data to challenge these narratives.[74] First, shifting cultivation of indigenous crops of pearl millet and sorghum, alongside tubers, legumes (e.g. cowpea), and the collection of wild shea nuts, was more than capable of supporting African communities in the region at the time of European arrival. Maize was virtually ignored until later centuries, suggesting that people had plenty to eat. Second, maize can prove incredibly costly for soil fertility, is vulnerable to pests, and can struggle under dry conditions, which periodically impact the seasonally dry forest areas of western Ghana, and did so at the time of European arrival.[75] Finally, maize became important more quickly along the coast. Here, African farmers profited from growing it in large quantities to meet increased demand from Europeans, who required it to feed themselves as well as the economic demands of new, capitalist business initiatives centred around human souls. As Amanda puts it, 'an obsession with Africa as a continent lacking in food security, and in need of surplus from introduced Eurasian and American crops, goes completely against millennia of African farming innovations that effectively managed environmental perturbations and that processes of coloni-alism have prejudiced over the last five centuries'.

Indeed, the introduction of new crops and animals has, as we have seen in the Canary Islands and the Americas, often left legacies on landscapes that are not entirely positive. In the case of Africa, a focus on monoculture crops such as maize can leave farmers more vulner-able to climatic change.[76] In China, the arrival of maize and sweet potato, with their very different climatic and soil tolerances when compared to Old World crops, offered the opportunity to expand agriculture on to new earthy horizons. Sweet potato arrived from the Philippines in the late sixteenth century, and provided a crucial food source for local farmers reeling from climatic-induced devasta-tion of their rice fields. Sweet potato and maize, which arrived via Portuguese-held Macao, also supported emigrating Chinese farmers moving as far south-west as Sichuan and as far north-west as the Gobi Desert. By the eighteenth century, agricultural capacity and popula-tions in these previously sparsely utilized regions sky-rocketed. Palaeoecological records and historical observations document the increasing deforestation and soil erosion that resulted.[77] The rich nutrients and carbohydrates provided by these new crops certainly, in part, contributed to China being the most populous country on the face of the planet. However, geomorphological study of ancient soil quality, in combination with archival references to soil instabil-ity, shows that these new crops replaced tree cover. In doing so, they contributed to the washing away of unprotected soils, the loss of nutrients, an increase in major flooding and, ultimately, malnutrition of over-extended rural populations that the country is still grappling with in many areas today.[78] The 'Columbian exchange' certainly operated in various directions between the now-linked worlds, both in terms of cultural influences and potential environmental and nutritional repercussions.

Although the discussion of the 'Columbian exchange' is often pop-ularly described as a product of nations or empires consciously 'transporting' and using different crops or animals (and I am also guilty of that in the above!), the examples of West Africa, the Philippines and China show that, ultimately, it was a network built upon the backs of a variety of merchants, food producers, travellers, writers and government officials, from all continents and cultures. Seeds and animals were simply not enough; their appropriate management, especially in the tropics, required different traditions of cultivation and herding that came in the form of people moving. To take one revealing example, between the sixteenth and nineteenth centuries, approximately 12–13 million Africans were forcibly removed from their homes and transported across the Atlantic as part of the transatlantic slave trade. We will see in the next chapter how this represented the onset of increasingly globalized, capitalist and racist exploitation of landscapes and labour in the tropics. However, it is also worth noting that the Africans who were forcibly shipped, and whose descendants now make up significant proportions of many nations in the Americas, brought with them yams, black-eyed peas, watermelons and plantains, as well as knowledge of how to farm them,[79] working small 'plots' at the edges of the plantations on which they were forced to work. Not only that, but as many of these individuals had already been farming in fields with maize and manioc back in West Africa from the sixteenth century, they also brought a cultural tradition of combining these different foods into new, sustainable systems and eclectic cuisines. The initiation of rice cultivation in North America has been suggested to have arrived in a similar manner from West Africa, rather than Asia.[80] These are far from being the only examples, but clearly show that often-neglected, marginalized groups were also key players in the formation of new post-Columbian ecosystems and economies.

European colonizers brought along their own belief systems and concepts of land use as well as new biological entities. Intense, direct control over the natural world, concepts of property, ideas of how a settlement should be laid out, and the priorities placed on maximizing output from a given portion of land, represented a new, frequently traumatic, experience for tropical landscapes and their occupants in what has become known as 'ecological imperialism'.[81] One of the more obvious examples of this new ideology of exploitation can be seen in the form of Iberian desires for Neotropical silver and gold. Seeing the rich use of ores by Indigenous populations, the Spanish set about fervent survey and extraction. Beginning in the mid-sixteenth century, after just 100 years Spain alone had extracted three times Europe's previous silver wealth. The scale of these operations can be seen in the size of mining cities that emerged, such as the high-altitude city of Potosí, Bolivia, which, by 1660, was home to 100,000 people[82] – more than Madrid and Rome at the time. Forced movement of enslaved individuals and the flocking of Europeans, among others, seeking to make their fortune, led to rampant deforestation to fuel both mining and construction. Similarly, when rich gold deposits were found in the current state of Minas Gerais in Brazil in the late seventeenth and early eighteenth centuries, the region's dry tropical forests were destroyed to make way for mines, pastures, crop fields and villages to support the thousands of people seeking quick enrichment.[83] In the case of silver, more insidious environmental impacts also came about as a result of the use of mercury to more efficiently extract this precious metal from the sixteenth century onwards. Mined at locations such as Huancavelica in Peru, the release of mercury into soils, water systems and the atmosphere, not to mention the lungs of forced labourers, had lethal, lasting consequences visible in the chemical contents of the soils of the exploited regions to this day, with ongoing impacts on the health of contemporary inhabitants.[84] Some of these effects have been documented in pre-Columbian periods in sediment cores. However, the invading Iberian powers undoubtedly ramped them up to whole new levels to support expanding colonial infrastructure, warfare, and their economic and political designs back in Europe.

Colonial Iberians also brought their ideas of where and how to live to their new tropical colonies. In many cases, the Spanish built their administrative centres on top of the Indigenous cities they so admired. However, they often neglected the traditional knowledge that had dictated landscape use prior to their arrival. Hernán Cortés famously refused to abandon Tenochtitlan despite the fact that, without Indigenous knowledge of water and field management, the colonial residents were vulnerable to stagnant water and seasonal flooding.[85] Similarly, in the Philippines, many towns and churches built by the Spanish in river valleys were later abandoned due to flooding.[86] In other cases, the Spanish, and their Portuguese counterparts, made their own towns and cities, often laying them out around central plazas, with key, wide streets, and specified sectors for craft activities, just as they had back home. Environmental historian Shawn Miller has noted that amazingly nearly 50% of people in Spain's American colonies lived in cities by 1600, something not achieved in England until the mid-nineteenth century.[87] Although this figure is likely problematic, given that many Indigenous people deliberately fled into the hinterlands and rural areas to avoid census recording,[88] it does show the emphasis that the Spanish placed on attaining their urban ideals. Across the wider landscape, analysis of historical records shows that the Spanish also tried to organize the populations in their new tropical realms for maximum political and economic control. This included the system of *reducciones* (or *aldeias* in the case of Portuguese realms) from the sixteenth century, which forcibly relocated many Indigenous people into settlements. This enabled easier taxation, census counting, enculturation and recruitment of Indigenous populations through *encomiendas*, a royal grant of effectively enslaved labour to chosen Spanish elites.[89] In the case of the Philippines, this involved centralized *cabacera* towns that were frequently built around a church and town hall, with the former holding immense power over the labour and disciplining of local Indigenous populations.[90] These systems were often also taken advantage of by Indigenous *alcaldes* (mayors) and *gobernadorcillos* (low-level governors) who sought to gain influence.[91] Still visible in the building plans and architecture of key cities such as Vigan and Taal today, these novel forms of settlement organization could all impact the landscape, demanding wood for construction, more intensive cultivation of fields for tribute and for larger, urban populations, and greater intensity of domesticated animals in a given locality. European ideologies as to how people should live were thus transplanted from the other side of the world into the tropics.

While the effects of mining and new types of settlement and

organization could leave deep scars upon local tropical landscapes, more pervasive changes occurred as a result of the imposition of plantation agriculture. The Spanish realized that many economically seductive crops could be more productively grown in their new tropical lands, away from native pests. Prime among these examples is sugar, which originated in the tropics of New Guinea. As early as 1493, on just Columbus's second voyage to the Americas, the Spanish brought sugar for cultivation in the Caribbean and, by the sixteenth century, it was also being grown across the Canary Islands, Mexico and Central America. The Portuguese, meanwhile, had introduced sugar to the Atlantic islands of Madeira and São Tomé and, by 1516, the shores of Brazil, planting it directly within former Indigenous field systems as a morbid symbol of what was to come. While not always grown in a strict 'monoculture' at this early stage, sugar sucked nutrients from soils and led to vast deforestation for both space and the fuel for its refinement. By 1550, Madeira's forests were in tatters and sugar production had to be abandoned, and historian Jason Moore has estimated a cumulative deforestation total of around 155 square kilometres in just over a century.[92] By the 1600s, the forests of the Canary Islands and Brazil's Atlantic coast were facing similar pressures.[93] The same type of over-exploitation can also be seen in the Spanish focus on 'ranches' of livestock. Here, herds of hundreds and thousands of animals were used for profit rather than subsistence. These models of land ownership, land use and demand-driven operation were new to these landscapes, and still haunt them to this day. This is not to mention new, market-driven patterns of exploitation of wild animals such as marine mammals off tropical coasts, the hunting of civet cats for their skins and scent glands, and the capture of parrots for their feathers undertaken by European merchants.[94]

These changes in perception of the natural world, ownership of the land, and intensity of land use to feed wider, global tastes and demands undoubtedly had massive pan-tropical implications. However, once again, as the historians Amélia Polónia and Jorge Pacheco have highlighted, it is important to balance the global picture with the local result.[95] As we have seen, the ultimate outcomes of colonial Iberian tropical endeavours could be shaped by environments, existing Indigenous populations and traditions, and different forms of European rule. Ecologies, soils, and histories of use moulded the nature and impacts of sugar production in different tropical island and mainland contexts. In contrast to the Americas, gold mining in Africa followed Indigenous patterns of exploitation until the nineteenth century, as the Portuguese, mainly located on the coast, were unable to interfere directly in the mines. Similarly, different patterns of *reducción* and *aldeias*, seen in historical and archaeological records of colonial period settlements, shaped demography, and the spread of disease, to variable extents between the Caribbean, Mexico, the Philippines, religious missions in Brazil and traditional settlements in Africa.[96] Finally, Indigenous populations, in contrast to being 'uncontacted' and 'isolated', could be highly active in participating in and shaping these new economic systems. More traditional approaches to food production often fed the workers and enslaved people working on new agricultural field systems; meanwhile, historical texts state how hunter-gatherers in India and Brazil brought honey and jaguar skins to the worldwide market, respectively.[97] Not only that, but a number of European invaders inter-married with Indigenous populations. Although this involved its own form of power dynamics, with European men often marrying the daughters of local elites, it did eventually result in the forming of new colonial identities and cultural mixes, or, in the case of some Europeans, the abandonment of their previous ways of life and the joining of new families in tropical forest environments.[98]

The Philippines provide a particularly elegant example of some of these more local colonial experiences. In the present-day capital of Manila, although the Spanish erected a stone-walled town to develop a clear plaza and compact urban system that was more familiar to them, they were repeatedly distressed by the appearance of a sprawling 'second' Manila outside its walls. This so-called *Extramuros*, or 'city beyond the walls',[99] began developing as early as 1583, and included a district, known as the Parián, which was inhabited by Chinese merchants and their families eager to access silver from Acapulco. It eventually dwarfed the official city, and the Spanish tried several times to remove and destroy it. Similarly, as Filipino archaeologist Dr Grace Barretto-Tesoro of the University of the Philippines puts it, 'the enculturation of Indigenous populations occurred at a different pace and to different extents across the Archipelago, and was often dictated by local populations'. For example, Indigenous populations on Mindanao known as the *Lumad* took advantage of Spanish presence to counteract the powerful Muslim groups of the southern Philippines. Back on the terraces of Ifugao, growing, politically powerful populations actively traded with the Spanish, while also raiding and resisting them.[100] The pre-colonial environmental and cultural situation of the Archipelago also fashioned Spanish interactions with the environment. As environmental historian Greg Bankoff has elegantly noted, endemic termites or 'white ants' led to there being only 30 to 40 species of tree which could be profitably felled for forts, shipbuilding and construction by Indigenous populations and Spaniards alike. This meant that although certain key species became threatened, archival reconstruction of forest cover through time shows that blocks of overall forest remained until nineteenth-century commercialization.[101] These local colonial ecological and cultural experiences occurred across the new Iberian realms.

The decapitation and removal of statues of Christopher Columbus, up and down North America, in 2020 shows the degree of controversy that his legacy continues to cause.[102] More and more Indigenous activists and institutions are calling for Columbus-reverence to be replaced by a celebration of pre-colonial Indigenous societies and an acknowledgement that the actions of European 'explorers', and their consequences for tropical cultures, populations and landscapes, are by no means cause for fanfare.[103] The rampant advance of European diseases, alongside active enslavement, murder, abuse and invasion, meant that the previously significant population of the Americas did not reach its former levels until the nineteenth century. The introduction of new plants and animals often permanently reconfigured ecosystems, leaving lasting legacies and conservation challenges to this day (as we will see in Chapter 13). Yet the agency of Indigenous and other marginalized groups, tropical diseases, and difficult geography could still dictate, or even prevent, Iberian expansion into much of the Amazon Basin, the Caribbean, North and Central America, West and Central Africa, South Asia and Southeast Asia between the fifteenth and seventeenth centuries. In many cases, local elites, some of whom were subordinate to existing tropical empires, grappled to make use of incoming European soldiers to further their own political advantage. Pre-colonial exchange systems spanning continents, as well as trades in enslaved individuals, already existed in many parts of the pre-colonial tropics,[104] meaning that Europeans often tapped into established economies to apply their new ideas of towns, mining and profit-making agricultural and pastoral extraction. However, it was ultimately to be Europeans that possessed the keys to this new globalized world. Following decimation of Indigenous populations across the Americas, and with a monopoly on naval power and technologies, it was Europeans alone that routinely and regularly plied the world's oceans. With an intent to maintain global trade and dominance, they were left in a position to exploit the resources, labour and landscapes of the tropics on an unprecedented scale.

Although little discussed, between the fifteenth and seventeenth centuries, to some extent, Iberian royal monopolies and administrative flaws could actually limit the transformation wrought on tropical landscapes by colonial enterprises. In Brazil, the Caribbean and Mexico, kings and queens reserved forests and certain tree species for state interests in building ships or fortresses.[105] The King of Portugal controlled sixteenth-century whaling activities off Brazilian shores. Indigenous enslaved labour in new Spanish plantations was also initially regulated by royally decreed *encomiendas* to chosen nobles.[106] Meanwhile, in the Philippines, while the church was highly active in the settlement of the archipelago,[107] the Spanish monarchy did everything it could to control the 'galleon trade' which sent silver and gold from the Americas to the shores of the Philippines and China and saw silk and porcelain return in the opposite direction.[108] Indeed, you might have noticed that the Iberian Empires themselves have also had something of their own monopoly on this very chapter. This is because it was the expansion of the Spanish and Portuguese Empires between the fifteenth and seventeenth centuries that, with some limited exceptions, primarily initiated, and sought to control, novel economic and cultural exchanges into and within the tropics. These systems would have been unrecognizable to the supposedly 'free' market consumers of the twenty-first century. As we will see in the next chapter, however, from the seventeenth century onwards, other hungry European states, institutions and private individuals, grappling for status and power on their own continent, and seeing the riches arriving on the shores of the Iberian Peninsula, made their own, increasingly aggressive advances into the tropics.

The Dutch Empire of the sixteenth to eighteenth centuries sought a foothold in the Caribbean and South America as well as in South and Southeast Asia. The British Empire became the largest empire in history between the sixteenth and early eighteenth centuries, ruling territories on every tropical continent across an area where the sun was said never to set. By the nineteenth century, the French had also established a tropical colonial empire in the Caribbean, Africa, South Asia and Southeast Asia, and the Belgians had done the same in Africa. These jostling competitors broke down former Iberian domination, and benefits became distributed across an increasingly wealthy European elite. The resulting new trading opportunities were nevertheless not driven by homogenous empires, but rather by lots of individuals, including mobile merchants, gambling chancers and food producers of all cultures. The seventeenth to nineteenth centuries saw the laying of the framework for the system of competitive, global flows of capital and labour we know today, as Europeans expanded existing inequalities and systems of exchange into the global sphere. A consolidation of profit-driven approaches to tropical environments and to people from the tropics saw haunting commercialization and racialization of enslaved labour in the form of the transatlantic slave trade. Intensifying plantation agriculture stripped forests on new scales to produce 'cash crops' as cheaply and as numerously as possible, simultaneously shaping and being driven by European and northern North American consumer demand. The result was a wealth imbalance between Euro-America and the tropics. The result was tropical nations plagued by issues of limited economic development, lagging infrastructure and conservation challenges. The result was racial discrimination and violence that continue to afflict twenty-first-century societies. The result was a situation where we all, whether we like it or not, are responsible for what happens in tropical forests in the twenty-first century.

11.   Globalization of the tropics

Every day we turn on the TV or open a newspaper to new international political squabbles, economic difficulties and catastrophic climate change. Many tropical nations, across the Americas, Asia, Africa and the Pacific, are on the frontline of these issues as they try to balance needs for economic growth with the burning of fossil fuels, desires to alleviate poverty with biodiversity conservation, and nationalism with the need for international communication and global solutions. It is in the tropics where islands are already beginning to disappear underwater. As global temperatures increase, ice-caps melt and sea levels in the Indian and Pacific Oceans rise at a rate of around 4 millimetres every year.[1] It is in the tropics where increasingly unpredictable climates are leading to both droughts and floods in the same regions in the same years. It is in the tropics where, should all tropical forests become disturbed by human deforestation and infrastructural development, a mass extinction event is looming for plants and animals. Some predict declines of as much as 30% of all tree species and 65% of ant species,[2] and bushmeat hunting and climate change provide further challenges for the disproportionate biodiversity of these regions. It is also in the tropics where many of the fastest-growing but also the poorest populations reside, some living on less than one US dollar a day, placing increasing pressures on the public health, economies, soils and forests of these nations.[3] The frequency and urgency with which these shocking stories are conveyed to us, as well as their occurrence often many thousands of kilometres away, might allow us to think that they represent a recent, emerging problem that is detached from the pasts and behaviours of the rest of the world. Yet no matter how hard some of our politicians try to convince themselves, and us, of this, nothing could be further from the truth.

We have seen in Chapter 10 how the dramatic expansion of

European powers, primarily from Iberia, heralded a period of new pan-tropical and extra-tropical global contacts and exchanges between the Americas, Asia, Africa and Europe during the fifteenth and seventeenth centuries. Some recent treatments have stopped the story there, suggesting that early interactions and geographical 'chances' wholly determined the rest of history as we know it.[4] However, this ignores the ongoing colonial, imperial and capitalist processes that were only just beginning. Between the seventeenth and twentieth centuries, a variety of European and, later, northern North American, leaders, landowners, merchants and investors increased their attempts to control and exploit the environments and people of the tropics. During this period in the tropics we see some of the earliest, but also worst, impacts of unchecked, unregulated flows of capital on human bodies, cultures and landscapes in world history. We see the consolidation of a global, racialized trade in forced human labour.[5] And we see the appearance of the first documented recognition that human impact on the environment could have regional consequences for the climate.[6] These economic and political processes were resisted and exploited by societies and individuals from the tropics.[7] They also eventually led to some of our modern laws on human rights and international property.[8] However, in the end, they created a new and monstrous redistribution of wealth, away from the tropics and into the hands of the western half of Europe, the United States of America, and Canada. They altered tropical landscapes beyond all recognition, leaving conservation challenges for many nations today. And they sprouted the roots of racial discrimination and violence that continue within many Euro-American societies in the twenty-first century.

We will now turn to the latest historical, archaeological, palaeo-ecological and anthropological research that has explored how tropical forests became key witnesses to this new global order. They were home to millions of Africans forced from their homes in West and Central Africa as local raids and European demand for labour drove the transatlantic slave trade. Tropical forests were also the destination of the transported enslaved individuals, cut back as the increasingly monoculture sugar plantations, particularly in the Caribbean and

South America, had violent, expanding impacts on Neotropical environments. Later in time, as global markets for varied tropical crops solidified, tropical forests watched as Brazil had the crucial resource of rubber taken from it to be planted elsewhere, as India had its own cotton sold back to it, and as farmers in Myanmar cleared entire river deltas to plant sweeping fields of rice to fuel the Western world and an ever more unequal distribution of global wealth. Landscapes previously home to tropical forests provided the tea, coffee, potatoes and bananas that began to stock increasing numbers of Euro-American kitchens. This is the story of how the world we know today emerged from the tropics, alongside the stories of the people and plants that watched it happen. It is the tale of enslaved individuals, labourers, merchants and empires experiencing and manoeuvring a rapidly changing world of global economics and cultures, and of the tropical plants that now line our cupboards and wheel our cars and bikes. It forces us to confront the origins of globalized issues in sustainability, climate change, biodiversity loss, political disputes, warfare and racism that confront not just tropical nations, but our entire world, in the twenty-first century.

At the close of the twentieth century, archaeologists excavating at the Spanish colonial Hospital Real de San José de los Naturales (now San José de los Naturales Royal Hospital) in Mexico City, in preparation for the construction of a new subway line, uncovered a mass burial of human remains. Many of the discovered individuals were Indigenous people, buried after succumbing to the rampant spread of diseases that arrived with Europeans. However, scientific research published in 2020 showed that three of the individuals, dating to the very earliest phase of the hospital in the sixteenth century, had a very different origin altogether. Using analysis of their preserved DNA, strontium isotope analysis of their tooth enamel (which reflects the geology where an individual lived during the formation of the tooth), and studies of deliberate cultural dental modifications, they were shown to all be men, all be between twenty-five and thirty-five years old at death, and to all have come from sub-Saharan Africa,[9] over 10,000 kilometres away. At another burial ground, near the seventeenth-century Dutch

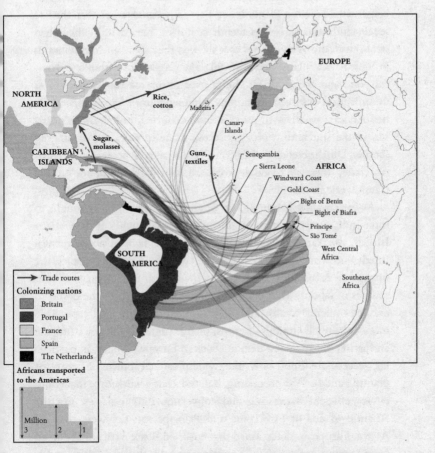

Map of the transatlantic slave trade between 1500 and 1870. The colonized areas of
Britain, Portugal, France, Spain and the Netherlands are shown in the Americas,
as well as key locations along the Atlantic coast of Africa. The number of Africans
forcibly abducted and transported to the Americas along different routes is shown.
Arrows show the trade routes of different products and materials, with a focus on
Britain, as discussed in Chapter 11.

capital of Philipsburg on the Caribbean island of St Martin, archaeologists uncovered individuals with genetic affinities to people living in tropical Cameroon, Nigeria and Ghana.[10] Likewise, at British-constructed Newton Plantation in Barbados, dated to between the late seventeenth and early nineteenth centuries, buried individuals had strontium isotope signatures that showed they spent their childhoods in West Africa.[11] How and why did these people move from one tropical continent to another? The answer is one of the largest, most inhumane set of forced migrations in human history. From some of the first Africans to arrive in the Neotropics courtesy of Iberian colonialism in the sixteenth century to those made to work for expanding Iberian, Dutch, French and British global economic powers two centuries later, these individuals were all first-hand witnesses to the transatlantic slave trade.

As we saw in Chapter 10, although Europeans initially sought to force Indigenous populations, including the Guanches of the Canary Islands, the Taíno of the Caribbean and the Tupí in north-eastern Brazil,[12] to work their newly conquered lands in the Neotropics, the impact of devastating diseases left them with a demand for other sources of labour. This was particularly the case given the desire to extract mineral wealth and to transform tropical landscapes into monoculture plantations of crops, such as sugar, to cheaply stimulate and satisfy the growing demand back in Europe.[13] Sugar, in particular, necessitated not only the widespread clearance of trees for planting and fuel for processing, but also a large workforce that could ensure efficient harvesting and conversion into molasses. On the Atlantic islands of São Tomé and Príncipe, off the west coast of Africa, Europeans determined that enslaved Black Africans abducted from tropical portions of the continent could, thanks to early exposure, resist tropical diseases like malaria and yellow fever that were creeping into the Americas and destroying European migrant workers and Indigenous populations alike.[14] Although enslaved Africans had been arriving on the shores of the Americas since shortly after Columbus's arrival on Hispaniola,[15] the expansion of sugar plantations across the Caribbean and Neotropics throughout the sixteenth and seventeenth centuries, and the unification of the Spanish and

The real drivers of the 'industrial revolution'? Black labourers working on a sugar plantation in the West Indies around 1900. Although after the formal date of abolition, many former enslaved communities were forced into multi-year 'apprenticeships' that included working for long hours with no pay. Some of the workers are children, harvesting under the watchful eye of a white supervisor.

Portuguese thrones in 1580, saw African labour become progressively more demanded and extracted.[16] Between 1550 and 1650, Iberian ships had already transported over half a million enslaved Africans to colonies in the Americas.[17] Although slavery had existed in various forms in societies around the world up to this point,[18] this new, European-driven trade saw two significant shifts. First, enslavement became a global endeavour, based on a violence against Black African bodies and violence against tropical landscapes. Second, although it drew upon existing economic structures and political conflicts, the profits and flows of wealth resulting from the increasingly dispassionate, mechanistic, profit-driven exploitation of this expanded trade in enslaved labour were primarily in the hands of Europeans.

Working with sugarcane was notoriously brutal, requiring constant attention, heavy labour, tropical lumber, and dangerous refining through boiling and distillation. This is even before the abuse, malnutrition and insanitary conditions forced upon enslaved individuals by plantation owners are considered.[19] The demands placed on the new arrivals to Neotropical shores saw many enslaved peoples worked to death within a matter of years. Yet this trade was only in its infancy. By the seventeenth century, sugar plantations established along the Portuguese-controlled coast of Brazil, such as those at Pernambuco and Bahia, were producing most of the sugar exports to Europe.[20] Something that other European nations sought to change. The Dutch West Indies Company, after a failed attempt to take control of this emerging 'sugar coast', worked their way into the lucrative sugar business by other means. Namely, they invested in the dedicated shipping of enslaved individuals from Africa to the Caribbean to work on the plantations of others.[21] By the end of the seventeenth century, the British Empire had colonized a number of islands in the Caribbean, including Jamaica, Antigua, Saint Kitts and Nevis, and Barbados, which had been largely neglected by the Spanish due to their lack of ore resources. Here, the British established colonies, not to settle or to control, but rather to extract, rolling out the sugar plantation as a concept based on chattel slavery across tropical landscapes to make maximum profit.[22] The French founded a colony on Saint-Domingue (now Haiti), based on a similar economic system.[23] Although British people launched slave raids in West Africa as early as the sixteenth century, in the seventeenth and eighteenth centuries the British Royal African Company, as well as a number of French companies, sought greater control of plantation workforces and challenged Iberian and Dutch monopolies by increasingly specializing in a trade in enslaved Africans. At the same time, racist theories, supported by prominent philosophers, served to make associations between concepts of enslaved labour and Black African bodies particularly rigid in English social and cultural discourse.[24] By the eighteenth century, the 'triangular trade', as it became known, had not only enveloped the entire continent of Africa in its demand for labour, but also saw African individuals shipped to European colonies and business ventures in the Americas, South Asia and the Pacific[25] as part of economic extensions spread well beyond sugar plantations (e.g. mining and other crop monocultures) and the

Atlantic Ocean.[26] All to meet an increasingly coordinated, global flow of investment that had become disturbingly mechanical.[27] Other European countries, including Denmark and Norway, became involved. So did an independent Brazil in 1822, as well as other American-based trading polities.[28] Smaller private merchants, alongside the wealthy elite and large companies that had thus far dominated coordination of the trade and ownership of plantations, also played an increasingly large role. These merchants ensured a flow of wealth into cities like Liverpool, London, Bristol, Birmingham, Glasgow and Nantes, managing the shipping of cloth to African merchants in exchange for enslaved individuals, the shipping of enslaved individuals on to the Americas, and the return of sugar, as well as cotton, tobacco and other plantation goods.[29] The human cost of these increasingly formalized and racialized flows of capital is shocking. By the time an independent Brazil became the last country in the Americas to ban slavery in 1888, it is estimated that anywhere between 10 and 20 million Africans had been forcibly uprooted from their primarily tropical homes and families and transported across the Atlantic.[30]

Many of them died before they even stepped ashore. The eighteenth-century Portuguese shipwreck of the *São José-Paquete de Africa*, uncovered by underwater archaeologists off the coast of Cape Town, is a haunting graveyard of iron bars and copper nails that shackled more than 400 trapped African men, women and children on their way to Portuguese-controlled Brazil until stormy waters took them to their deaths.[31] First-hand written accounts by freed Black African enslaved individuals such as Olaudah Equinao (Gustavus Vassa) and Ignatius Sancho highlight both the terrors faced in capture and transport and the racist and capitalist treatment of Europeans selling people as commodities. As Sancho wrote, 'Look around upon the miserable fate of almost all our unfortunate colour . . . see slavery and the contempt of those very wretches who roll in affluence from our labour.'[32] Although the degree remains debated, the transatlantic slave trade certainly impacted West and Central African societies and polities over the long term. African historians such as Walter Rodney and Joseph Inikori have argued that it was a major factor, alongside droughts and famine, in changing demographics in sub-Saharan Africa between the seventeenth and nineteenth centuries, using historically recorded and estimated numbers of captured slaves to show how depopulation and severely imbalanced ratios of men to women resulted.[33] Similarly, the increasingly active meddling of European powers in African politics and sponsoring of local slave raids to obtain captives for trade played an important role in further stimulating a long history of ethnic and religious violence in West and Central Africa, leading to the movement of populations further into mountain or forest settings, preferences for isolated villages as a key form of safety from attack, and the creation of political and cultural fault lines that continue to be exploited by modern terrorist groups such as Boko Haram in the tropical forests of southern Nigeria and northern Cameroon.[34]

Traditional approaches to the transatlantic slave trade have often framed Africa as a land of passive victims, perpetuating colonial stereotypes that still depict Africa as an isolated, impoverished continent. Much of this is a product of the fact that most of the historical records used to understand West and Central Africa at the time of European involvement were written by Europeans[35] or Arabic traders and scholars. Meanwhile, as seen in the recent controversy of the 'Benin bronzes', preserved material culture that highlights the diversity and wealth of pre-colonial African states was removed from the continent by European colonizers.[36] Nevertheless, the last thirty years of historical revisionism have presented a more active African role in an increasingly globalized world. Like Europe and the Mediterranean,[37] capture and trade of enslaved individuals was prevalent in many parts of Africa at the time of European arrival on its shores.[38] Trade across the Saharan caravan routes and through the Red Sea or Indian Ocean saw many enslaved peoples traded with Muslim powers from the first millennium AD.[39] As we saw in Chapter 10, from the beginning the Portuguese relied upon African political goodwill and traditional marriage networks to access enslaved individuals who were already being captured and traded throughout the region as part of warfare and regional demands for labour and prestige.[40] In the fifteenth century, thriving local kingdoms, such as those of Kongo and Benin, actively exploited or resisted new European contacts, setting prices, limits on the numbers of enslaved people taken, and making economic and religious demands. Indeed, through the transatlantic slave trade, European merchants and captains were only likely to be successful if they tailored the goods they brought with them (e.g. weapons, cloth, pipes) to the luxury tastes of their regional partners, with whom they often formed close, personal relationships.[41] Africans meanwhile incorporated and used Europeans in their geopolitics, as can be seen in the fact that European traders built their houses alongside African elites and merchants at sites such as Elmina, Ghana and Savi, Benin.[42] Even when Europeans did actively take part in raids in the interior, they were reliant on allies and could quickly succumb to a changing of political allegiances. This was amply demonstrated when Queen Nzinga (Njinga Mbande) of the Ambundu Kingdoms of Ndongo and Matamba (modern-day Angola) shifted from being a Portuguese Ambassador in the region to being their staunch enemy after aligning with the Dutch to further the independence and stature of her own realm.[43]

Just as we should not reduce Africa, and its diverse states and polities, to a passive clay that was moulded by Europeans, we should also not forget the individual examples of resistance to the transatlantic slave trade. Increasingly recognized numbers of escapees, slave mutinies on Atlantic voyages, and rebellions across the Americas,[44] including notably in Haiti, highlight the ongoing powerful, determined resistance of Africans and their descendants to horrendous conditions between the sixteenth and nineteenth centuries.[45] Free 'maroon' settlements of former enslaved peoples, and sometimes also Indigenous communities, developed across the Caribbean, Central and North America, and South America,[46] and, as in the famous case of the *Quilombo dos Palmares* of Brazil, could provide staunch militant resistance to imperial powers.[47] Away from these more dramatic events, Dr Alicia Odewale, of the University of Tulsa, has also analysed historical records and the material remains left behind in the ground to study the everyday experiences of enslaved communities living in the Danish-controlled, Christianized and sugar-oriented urban centre of St Croix in the Caribbean. These communities negotiated and resisted their condition through their daily activities and use of mundane items such as pottery.[48] At the tropical urban site of St Croix, enslaved communities made significant use of undecorated and hand-made ceramic wares, alongside European pottery, as part of a growing reliance on skills developed within the enslaved community for daily survival.[49] At St Croix, and elsewhere in the Caribbean, enslaved communities also coordinated informal trading systems that saw tobacco, pipes, pottery and glass exchanged between communities as part of the negotiation of new identities and connections.[50] As Alicia puts it, 'even under the immense structural power of imperialism and slavery, enslaved communities still expressed their own power and transformative capacity'.[51] Beyond slavery, numerous diasporic African voices are also now beginning to be heard as merchants, voyagers, investors and crucial workers in the formation of globalized, and later industrialized, cultures across Africa, the Americas and Europe.[52] There were far more Africans who were captured and brutally transported to the Americas than there were Europeans who arrived on these shores prior to the nineteenth century. In many parts of the Caribbean and South America, they outnumbered Europeans by as much as 25 to 1, and people of African heritage still make up a considerable portion, or even the largest portion, of populations in many Caribbean and South American countries today.[53] It is no exaggeration to say that the mixed cultures, demographies, economies and societies of the new globalized world were built upon their shoulders.

Nevertheless, it is impossible to ignore the global wealth and status discrepancies that emerged as a product of the transatlantic slave trade. Historians, exploring records of income, trade and demography, have powerfully argued that the capital gains made by European and northern North American plantation owners, merchants and families[54] that were directly and indirectly involved in the trade provided the key financial bases for the industrial revolution and the development of wealth and infrastructural imbalances that went on to underpin the modern world economy as we know it today. Technological innovations seen in a number of sugar colonies have, themselves, been seen as a part of the dawning of an 'industrial age'.[55]

Analysis of historical nutrition has also shown that cheaper, more readily available energy-rich plantation sugar between 1600 and 1850 increased public health across British society to such an extent that it spurred on the formation of an industrial working class.[56] Even following the abolition of slavery in the United Kingdom in 1807, and in its colonies in 1833, the British government paid out massive amounts of money to companies, landowners and individuals with investments in the slave trade in repayment for losses of 'free captive labour'.[57] Europeans did not *need* to take part or invest in this trade, and there was opposition back home.[58] Yet many individuals such as Edward Colston, whose legacy has left a mark on a number of institutions across the city of Bristol today, joined initiatives such as the Royal African Company and invested in the transport of enslaved individuals between continents. In Europe and northern North America, as we go about our daily routines, it can be easy to forget just how much of our streets and daily lives the resulting profits have touched, and how many of our predecessors could so quickly take part in this flow of forced labour and capital. Capital accumulated from the transatlantic slave trade has flowed into our hospitals, schools, churches and universities through charitable donations. The people who earned wealth from, and took part in, the trade are honoured in street and building names, as well as statues. Many innovations in insurance that characterize the financial sector of our economies today can also find their disturbing origins in the transatlantic slave trade, as investors sought financial safety nets for both cash crops and human cargo.[59] The racialized abuse of Black African bodies that was formalized in the transatlantic slave trade also undoubtedly left an ingrained legacy of inequality and discrimination that continues to influence North and South American, and European, societies and political conflicts into the twenty-first century.

Back in the tropics, although some African elites and merchants made profits from the trade, the benefits of sugar and the global flow of capital did not reach African shores in the same way. As we have seen, it disrupted the demographic and political situation in much of the continent with lasting effects. From the eighteenth and nineteenth centuries, subsequent, sustained conquest of different parts of West Africa also resulted in brutal warfare, the dismantling of traditional land use and social systems, and the robbing of heritage which remains locked up in European and North American museums. Profits were also not seen on the sites of many of the industrial plantations where enslaved African labour ended up. Places such as São Tomé and Príncipe, and a number of Caribbean islands, were left as some of the poorest countries anywhere in the world following eventual abandonment by Europeans, with landscapes brutally shaved of their tropical forests.[60] On the Caribbean island of Nevis, nearly all the island's plants and animals are now introduced or invasive species, with the majority of their native counterparts now being extinct, as the island was nearly deforested within the first fifty years of sugar plantation being established.[61] While roads might have been built to help support the movement of this valuable crop, the British made no attempt to introduce critical infrastructure such as schools or farms for nutrition.[62] Returning to the island of St Croix, Dr Justin Dunnavant of Vanderbilt University is exploring how the ecology, and even the shoreline, has changed as a result of these processes and their legacies. Widespread clearance of trees began with sugar plantations, and there is clear evidence of lasting soil degradation. Meanwhile, the island's coral shorelines have been literally, and visibly, mined away, not only for building materials to support the forced labour of enslaved individuals, but also to clear a pathway for the tax-free smuggling of rum.[63] The expansion in global production of this popular beverage, enjoyed straight and in cocktails, was often associated with existing or former sugar plantations in various parts of the Caribbean.[64] The unequal global inheritances left behind as a product of flows of capital during the transatlantic slave trade are something more of us should take heed of, as we all too often watch environmental degradation, poverty, political turmoil, discrimination and ongoing discrepancies in pay and opportunities, both at home and in the tropics, from the comfort of our armchairs.

Sugar was, of course, not the only high-value product at the heart of the transatlantic slave trade. Nor was it the only crop that saw the tropics increasingly caught up in a globally imbalanced exploitation of landscapes and labour between the seventeenth and twentieth centuries. Today, cotton quite literally puts the shirts on our backs. In the seventeenth century, however, cotton cloth from India provided a major luxury good that was given by Europeans to African merchants and rulers in exchange for enslaved individuals. In turn, expanding growth in extensive plantations in the southern United States was supported by the brutally extracted labour from these people once they had been forcibly transported from their home continent.[65] Cotton's impact on labour and environments in the tropics did, however, reach still further than this new Atlantic economy. Between the early sixteenth and the early eighteenth centuries, the Mughal Empire, which extended across India, Pakistan and Bangladesh, was perhaps *the* main global centre of cotton textile production.[66] As writer Meena Menon states, 'farmers across the Indian subcontinent grew many different varieties of cotton as part of mixed plots alongside food for subsistence'. These farmers were part of an intimate system, with ginners, spinners and weavers, that produced high-quality cotton textiles sought after as far afield as West Africa and Japan. This relationship was broken by eighteenth-century British imperialism and the dramatic change to cotton being grown as a cash crop for global export. Although the Dutch had already been importing Bengali textiles, the consolidation of the British East India Company in South Asia greatly increased the availability of these products in Europe in the eighteenth century, allowing it to replace wool as the wardrobe-maker of choice.[67] Following the conquest of Bengal in 1757, British agents used their political and economic power to transform the international cotton markets. Investing India's existing cotton wealth in English domestic development and technological advances that outstripped Indian hand-spun cotton, and restricting Indian cotton imports to Europe, British factories were catapulted to the top of the global textile market during the industrial revolution. Attempting to make up for a loss of reliable raw cotton from North America following the War of Independence, the British set up experimental farms in India to grow American cotton varieties which were suited to the British textile mills. Demand from these textile mills led to British encouragement of cotton monoculture instead of more sustainable mixed cropping, taking valuable productive land away from food production, forcing huge numbers of Indian cotton spinners and weavers to seek alternative incomes, and sending the diverse indigenous short-staple cotton varieties into decline. This resulted in a major crisis of food during famine years. Decline in local production and competition from European exports also saw Indian consumers buying finished cotton textiles, and even yarn, back from Britain, sometimes made using the very cotton exported from India with a significant price mark-up![68] India is today the second largest exporter of raw cotton anywhere on the planet and the largest cultivator of cotton. It remains renowned for the creation of muslin, printed and dyed traditional textiles, but its handloom industry has been marginalized. The case of cotton is a further example of how unequal global flows of capital and economic infrastructural developments between Europe and the tropics imposed themselves on both landscapes and labour.[69] Strikes within the Indian textile industry as part of the Independence movement of the early

Hand ginning of cotton in India by a Ponduru woman.

twentieth century further illustrate the role this crop had in the manifestation of colonial power dynamics and economic inequalities.[70]

Tea is a go-to feel-good drink across much of Europe, Asia and North America today. It is also another crop that had a big impact on tropical Asian forests and their inhabitants between the seventeenth and nineteenth centuries. Originating somewhere at the intersection between South, Southeast and East Asia, extensive tea cultivation began in China around the eighth century AD. It reached Europe only in the seventeenth century, courtesy of Dutch and Portuguese ships. Here, the ruling classes quickly gained a predilection for the drink, particularly in England, and it became the pre-eminent import from China. In South Asia, the Singpho people of Assam and Myanmar had, however, also already been growing their own varieties of tea. The expansion of British capital into India, following conquests and political negotiation with local rulers throughout the nineteenth century, drew on this local knowledge and planted tea widely across the state of Assam[71] and, later, other states including Bengal and Orissa (now Odisha). Widespread production and easy access through imperial connections soon made tea cheaper and more easily available to all social classes in England, at the same time as it was impacting tropical landscapes and farmers in South Asia. Tea became a major plantation crop in Sri Lanka, for example. Between 1883 and 1897 an estimated additional 20,000 acres were covered by tea plantations every year in the Sri Lankan highlands.[72] Rural villagers lost vast swathes of agricultural land, and their traditional, mobile swidden approaches to farming among the lowland and misty montane tropical rainforest in the region were vastly curtailed, as were the extents of these forests and their ecosystems. Similar changes occurred back in India, where workers on tea plantations had few rights and were treated poorly within the social systems of racial stigma under the British Raj.[73] The Indian and Sri Lankan tea industries remain strong international and domestic sources of income for those nations today, with much of Indian- and Sri Lankan-produced tea now consumed by their own citizens and with production increasingly in the hands of smallholders. Nevertheless, the degree to which foreign interests have impacted the wealth, forests and lives of local farmers, past and present, should not be forgotten.

The coffee that wakes us up in the morning and keeps us going through days of constant 'Zoom' meetings has no happier history. Originally native to Ethiopia, cultivated coffee arrived in Europe, via the Middle East, during the seventeenth century – warming lawyers, politicians and philosophers in London, the nobility of Germany, and the financial sector of Amsterdam. The Dutch, recognizing that coffee plants thrived in the tropics, from 1699 cleared rainforests in Java to make way for its growth – a region where it is still grown commercially to this day. Coffee was soon exported for growth in the Caribbean. Supported by enslaved labour, the tiny French colony of Saint-Domingue (now known as Haiti) came to dominate the global coffee market in the eighteenth century.[74] However, by the time the Haitian Revolution of 1791–1804 saw self-liberated enslaved people create their own independent nation, no one would have known it. Thanks to tropical deforestation and soil erosion, the French left it one of the poorest nations and most degraded landscapes anywhere in the world. Although coffee continued to be produced on the tropical Caribbean islands of Cuba, Puerto Rico and Jamaica, by the middle of the nineteenth century Brazil dominated coffee production. Again, supported by a trade in enslaved individuals, one that continued up until the late nineteenth century, Brazil was able to produce coffee in massive quantities and at a low cost. Plantations also expanded across Dutch-held Sumatra, Bali, Sulawesi and Timor in the late nineteenth century. Soon coffee, previously the preserve of the wealthy, was being regularly drunk across the working classes of northern North American and European society.[75] The local benefits of growing coffee have been heavily debated. On the one hand, it has been argued that profits aided Brazil's emergence as a large, independent economy. On the other hand, it has been suggested that long-term continued use of slavery, the exploitative use of cheap labour, including children and marginalized Indigenous groups, inequality, poverty, the buying up of smallholdings by larger international companies, and deforestation, have offered few sustainable benefits to coffee-growing regions in the tropics[76] and have presented problems for tropical environments extending into the twenty-first century. Something to think about next time you take a sip . . .

Rubber is one of the most important tropical products around the world today. From balloons to the tyres of your next aeroplane holiday flight, from condoms to waterproof shoes, from rubber balls to supermarket conveyor belts, from diving gear to hospital tubing – you can barely go anywhere or do anything today without coming across this remarkable material. Natural rubber is made from latex, which oozes from certain plants when they are wounded. The only commercially viable natural sources, however, are the tropical forest plants of *Hevea brasiliensis*, a tree that grows wild in the Amazon Basin, and *Landolphia owariensis*, a wild vine that grows in West and Central Africa. The former was used by Indigenous populations across the Neotropics to produce clothing, boots, balls, storage carriers and toys long before any Europeans got there. In fact, Europeans did not realize the benefits of the material until the late eighteenth century, and economic production of rubber products began in the nineteenth century following the patenting of the vulcanization process. The rise of automobiles and other rubber-reliant inventions drove the subsequent 'rubber boom' which, between 1879 and 1912, spurred on an expansion of profit-making enterprises across the Amazon Basin of Brazil and Bolivia to feed growing global markets, particularly in Europe and northern North America. In the Upper Amazon, the resulting wealth galvanized the development of cities such as Manaus, where per capita income became the highest in Brazil. It had the first electric street lights anywhere in the country.[77] Native pests meant that attempts to grow rubber in plantations in Latin America repeatedly failed,[78] leading to a scramble of entrepreneurs seeking to make profits in an industry reliant on the tapping of wild dispersed trees in the Amazon Basin. Indigenous groups often became caught up in the race for profit. Some, such as the Mundurukú, actively engaged with the boom, moving in search of wages and European products. Others, however, were forced into labour agreements that effectively amounted to slavery and were killed and abused by so-called 'rubber barons'.[79] A study of living gigantic Amazon nut (Brazil nut) trees in the vicinity of the city of Manaus, by

Victor, whom we met right at the beginning of this book, has even shown that these trees grew more poorly as Indigenous populations and their traditional management effectively disappeared in the face of the pressures of rubber extraction in the nearby rainforests.[80]

The hands of Europeans and North Americans went on to shape the expansion of rubber as the commercial product we know today, as well as its impact on labour and environments in the tropics. In 1876 Sir Henry Wickham watched over the transport of 70,000 seeds of *H. braziliensis* to Kew Botanical Gardens in the United Kingdom. From there, the British Malayan and Dutch Sumatran rubber cartels promoted the planting of rubber in Malayan tea plantations in Southeast Asia, eventually securing their own direct access to the resulting rubber and profits in the early twentieth century. Away from native Amazonian pests the trees excelled and plantations were rapidly expanded. Although the addition of rubber into smaller plots could benefit local smallholders and entrepreneurs, intensification of plantation land use, and growing pressures from European and northern North American companies (particularly those with interests in the rapidly expanding automobile industry, such as Michelin), resulted in a widespread loss of local biodiversity and, frequently, the exploitation of Indigenous, as well as indentured migrant, labour under conditions that were little better than enslavement.[81] Today, Thailand, Indonesia, Malaysia, India and China are the top five leading global producers of rubber, using a tree from the other side of the tropical world and far surpassing Brazil.[82] Though the economic benefits are clear, monoculture growth of rubber has undoubtedly left its scars on the Southeast Asian landscape.[83] A more horrific European intervention in the global rubber market occurred when Leopold II of Belgium, seeking his own colonial empire, established the 'Congo Free State' in what is now the Democratic Republic of the Congo in the late nineteenth and early twentieth centuries. Seeking wealthy returns to rival other European entities during the rubber boom, Leopold's state forced local populations living in the rainforest into service, killing them, razing their villages, and cutting off their hands if they refused, to exploit the local latex-producing vine of the region. The atrocities observed and documented by appalled

European observers[84] almost certainly contributed to the collapse of the local population in just a few decades. Based on records of rubber production and the mortality rate, it has been estimated that one person in the Democratic Republic of the Congo died for every ten kilograms that was exported between the 1880s and 1923.[85] With the rise of electricity and two world wars, Western nations and business interests continued to grapple for access to rubber throughout the twentieth century, with little regard for local working conditions. Not for nothing has historian Professor John Tully of Victoria University called rubber 'the Devil's Milk'.[86]

In the above I have attempted to distil incredibly complex, and often contentious, processes in order to highlight a few key examples where crops and products we take for granted today were at the heart of the rise of global markets between the seventeenth and twentieth centuries, with major impacts on people and environments in the tropics, and economic and political echoes that still reverberate today. Even while Western imperialism and control over new worldwide flows of capital framed the all too often unequal outcomes, it is important to remember the huge variety of ways in which local governments, merchants, smallholders, consumers and Indigenous populations actively negotiated and challenged the introduction and exploitation of these economically valuable products. One example serves to highlight this local agency. Oil palm is currently one of the most prominent tropical plantation crops. As we saw in Chapter 7, it was long consumed across West and Central Africa and, between the seventeenth and nineteenth centuries, was produced to feed captives sold into the transatlantic slave trade. With the abolition of the trade in 1807 in the United Kingdom (and its colonies in 1833), however, it increasingly rose to prominence as an export crop. Nevertheless, in many parts of the continent its entire production process lay in the hands of African producers, and the palm crop was ill-suited to models of plantations built on enslaved labour.[87] Attempts to introduce mechanization repeatedly failed to boost productivity, with farmers in Sierra Leone, Nigeria and Senegal seeing no advantages. Local ecological knowledge, and the benefits of growing the crop within mixed subsistence plots, meant that groups with little wealth and their own labour could easily produce for export, including a group of enslaved individuals near Lagos who ultimately raised enough to buy their freedom.[88] A similar situation is evident in the failure of the Portuguese to establish oil palm as a commercial crop among the Atlantic coast rainforests of Bahia, Brazil. Instead, oil palm, until today, is grown here within 'subspontaneous' polycultural groves that contribute to subsistence and local economic wellbeing and are underpinned by centuries of African agroforestry.[89] Continued European attempts to control the palm oil trade eventually stimulated colonial conquest of many African states, and also saw it exported to Southeast Asia on a plantation-based model. Yet, ultimately, smallholders across the tropics retained some share of profits throughout. Interestingly, today, oil palm is still heralded as a solution for rural, smallholder development in Africa, South America and Southeast Asia, though, as we will see in Chapter 13, not without its challenges.

It was not just these more classically 'luxury' crops that became part of a growing desire to control the plants, workforces and environments of the tropics within increasingly unequal, globalized economic systems between the seventeenth and twentieth centuries. In Europe, growing subsistence reliance on and monoculture planting of the potato between the eighteenth and nineteenth centuries were to have major economic, demographic and environmental consequences some distance from the equator. Arriving from the cool montane forests and grasslands of the Andes, this tropical tuber was initially ignored in favour of traditional reliance on grain. Yet poor harvests in the face of the Little Ice Age, price fluctuations, frequent famine and public unrest gave the potato its moment to shine. Extolled by agricultural economists, scientists (notably Frenchman Antoine-Augustin Parmentier) and even rulers (namely the Prussian 'Kartöffelkonig', Frederick the Great), from the second half of the eighteenth century it became widely planted by smallholders and large landowners alike.[90] By the early nineteenth century, the potato was part of the rolling out of the European 'Agricultural Revolution' which involved the healthy application of fertilizer, such as seabird guano mined from its original home in Peru, and industrial ploughing methods.[91] The result was a significantly greater, and more reliable, European food base, one that has been credited with, at least in part, catalysing population increase and even the beginnings of the industrial revolution in England. The result was a monoculture *par excellence*. Unlike the original potato fields back in the Andes, this new form of food production exchanged genetic variability for uniformity, increasing vulnerability to disease with disastrous consequences. The Great Famine of Ireland (1845 to 1849) saw the decimation of this nation's new, homogenous food staple, thanks to the arrival of the potato blight from the Americas. Exacerbated by the callous negligence of the British imperial government in failing to provide alternative food sources, the resulting famine led to widespread suffering and a reduction in Ireland's population through death and emigration. Irish emigration stimulated by the Great Famine formed a significant part of the total of 10 million people of Irish ethnicity who left the island from the start of the eighteenth century, desperately seeking new labour opportunities in the face of often significant abuse. To this day the number of people living within the island of Ireland has still not recovered to pre-famine levels.[92] This catastrophe was a warning of the dangers of this new form of farming. One that heralded the beginning of the cycles of pesticide experimentation and application which most monoculture agricultural fields remain locked into around the world in the twenty-first century.

Back in the tropics, imperialist control over land use and a capital-driven demand for the more efficient growth of tropical subsistence crops also had consequences for local food security and environmental change around the equator. Taking the example of rice, historical ecologist and archaeologist Professor Kathleen Morrison of the University of Pennsylvania, alongside her colleague Dr Mark Hauser, has described the creation of imperial 'rice bowls' to serve growing globalized exchange networks and capital investment.[93] In the case of the Americas, as we saw in Chapter 10, the knowledge of how to grow rice, as well as rice itself, likely took initial hold as a product of African agency. However, during the eighteenth century the organization of its production changed significantly. Plantations in primary areas of production, such as South Carolina, Georgia and Brazil, created by European settlers and worked by enslaved African labour, became essential to providing food for Europe, northern North America, and, particularly, to sustain the valuable cotton, tobacco and sugar plantations of the Caribbean. This created inter-regional relationships of nutritional dependency, making local farmers and plantation communities less self-sufficient, and less secure in the face of natural hazards, such as hurricanes, that could destroy lives and wipe out vital arteries of food sourcing. In South and Southeast Asia, a lively market of regional rice exchange already existed in pre-colonial times. However, with increasing British political and economic influence, export-driven rice production increased significantly. This was particularly the case in Burma (now Myanmar) following British annexation in 1852. The entire tropical coastal delta of the Irawaddy River was transformed and deforested into a managed landscape focused on rice growth, for local consumption but also primarily for significant export to India and Europe. As imperial structures forced a move away from local food security towards wider markets of demand, local populations became increasingly reliant on global commodity prices, imported food and often unsympathetic governance.[94] The result was the destruction of local biodiversity, inequality, declining flexibility for smallholder land use, and potential famine. As Kathleen puts it, in the case of both the Atlantic and the Indian Ocean, 'the production and movement of rice around and outside of the tropics in the eighteenth to the twentieth centuries is one example of imperial imbalances of power, hunger, and risk' that have left an economic and environmental legacy in global economic relationships between the western half of Europe and northern North America and the tropical nations of the Caribbean, Asia, Africa and the Pacific in the twenty-first century.

The bananas we put in our baby food, smoothies, pies and ice cream, or simply enjoy raw, provide a more recent example of a tropical staple food turned from a local source of calories into an agro-industrial complex at the mercy of global consumer demand in the nineteenth and twentieth centuries. We saw in Chapter 7 how this tropical forest tree crop was first domesticated in New Guinea,

Woodblock images reproduced from Proctor 1888 in Morrison and Hauser[95] from Myanmar (formerly Burma). Upper left, men with bullocks threshing a rice paddy field; upper right, winnowing apparatus; lower left, hulling and decorticating devices; bottom right, image showing cultivators being paid. Following the British annexation of Burma in 1852, traditional methods of rice growth were often abandoned in the face of wide-scale clearance and mass production to support export to elsewhere in Asia and Europe.

being introduced into the Americas by the Spanish in the fifteenth and sixteenth centuries before spreading widely among local farmers in lowland tropical forests. In the late nineteenth century, however, it became a plantation crop. In the 1870s, United States railroad promoters in Costa Rica experimented with their widespread planting as a cheap, nutritious way to feed their workers. Realizing the potential for mass export of bananas to the United States of America, multinational corporations (such as the United Fruit Company, which formed in 1889) began to buy up land in nations such as Costa Rica, Panama, Honduras, Ecuador and Colombia. In the latter case, Colombian planters had actually already developed their own export networks to New York. Nonetheless, the United Fruit Company, buying land and railroads, forced local farmers to sell to them at low prices. Moreover, in order to spread risk in terms of the potential of diseases, soil erosion and climatic events hitting their crops, these multinational corporations frequently bought up and cleared a vast amount of tropical land simply so that it would be ready should plant-ing need to extend to new areas to maintain profits.[96] The result was the removal of local farmers from their smallholdings, the forcing of local farmers and planters into wage labour, and an increasing reli-ance on imported foodstuffs. The economic sway of these international corporations, and their tight grip on access to foreign markets, was so strong that they could even make and shape the politics and econo-mies of entire tropical nations. The 'banana wars'[97] also saw the government of the United States of America exert active imperialis-tic political influences on various nations in the Caribbean, Central America and South America, in order to secure access to tropical pro-duction and trade. The security of the United Fruit Company's operations in Honduras was backed up by US troops on more than one occasion, leading to long-term political instability. The seizing of Cuba and Puerto Rico from Spain in 1898 also saw the United States attempt to forcibly maintain beneficial economic flows of pro-duction and capital for themselves at the expense of entire nations.[98] Similarly, the United States of America intervened to support Pan-ama in its independence from Colombia, enabling it to secure permission to construct and control the Panama Canal (which opened in 1914), allowing it to dominate trade in the region.[99] Many coun-tries became almost completely reliant on certain key crops as a result of economic and political interventions by the United States of America, leading to unequal distributions of agricultural land, infra-structure produced solely to meet the needs of external capitalist enterprises, immense wealth inequality and low wages, degradation of tropical forest environments at the expense of plantations, and political conflict.

Away from crops, the increasingly intensive capital-driven pressure on tropical landscapes between the seventeenth and twentieth centuries is perhaps most vividly seen in the ways in which tropical timber and tropical land themselves became the objects of a 'rapacious capitalism'.[100] Mahogany furniture is a common feature of stately homes and historical museums in Europe and northern North America, venerated for its elegant craftwork and exotic, workable wood.[101] Yet the trade in this rainforest timber, driven by growing luxury Euro-American demand and profit-hungry merchants between the seventeenth and nineteenth centuries, involved the exploitation of enslaved African labour in 'woodcutting gangs'.[102] It also saw waves of 'harvesting bonanzas' crash upon entire ecosystems, from the Caribbean island of Cuba to the Central American coast of Honduras.[103] The 'commercialization of the forest' occurred a little later in the Philippines, intensifying as American imperialism replaced Spanish colonial rule and sought to determine how much wealth it could extract from the landscape. From 1898 up until the mid-twentieth century, a moving timber frontier, using railways and machines, began to cut lower-quality hardwoods indiscriminately, to meet Japanese, Chinese, North American and European demand.[104] Beyond being a product in its own right, timber could also be cleared to 'liberate' the land below for more 'profitable' uses. In Australia, from the nineteenth century onwards, intensified expansion across many parts of the continent by European settlers was driven by sheep and cattle pastoralism, and in 2010 Australia was the third largest dairy exporter in the world.[105] In the late nineteenth and early twentieth centuries, significant portions of the 'wet frontier' of Australia began to be cleared as farmers sought to exploit increased global and regional demand for animal products, particularly in Europe and other parts of Australia, exploiting British imperial connections. Expecting fertile soils must lie under these heavily forested regions, larger landowners, as well as small-scale farmers who rented contracts, undertook significant clearance of the semi-tropical eucalypt forests and wet rainforests of Queensland, including, notably, on the Atherton

Tablelands.[106] However, what these farmers found was often poor grazelands or wetlands, inevitable soil erosion, difficulty of exporting dairy from isolated settings, and a perilous existence at the mercy of government controls and global prices. Moreover, the murder and forced relocation of Indigenous traditional owners of these tropical forest landscapes, such as the Jirrbal, was not only morally unscrupulous but also led to a suppression of ecological knowledge and traditional land management which, as we will see in Chapter 13, still plagues the threatened biomes of this region to this day.[107] These examples simply scratch the surface of the increased deforestation and land conversion that took place across the tropics as governments, landowners and companies sought to maximize the benefits of growing regional and global markets of production and consumption.

New global circulations of capital and products, and demands for tropical land conversion, also continued to place pressure on workforces and migrants living in the tropics. The abolition of slavery did not necessarily mean that such movements became entirely voluntary, however. Even after formal abolition, enslaved Africans within British imperial territories could still be forced to enter multi-year 'apprenticeships' where they would work long hours for no pay.[108] When that practice finally stopped in 1838, landowners looked elsewhere for labour. In the early nineteenth century, many thousands of men from tropical South China and the Pacific Islands were coerced into contracts that nearly amounted to slavery, as they performed frequently fatal work on the guano mines of Peru that fertilized the 'Agricultural Revolution' of Europe.[109] From the early nineteenth century to the early twentieth, historical records demonstrate that more than 1.5 million labourers from southern India moved to plantations in British, French and Dutch colonies in the Caribbean, South Africa, Indian Ocean islands like Réunion, Mauritius, and Sri Lanka, Myanmar and Fiji. Although the contracts of 'indentured servitude' that formed the basis of this system were supposedly voluntary, they exploited sectors of the Indian caste system with low social status, offered poor working conditions, and often involved significant coercion.[110] In the case of Sri Lanka, the poor treatment of arriving Tamil-speaking labour on British tea plantations in the highlands, which local Sinhalese farmers had refused to work, may have contributed to ethnic conflicts that have since confronted this tropical island.[111] An often-forgotten part of Germany's colonial history is the role it played in the movement of nearly 200,000 people as indentured labourers from the Solomon Islands, New Guinea, and the Banks and Torres Islands, among other regions, to plantation work or other forms of labour in Australia, Fiji, Samoa, New Caledonia and other parts of New Guinea between 1863 and 1919. Potentially nearly one quarter of these individuals died within the duration of their contracts.[112] Migrating Chinese investors, landowners and labourers were a significant force in the twentieth-century planting of rubber in parts of Southeast Asia.[113] In the late nineteenth century, Italian labourers migrated to coffee plantations in Brazil, while in the early twentieth century[114] the arrival of Filipino migrants as fruit pickers in Hawai'i[115] formed the beginning of a labour export market for a tropical nation in which over 10 million of its 85 million population lives and works abroad, often facing heavy discrimination in the process.[116] Although this list does not come close to accounting for all the individuals, families and communities that moved, willingly or forcibly, as a result of growing global demands for production, capital and consumption, it illustrates the scale on which these processes have left their mark on demographic and cultural diasporas across and beyond the tropics right up to this day.

Through the seventeenth and twentieth centuries, the tropics, and their people and environments, were at the core of new, global flows of capital, demands for labour and exploitations of land. While the varied examples I have dealt with have, given their own unique temporal, social, economic, cultural and political contexts, all been the subject of their own in-depth historical, anthropological and archaeological studies, together they highlight how people, plants and landscapes from the tropics experienced, and also pushed back against, the formation of economic systems that majorly shape the world in which we live today. Throughout the seventeenth and twentieth centuries, and indeed even before, flows of investment were based on a 'taproot of imperialism'.[117] This was a European ideal that new lands and new sources of labour could be exploited to harness more profits and is what ultimately birthed the concept of 'capitalism', its critiques, and theories of modern global economics in the nineteenth century. From the racialized transatlantic slave trade to the exploitation of indentured workers and the migration of various, often stigmatized, ethnicities, labour from the tropics was sought out by landowners, companies, and even individuals seeking profitable investments as a cheap commodity. Meanwhile tropical landscapes were shaved so that they might be used to produce crops or resources that did not fuel local or regional need, but rather global flows of wealth and demand as part of a new 'world system'.[118] Societies and existing regional politics and economies in the tropics were involved, and individuals resisted in various ways. Overall, however, European and, later, northern North American control of the direction and output of investment on a global scale left legacies of inequality, cultural and political prejudice, racism, and land use, that saw certain sectors of Western society enriched and diasporic arrivals, as well as their tropical nations, impoverished. They also left a blatant mark on tropical environments. When the famous German naturalist Alexander von Humboldt explored the Spanish plantations of Lake Valencia in Venezuela in the nineteenth century, he noted, 'When forests are destroyed, as they are everywhere in America by the European planters . . . the springs are entirely dried up . . . '[119] In doing so, he was one of the first European observers to suggest a link between conversion of tropical land for new, profit-driven uses and a change in local and even regional climatic conditions.

It might be hard for us to see the origins of our own twenty-first-century economics and politics in this early capital-driven violence against landscapes and labour in the tropics. Yet, taking the example of the transatlantic slave trade and monoculture agriculture with which we began this chapter, we can track how a so-called 'plantation logic'[120] still continues to shape global economic markets, the regulation or un-regulation of the rights and treatment of workers, international politics, infrastructural developments, structural racism, and environments within and beyond the tropics in the twenty-first century. Dr Alex Moulton of the University of Tennessee, Knoxville and his colleagues have elegantly described how much of the Caribbean, for example, has witnessed three consecutive eras of agricultural management for capitalist accumulation.[121] The 'plantation era' refers to the imposition of chattel slavery as a 'productive force' which combined desire for biological control of tropical forest landscapes with brutal disciplining and control over enslaved bodies. The second era refers to the nineteenth- and early twentieth-century emergence of independent states such as Jamaica, Haiti and the Dominican Republic, which saw governments focus on the productivity of rural populations, deforestation and expansion of usable land, and the modernization of farming. Legacies of colonial infrastructure, however, often meant this involved succumbing to significant foreign debt as well as somewhat patronizing, even if well-meaning, perspectives that the 'Global South' was in need of 'development'. The third era, from the mid-twentieth century to the present day, saw some governments of tropical nations and the ongoing diffusion of global market capitalism promote internal competition between entrepreneurial farmers, seeking to out-do each other and leave behind traditional agricultural approaches. It has also seen European and northern North American leaders politicize 'help' and 'relief' to Caribbean communities in the face of disasters (made all the more frequent by climate change driven by the emissions from their own countries), praising the 'resilience' of local people while completely ignoring the colonial injustices that have left them the most at risk.[122] As Alex describes, 'these three phases highlight the ongoing threads of colonialism, racism, and capitalism operating on tropical landscapes and labour that have channelled a "plantation logic" into the present'. While this is just one part of the tropics, we should not kid ourselves that Euro-American histories can be in any way disconnected from the fate of the tropics, their inhabitants, and the inequalities that shape global politics and economics more broadly in the twenty-first century.

The twentieth century saw many tropical nations previously under the yoke of colonialism gain independence, broad improvements in living standards, and increasing investment in their economies.[123] Nevertheless, the global and national wealth disparities set in place by the processes seen in this chapter and Chapter 10, the growing economic power of multinational corporations, the need for infrastructural development following centuries of brutal extraction, and the continuation of industrial farming systems enforced under imperialism,[124] have shaped the current threats faced by tropical forests around the globe that we will look at in the remainder of this book. Between 1980 and 2000, 2,275,320 square kilometres, 2,590,620 square kilometres, and 8,520,310 square kilometres of tropical forest were lost from Central Africa, Southeast Asia and the Amazon–Orinoco Basins, respectively, in the face of logging, land conversion, infrastructural projects and expanding settlements.[125] A total of 17,364,780 square kilometres of forest cover was lost from the tropics as a whole during this period.[126] That is an area of nearly double the size of Europe. Rapidly growing economies in Latin America, South Asia and Southeast Asia increased their rate of fossil fuel burning,[127] which, alongside industrial activities in Europe, Canada and the United States of America that had continued apace since their beginning in the nineteenth century, as well as the feedback resulting from the loss of tropical forest cover, contributed to a rise in CO_2 emissions and a 0.6°C global surface temperature increase over the course of the twentieth century.[128] Habitat loss and climate change meant that the high levels of biodiversity found in the tropics suffered, dramatically effecting global species numbers. Between 1950 and 1987 it was estimated that as many as 50,000 species of plants, birds, mammals, reptiles, amphibians and insects were lost as a result of tropical extinctions.[129]

Of course, differences in colonial history, environment, demography, cultures, and political and economic structures and decisions meant that the twentieth century saw each tropical region and nation develop its own unique conservation challenges, priorities, and strengths.[130] Even within each region, different social groups were impacted differently as the growing grip of capitalism on the tropics drove increasingly stark inequality. The widespread appearance of *favelas* in Brazil as a product of the movement of rural populations to growing, industrialized urban centres during the 1970s is one example. Notably, the majority of the population living in these *favelas* are descendants of Black Africans, highlighting ongoing wealth disparities and discrimination that reach back to the colonial period and a racialized trade in enslaved labour.[131] Another example is the growing poverty and marginalization of Indigenous populations, and the difficult choices they must make between isolation and becoming enmeshed in new forms of settlement, wage labour, and political and social structures.[132] The racial fault lines and traumas set up by the historical processes seen above and in Chapter 10 also began to play out on European and northern North American shores in the twentieth century. The systemic racism, segregation, discrimination and violence facing Black people in the United States of America, a lasting legacy of the trade in human lives of the fifteenth to nineteenth centuries, and the racist ideologies surrounding it, resulted in enduring injuries made visible in the rise of the Civil Rights Movement of the 1950s and 1960s.[133] In the United Kingdom, from the 1940s, immigrants arriving from tropical colonies in the Caribbean to seek work were discriminated against in searches for housing, and they, as well as ongoing arrivals from the Asian and African nations of the Commonwealth, faced racial profiling, discrimination and poverty that resulted in their murder by white supremacists, precarious living standards, and retaliatory riots such as those in Broomielaw, Glasgow in 1919 and Brixton, London in 1985.[134] Similar issues were faced by Black people living across north-western and Mediterranean Europe in the latter half of the twentieth century. They still are.

The twentieth century also witnessed a growing understanding that the plight of tropical forests was not going to be just a regional problem, but a *global* one. Rampant tropical deforestation and losses of species diversity became symbols of environmental movements around the world, as many ecologists drew attention to the imbalanced amount of planetary biodiversity that would be lost should these habitats disappear.[135] Yet as we saw right from the beginning of this book, when we met the first plants and forests to take root on Earth in Chapters 1 and 2, tropical forests are not just globally significant in the context of their amazing roll-calls of plant and animal species. As we discovered then, and as we will explore further in the next chapter, they also play a crucial role in the regulation of precipitation, flow of groundwater, soil formation and stability, and exchange of gases with the atmosphere, on local, continental and planetary scales. Whether European and northern North American nations and their citizens are willing to accept the fact that modern conservation tensions stem from their colonial and imperial encounters with, and attitudes to, the tropics, or not – and whether they are willing to do something about it – will be explored in Chapters 13 and 14. Regardless, however, the exploitation of tropical forests that has expanded throughout the twentieth and into the twenty-first centuries has impacted not just environments and human livelihoods in the tropics, but also, increasingly, the climates and atmospheres of all of us. Determining when and how this process began may offer important lessons as we try to deal with environmental feedbacks of our own making today. To do so, we must reckon with the increasingly urgent concept of the 'Anthropocene' – the period when humans began to have a dominating impact on earth systems.

12. A tropical 'Anthropocene'?

Since the beginning of this book, the broad, ordered periods of time, or epochs, that we have been using (e.g. 'Cambrian', 'Carboniferous', 'Pleistocene', 'Holocene') to split up our journey through the tropics have been identified on the basis of physical rock characteristics and the presence or absence of certain fossil plant or animal communities.[1] Distinct boundaries between these periods are often related to major events such as climate shifts, volcanic and tectonic activity, or, as we saw in Chapter 4, an extra-terrestrial impact. However, the pace and extent of twenty-first-century human impacts on the Earth's surface, its rock layers and climate have led some scientists to argue that we have now entered a completely new era of geological history, the 'Anthropocene'. One whose boundary is set by us. Combining the Latin word for 'human' with the standard suffix for a period in geological time ('-cene'), this term signifies that we are now living in an epoch when human activities are playing an equal, or even dominant, role to natural forces in the operation of so-called earth systems. The Anthropocene has become an increasingly popular term,[2] not only thanks to artistic and philosophical musings on what the so-called 'end of nature' might mean for all of us,[3] but also because the Anthropocene Working Group (AWG) has been increasingly pushing for its official acceptance as a division of geological time by the International Commission on Stratigraphy in order to highlight the sheer force our species now wields in the context of the deep time of Earth's history. Geologically, such a status requires agreement on a clearly visible global 'spike', such as the biological and chemical fallout produced by the asteroid impact that doomed the reptilian dinosaurs at the K-Pg boundary. Attempting to do the same for the Anthropocene, the members of the AWG, headed by Jan Zalasiewicz at the University of Leicester, have narrowed in on plastic particles produced during an expansion of manufacturing of synthetic materials or radionuclides produced by the testing and dropping of the devastating atomic bomb during the twentieth century, which turn up in marine and lake sediments and caves around the world.[4] Yet while it is undeniable that human impacts on planetary systems since the twentieth century are on a scale never seen before in human history, to truly understand their origins and predict their outcomes we quite literally need to dig deeper than a single line of plastic or radiation.

As Professor Nicole Boivin, the Director of my own Institute in Jena, puts it, 'As a process that involves humans, it is essential that geologists include social scientists, specialists in the study of human-environment interactions, to properly come to terms with what the "Anthropocene" is and where its origins lie.' Our species has shaped the Earth for much of its existence, and to limit focus to post-industrial timeframes is to miss a long history of human relationships with the natural world. By including archaeological, historical and palaeoecological perspectives we can search for earlier potential roots of the 'Anthropocene' epoch, from the increased burning of fossil fuels following the industrial revolution of the late eighteenth and nineteenth centuries[5] to the origins of agriculture thousands of years ago.[6] We can also move past a desperate search for a single, formal epoch and place the geological timescale to one side. Using the anthropocene as a framework – with a small 'a'[7] – we are able to explore the long arc of human impacts on environments and the dynamic, and often imbalanced, human influences on earth systems that have led us to where we are today.[8] Tropical forests provide perhaps *the* ideal setting in which to do just that.[9] From our exploration of their origins in Chapter 1, we have seen that tropical forests, since their first arrival on our planet ~300 million years ago, have been a crucial part of earth systems. They built on the work of the first plant life to make the atmosphere breathable for life on land. They have regulated the distribution of the Sun's energy and also shaped where and how much rain fell in a given area. Their roots stabilized the planet's surface and created soils. They housed some of the first complex ecosystems to inhabit the planet and shaped some of the most marvelled-at periods in the evolution of life on Earth. They remain perhaps the most significant biomes at the heart of anthropocene-derived changes seen around the world in the twenty-first century.[10] As even local and regional human alterations to these environments can combine to have global feedbacks, what better place for us to look than tropical forests for the origins of human interactions with earth systems?

We will now turn to the latest ecological and earth science evidence that shows how tropical forests remain keystone environments for various earth systems today. They are host to the planet's most varied plant and animal communities and protect them from natural hazards, they are key to regional and global precipitation, they curate the soils of entire river basins, and they interact with the atmosphere.[11] Using the archaeological and historical background developed throughout the book to this point, we will then investigate the changing interaction of human societies with tropical forests, and the potential scale of its consequences. We will explore how the expansion of water buffalo and rice in the tropics 6,000 to 3,000 years ago may have resulted in visible greenhouse gas emissions, how the spread of farming and associated tropical deforestation in Africa might have influenced regional and continental soil erosion, and how the murder and disease inflicted upon huge Indigenous populations across the Americas at the time of European contact may have resulted in a regrowth of forests so significant that it changed the amount of CO_2 in the atmosphere. We will then move to the colonial and imperial land use that followed, and its legacy for perspectives on the 'Anthropocene', and indeed conservation, which can too often fail to acknowledge the wealth and power imbalances that have produced current sustainability tensions in the tropics. This chapter in no way suggests that our current impacts on tropical forests are 'precedented' or 'OK'. Rather it highlights how our actions diverge in both scale and severity, how we have become depressingly detached from the consequences of our choices, and how politicians and scientists must bear this long history of planetary influence and social exclusion in mind if they are to design effective conservation and policy.

For those of you who regularly watch nature documentaries on television or make wildlife encounters a key part of your summer

holidays, it will not surprise you to learn that tropical forests are ecological hotspots. They are home to over half the world's plant and
animal species.[12] Their lofty trees and bustling undergrowth, as well
as their vast array of crawling, walking, swimming and flying life-
forms, make up a staggering third of the Earth's land-surface
biological productivity.[13] To put this into perspective, more species of
ant can be found scurrying on a single tree in the Amazon Basin than
across the entirety of the British Isles.[14] Similarly, tropical forests can
have up to 1,000 species of tree per hectare,[15] as opposed to around a
dozen for temperate forests.[16] This plethora of biological life is amplified by the fact that the huge diversity of tropical, and sub-tropical,
forests we first met in Chapter 2, ranging from humid lowland evergreen rainforests to semi-deciduous tropical forests with a pronounced
dry season, from acidic peat swamp forests to cool montane tropical
forests, means that each and every tropical forest ecosystem has its
own characteristic roll-call of plants, insects, mammals, amphibians,
reptiles and birds that are often strung together in vital, intimate relationships.[17] Certain key species can have massive influences over
forest regeneration and health, for example the critical role of seed-
dispersing primates, bats and birds, which we gained a deep-time
perspective on in Chapters 3 and 4 and which is documented the
world over. As world-renowned tropical forest ecologist Professor
Yadvinder Malhi of the University of Oxford puts it, 'in the context
of the biosphere, tropical forests, in many ways, have the most to
offer. However, they also have the most to lose.' These environments
are currently facing some of the worst habitat loss and threats to their
wildlife in the twenty-first century. Given the huge proportion of
planetary biodiversity tropical forests host, such losses will undoubtedly echo around the planet.

Lord of the Rings aficionados will be delighted to know that tropical
forests are also gaining something of an Ent-like reputation[18] in the
way they manage and secure the land on which they grow. Below
ground, grasping root systems anchor soils in place and shape the
flow of water through the ground around them. Above ground, significant canopies and leaves regulate the flow of rain to the floor,
further stabilizing the earth.[19] Together, these attributes enable

tropical forests to quell rushing floods and to stem the movement of land. Coastal mangroves and forest plantations were even shown to have protected parts of Sri Lanka against the full force of the 2004 Indian Ocean tsunami. Elsewhere, in their absence, the impacts were much worse.[20] Tropical forests additionally host unique microorganisms that rapidly break down dead organic material, processing almost the entirety of falling leaf litter within a year, to keep up an important supply of nutrient-rich soil.[21] These microorganisms, roots and decaying vegetation even filter the water passing through and over them, removing contaminants for their inhabitants. A team of scientists, led by ecological modeller Dr Nadja Rüger of the German Centre for Integrative Biodiversity Research, working on an island in the middle of the Panama Canal, has shown that long-lived pioneer tree species such as mahogany and Amazon nut (Brazil nut) can represent nearly half the biomass in a given tropical forest.[22] Meanwhile, studies of large-diameter trees in tropical forests more widely show the major control they have on carbon storage and cycling in forests.[23] As a result, 'many of these services to the geosphere, and a series of earth systems, can be attributed to the tropical "Treebeards" out there', as Nadja puts it.

The contributions of tropical forests to earth systems also reach far into our skies. Their large leaves and extensive canopies provide the perfect transfer platforms for water back into the atmosphere. Tropical forests are responsible for one-third of the evapotranspiration (evaporation of water from leaves) around the globe at any one time.[24] Once in the air, the pollen, fungal spores and compounds unleashed by tropical forests bind this moisture together to form clouds and, thus, life-giving rain.[25] Climate scientists, using the sandbox technique similar to that of Tim Lenton, whom we met when exploring the first planetary impacts of plants in Chapter 1, have shown that this evapotranspiration is incredibly important to the water cycle (or hydrosphere) on both local and regional scales. If one were to completely deforest the Amazon Basin, precipitation could decrease by an average of 324mm per year locally (or as much as 640mm in some estimations, which would be more than the entire average annual rainfall in and around Newcastle-upon-Tyne[26]), while also leading to reductions more widely across South America.[27] Similarly, deforestation of the Congo River Basin in Central Africa would not only reduce rainfall by ~50% in the immediate region, but could also lead to less precipitation falling in West Africa along the Guinea Coast.[28] Modelled impacts of tropical deforestation even extend beyond the tropics, thanks to global climate circulation systems. Complete deforestation of the Amazon Basin could potentially lead to declines in rainfall in the mid-western and north-western portions of the United States of America.[29] Meanwhile, models of deforestation in the Congo Basin simulate reduced rainfall in Mexico, the United States of America, the Arabian Peninsula, southern Africa, and southern and eastern Europe.[30]

The cloud-forming canopies of tropical forests are not just significant for their role in the water cycle, they can also quite literally act as parasols in the regulation of local, regional and global temperatures. Due to the fact that the tropics are exposed to energy from the Sun more prominently than any other parts of the planet, the shade and moisture provided by the water-filled air skirting above these forests not only cool the ground beneath them, but actually impact the distribution of temperature across the Earth more widely.[31] That is not to mention the fact that the forests of the tropics store nearly one quarter of the world's land-based carbon, with peat swamp forests acting as particularly bulky carbon 'sinks'. As tropical forests perform a staggering 34% of the planet's photosynthesis, if they disappeared, not only would this trapped carbon be released into the atmosphere, but less CO_2 would be absorbed into the biosphere in their absence.[32] Together, these greatly increased emissions would trap more of the Sun's energy and exacerbate global warming still further. Although interactions between changes in land cover, 'albedo' (how much of the Sun's rays are reflected back into the sky, determining how much the surface warms), and CO_2 shifts are highly complicated, a series of modelling studies, as well as direct observations,[33] have highlighted the potential impacts deforestation could have on regional and global temperatures. Complete deforestation of the Amazon Basin, Central Africa and Southeast Asia is expected to produce regional warming of anywhere between 0.1 and

3.8°C, 0.5 and 2.5°C, and ~0.2°C, respectively.[34] Meanwhile, pan-tropical deforestation would also lead to an increase in average temperatures globally of as much as 0.7°C, a total that would double the warming observed since the mid-nineteenth century and intensified burning of fossil fuels.[35] This would undoubtedly lead to further polar ice melt, climatic instability, and the demise of various species of plants and animals within and beyond the tropics. This is not to mention the fact that most of the coal already burned since the start of industrialization, as we saw on our search for some of the first trees on Earth in Chapter 1, originated from the earliest tropical forests in the Carboniferous period in the first place.

Tropical forests are also often described as the 'lungs of the Earth'. Although it is not actually true that they supply 20% of the planet's oxygen,[36] they remain entwined in regional and global air quality. Without forests, landscapes become dustier.[37] Meanwhile the loss of their contributions to evapotranspiration increases the likelihood of fires.[38] These impacts are only exacerbated should their deforestation, or the draining and removal of tropical peat, occur as a result of deliberate burning to begin with. In 2019, Southeast Asian forest fires showed how widespread burning can have drastic impacts on the air quality across a sizeable area, with blazes in Indonesia being argued to be behind smog descending on Malaysia, Singapore and Vietnam and even forcing school closures in these areas – although it should be noted that Indonesia disputed that it was the only country at fault.[39] A lack of air filtering from forest cover,[40] increasing airborne dust and rising fire frequency, alongside deforestation-linked emissions of CO_2, nitrous oxide and methane, will eventually impact the air we breathe, and perhaps not only in densely populated tropical areas. If not our own breathing directly, alterations to tropical forests will certainly interfere with our food supplies. Human populations in Africa, Asia, the Pacific and the Neotropics are already facing challenges of food security as increasingly unpredictable climates impact their harvests. Furthermore, given that those of us in the Western world often rely on vast international networks, thanks to the colonial processes reviewed in Chapter 11, tropical and sub-tropical climate changes will also soon affect the coffee we import from

Brazil and Indonesia and the rice we obtain from Asia. The potential global climate changes wrought by dramatic declines in tropical forests could even damage agricultural growth of staple crops across the USA, China and India, with disastrous impacts for the large human populations of these regions.[41] Tropical forests therefore provide perhaps the clearest case of twenty-first-century human entwinement in earth systems. We can see how we are all reliant on these environments and we are all, in some way, linked to them. So, when did this begin? And how have past interactions with tropical forests shaped the potentially perilous position we now find ourselves in, where the loss of these environments will be felt all over the world?

Chapter 6 demonstrated how tropical forests were repeatedly occupied by late Pleistocene humans, as early as potentially 200,000 to 100,000 years ago, and certainly by 80,000 to 45,000 years ago.[42] But could these early hunting and gathering groups have somehow left lasting impacts on these environments and, thus, on earth systems? One way they might have done so is through their removal of their megafauna inhabitants. Conventionally classed as animals heavier than 44 kilograms, many of these giants faced extinction around the world during the late Pleistocene, at a time when *Homo sapiens* was occupying new areas of the planet. In South America, including the Amazon Basin, a staggering array of weird and wonderful megafauna disappeared during this time, such as giant sloths, the elephant-like gomphothere *Stegomastodon*, and the rhino-sized *Toxodon*.[43] Similarly, in Chapter 8, we saw how humans have been linked to later megafaunal extinctions in the Caribbean and Madagascar. Phylogenetic analysis of the late Pleistocene and early-middle Holocene extinction of a series of unique, large mammalian lineages, notably within South America, has been argued to show that this alone represented the demise of 2 billion years' worth of evolution.[44] Not only that, but because megafauna are so crucial to tropical forest ecosystems, through the dispersal of large seeds and fruits and the disturbance of forests through the swinging of their bulk, some scientists have suggested that if, indeed, humans brought about their downfall, this could be the first example of our species impacting not only the biosphere but also the carbon cycle, as the regrowth of large trees was hindered in their absence.[45] Other cases of potentially early human impacts on earth systems come in the form of deliberate burning and maintenance of forests in Borneo, Southeast Asia, the New Guinea highlands,[46] Near Oceania and in Queensland, Australia.[47] Ethnographic study of Indigenous populations demonstrates how widespread and long-lasting such activities can be in altering and managing tropical ecosystems, through the maintenance of forest boundaries, open patches, and clear forest floors.[48] Nevertheless, it can be difficult to definitely and directly attribute human activity these changes and, in the three 'megafaunal' cases mentioned, it seems unlikely that our ancestors were the sole factor in their tropical decline.[49] Moreover, in general, the overall earth systems influence of these activities would, while visible, also seem to be relatively localized.

More continental and global-scale human impacts on tropical forests and earth systems may have begun with the emergence of farming practices. We saw in Chapter 7 how a number of our favourite domesticated plants and animals have their origins in the tropics, including as part of some of the earliest examples of cultivation in the archaeological record. Meanwhile, we witnessed how the entrance of agricultural systems from extra-tropical areas could have major environmental repercussions. One of these latter cases has actually even been implicated in possible globally recognizable pre-industrial 'greenhouse gas' emissions. We are used to thinking of the rear ends of cows as significant methane producers, but perhaps less so rice. Nevertheless, where rice grows in flooded fields, or rice paddies, the water stops oxygen entering the soil, leaving it a hotbed for methane-producing bacteria. When William Ruddiman became the first scientist to suggest that the beginnings of the Anthropocene could actually be found with the origins of agriculture ~8,000 years ago, he suggested that, combined with CO_2 emissions from rampant agricultural deforestation in Eurasia, methane emissions due to prehistoric rice expansion across Asia produced a greenhouse gas effect so large that it actually prevented the onset of a glacial cycle.[50] Other scientists have linked the expansion of ruminating livestock into sub-Saharan Africa, monsoonal India and central China, as well as the expansion of irrigated rice and water buffalo throughout South-east Asia, to later, significant methane emissions between 6,000 and 3,000 years ago.[51] These models do not even yet include detailed estimates of the potential CO_2 emissions caused by prehistoric defor-estation across the entire tropics, as a product of both locally developed and externally introduced farming innovations. Although precisely and directly linking human activities to global atmospheric records remains challenging, and reconstructing the resulting climatic feed-backs even more so, these examples illustrate how prehistoric farming activities in the tropics may have already interacted with planetary systems.

The impacts of past tropical food producers on regional biospheres and geospheres can be a little easier to discern than their planetary counterparts, particularly in island ecosystems. The introduction of the domesticated pig and the Pacific rat as humans expanded across Micronesia and Polynesia are cases in point, resulting in observable soil erosion and the mass extinction of thousands of bird species, respectively (see Chapter 8). On another Pacific island, Guam, scien-tists believe that the arrival of swidden farming led to a complete environmental transition from forest to 'savannah'.[52] Such changes did not only occur on islands, however. In the Amazon Basin, culti-vation of indigenously domesticated plants not only involved significant landscape modifications to produce raised fields, but the active selection and promotion of economically useful species by past human societies has shaped the modern genetics and distribution of rainforest trees across this vast region to this day.[53] Working at Lake Caranã within a forest reserve in the Brazilian Amazon, palaeoecolo-gist Dr Yoshi Maezumi of the University of Amsterdam and her colleagues were able to demonstrate that 4,500 years of forest man-agement not only showed an enrichment of edible plant species represented in the sediment record of the lake, but had also left its mark on the composition of the surrounding forests.[54] Nonetheless, as Yoshi puts it, 'the widespread role of humans in ancient deforesta-tion and burning associated with these early tropical farming systems, as well as the legacy of human land use on biodiversity, soils, rainfall, and temperature, remains to be better resolved through the comparison of detailed, well-dated multiproxy records'. The same is true across Southeast Asia, southern China, much of tropical Africa, and most of the island environments we have visited on our tropical tour, where current investigations are often limited to local comparisons and the integration of archaeological and past environmental data into regional earth systems models remains a pressing area for future research.

The impacts of food-producing societies on tropical environments and earth systems were, however, only to increase and, as Chapter 9 showed us, between 2,500 and 500 years ago, a variety of populous urban settlements, states and empires had emerged in many of these regions. LiDAR and other remote sensing technologies have proven particularly powerful at revealing the sheer scale of the past populations buzzing across the tropics. In the Neotropics, in North and Central America, the Classic and Postclassic Maya occupied extensive urban landscapes, planted crops and manipulated forests. Recent assessments suggest as many as 11 million people may have been present in the region at 800 AD.[55] Estimates of the deforestation associated with these populations have now been factored into climate models to suggest that Mesoamerican urban societies reduced precipitation by 5–15% in the region, amplifying droughts, and potentially contributing to their own issues of sustainability in dry, lowland regions.[56] This is not to mention the vast hydrological and agricultural modifications made by the Triple Alliance (Aztec Empire) across the Valley of Mexico and out into their area of imperial control. Turning further south, the Inka Empire ruled over ~20 million people, building massive terracing and irrigation networks that deforested mountains and hillsides. Inka rulers were continuously combating soil erosion through the planting of trees and regulating the use of wood.[57] Meanwhile, recent demographic estimates of another ~8–20 million people among the 'garden cities' of the Amazon Basin,[58] through significant earthworks, occupation mounds and networks of causeways, were, by 500 years ago, permanently shaping forest composition, dynamics of regrowth and succession,[59] and perhaps even overall forest cover above ground, as they simultaneously changed the nutrient contents and fertility of soil below.[60] Although evaluations of their population sizes based on census data, later colonial records and archaeological records remain hotly debated, these pre-colonial Neotropical societies, additionally including the pre-colonial populations of the Caribbean, left their mark, not just on tropical forests, but also on the regional and continental biosphere, geosphere, hydrosphere and, perhaps, as we will see shortly, the Earth's atmosphere. Notably, however, these populations, even in urban systems, were often dispersed across the landscape, practising a combination of crop cultivation, tree management, and the use of wild animals on land and in rivers, reducing the overall per capita pressure on each parcel of land at any given time. As a result, despite some local and regional challenges, a population approaching that of Europe was broadly being successfully sustained in the Neotropics by the time of contact with Europeans.

Turning to Asia, the world's largest pre-industrial urban centre had documented impacts on the dry tropical forests of Greater Angkor in Cambodia. Studies of lake cores and the urban infrastructure itself have shown that human deforestation across this vast artificially constructed drainage basin gradually led to soil erosion that contributed, alongside climatic variability, to rulers eventually abandoning the region as their capital.[61] Limited palaeoenvironmental investigation means that we have almost zero understanding of how the significant states and empires of West and Central Africa, southern China, elsewhere in Southeast Asia, and Sri Lanka and the Indian subcontinent impacted their surrounding tropical forest environments, and thus earth systems, though current representations[62] are almost certainly under-estimates. In contrast to perspectives often put forward in conservation, ecology and earth science circles, then, these environments were not a blank canvas for what came next. Instead, many millennia of increasingly intensive manipulation meant that human societies were already locked into earth systems feedback mechanisms with plant and animal communities, soils, and perhaps even local and regional climates. Some scientists have suggested that such examples are, to some extent, anecdotal and incomparable to what happened next, and they might be right. Not necessarily only because they were on a smaller scale or because they were inherently less impactful when compared to the dominating impacts of humans on the tropics and earth systems today. Instead, it is also clear that while pre-colonial tropical societies modified their environments, often with far-reaching consequences, they retained a certain flexibility to adapt. We have seen that, with some notable exceptions, elites, merchants, and, most particularly, the farmers and horticulturalists on which these societies were built could see what was happening around them, change what and where they planted and exploited, modify landscapes, shift their political centres and structures, or even simply move on. The vice-like grip of proto-global and, later, global economic, political and social structures that followed increasingly swept these options from the board, particularly for marginalized, oppressed groups. It was this period that set us on a course for the present day, where some of the people most impacted by anthropocene-related changes to tropical forests are also the least able to change their practices and protect their livelihoods.

The size of human populations, and the extent of their modifications, across tropical forests by 1491 is dramatically illustrated by what happened as their worlds collided with Europe. In the Neotropics, we saw in Chapter 10 how Alexander Koch and his colleagues have used models of existing population estimates from historical documents, settlement areas and other sources, to estimate that within 150 years of European arrival staggering numbers of Indigenous people had died across the Neotropics as a result of diseases arriving from across the Atlantic. However, he and his team did not stop their research there. Assuming a given area of land use by Indigenous populations in different parts of the Neotropics, they sought to work out what would happen if these human societies were suddenly largely removed from tropical landscapes. They found that the estimated additional carbon uptake, which would have occurred as new forest grew rapidly across the Americas in their absence, made up a significant amount of the 7 parts per million decline in global atmospheric CO_2 concentration recorded in polar ice cores between the late sixteenth and early seventeenth centuries.[63] This decline reduced global temperatures by 0.15°C. This might not seem like a lot, but it actually

Ice-skating Near a Village. Dutch painting by Hendrick Avercamp showing icy conditions on Dutch canals and lakes during the 'Little Ice Age'.

contributed to the coldest part of the Little Ice Age ~1600 to 1650 AD when harvests failed across Europe and ice-skating on frozen lakes became a source of artistic inspiration. More detailed archaeological, palaeoecological, and historical based models of changing land use are necessary to confirm the scale of this 'reforestation'. However, for Alexander, the current balance of evidence suggests that 'the "Great Dying" of the Indigenous peoples of the Americas, particularly within the Neotropics, initiated by European colonization led to a human-driven global impact on earth systems, with serious feedbacks occurring far away from the tropics, long before the industrial revolution'.

Earth scientists Simon Lewis and Mark Maslin have even suggested that the 'Great Dying', and the earth systems feedbacks that ensued, represent the most significant global marker for the 'Anthropocene' as an epoch – separating what went before from the beginning of the modern 'world-system'.[64] Certainly, from a biosphere standpoint, the first global trade networks that joined the American tropics to Africa, China and Europe did not just transfer microbes on a scale never before seen, but also plants and animals. Lewis and Maslin, for example, in a prominent paper published in the journal *Nature*, have suggested that the appearance of pollen from maize in European lakes and marine sediments, and phytoliths from bananas in lake records in

Central and South America, from 1600 onwards, represents clear stratigraphic markers for the globalization of human foods and cuisines initiated by the so-called 'Columbian exchange'.[65] The same could be said of archival and archaeobotanical records of sweet potato in the Philippines, manioc in Africa, wheat in Mexico, potatoes and tomatoes in Europe, and sugarcane in the Caribbean, as the tropics sat at the centre of a process of changing landscapes and mealtimes across the world. The appearance of the bones of horses, cows, goats and pigs, as well as less intentionally moved black rat stowaways, for the first time in archaeological sites in many parts of the tropics offer further potential markers preserved in sediments. These animals could have major impacts on tropical biodiversity, vegetation cover, plant community structure, and the stability of soils as part of a 'swift, ongoing, radical reorganization of life on Earth without geological precedent'.[66]

This 'reorganization' also included new ways of appropriating, using and perceiving tropical landscapes that arrived with Europeans. We saw in Chapter 10 how, by the second half of the seventeenth century, after just a century of mining in the region, Spain had already received 16 million kilograms of silver from the Neotropics – tripling the amount that had formerly existed in Europe! This could leave deep scars in regions such as the silver mines of Potosí, where labour was forced into an intensive extractive exercise. Here, by 1600, trees were stripped, domesticated and wild animals disappeared, and the soils of entire mountain slopes had turned to loose gravel.[67] Meanwhile, mercury seeped into the soils and air. Likewise, excavation was so intense at the gold deposits in the Brazilian highlands that when scientists visited the mining zone in the early nineteenth century they assumed that the formerly lush, forested area had always been a barren plateau.[68] A similar disregard for sustainability in favour of profit is visible in the form of increasingly widespread plantation agriculture. Europe's insatiable lust for sugar had already eaten up ~5,000 square kilometres of Brazil's now threatened Atlantic rainforest by the middle of the seventeenth century[69] – that's nearly the same size as the entirety of Trinidad and Tobago. We find a comparable pattern on British, Dutch and French-owned sugar plantations in the Caribbean,

Dutch spice plantations in Southeast Asia, and French vanilla plantations in Madagascar. Beyond cash crops, globalized economic systems
even began to appropriate tropical landscapes for their own everyday
food and materials. For example, running out of wood on Mediterranean shores, the Spanish and Portuguese increasingly turned to the
lumber resources of the Philippines and Cuba, and south India, respectively, to build their fleets.[70] The scale of tropical deforestation initiated
by this lust for a 'cheap nature',[71] to be used and profited from, currently remains poorly known. However, given that, as we saw in
Chapter 11, concerned observers had already postulated a connection
between Neotropical plantation agriculture and aridity as early as
the nineteenth century, significant earth systems feedbacks seem
highly likely.

The expanding tendrils of this 'age of capital' have also been implicated in one of the most prominent anthropocene processes identified
in human history, the onset of the industrial revolution in north-
western Europe.[72] While this happened away from tropical landscapes,
the source of capital for investors, entrepreneurs, factory owners and
merchants that drove it was, in many cases, the wealth, productivity
and human lives that were forcibly taken from, or moved across, the
tropics at an increasingly merciless pace between the seventeenth and
twentieth centuries. Privileged, controlled access to energy-rich
sugar grown using chattel slavery, rice planted across entire tropical
river basins for the main purpose of export, and potatoes and fertilizer to develop local agricultural efficiency, meant that the tropics
also experienced a re-routing of calories. Opportunities to grow food
were taken away from the inhabitants of the tropics to fuel an increasingly nourished society in first Britain, then other nations in western
Europe and, later, northern North America. As such, the western
European imperial appropriation of tropical plants, land and labour
can be seen as a significant factor behind the onset of mass factory-
based, capitalist industrial production, expansion of effective transport
infrastructure of railways and steamships, and insatiable burning of
fossil fuels that left significant, globally recognizable increases of
atmospheric CO_2 in Arctic ice cores, and records of methane, nitrates,
ash, and remnants of burning in ice and lake records, from the

nineteenth century onwards.[73] Suggested by some as possibly representing the true beginnings of the 'Anthropocene'[74] as a formal epoch, industrial- and transport-based emissions remain some of the most significant culprits behind human-induced global warming in the twenty-first century.

The processes described above are not only important for our search for the gradual intensification of anthropocene phenomena over time. They have also been important for deconstructing the idea of a definitive, unifying, geological, scientific 'Anthropocene'. Studying the colonial and imperial processes that swept across the tropics, anthropologists, historians and historical geographers have suggested that searching for a single spike, either in the industrial revolution or in later twentieth-century events in Europe and northern North America, obscures the long-term processes of inequality on which twenty-first-century impacts on the natural world are founded. Global history between the fifteenth and twentieth centuries shows that the 'Anthropocene' is not a product of humanity as a whole, nor has it impacted all of humanity equally.[75] Instead, as Janae Davis of Clark University and her colleagues have put it, 'it is the product of interconnected historical processes driven by a small minority that provided the conditions for global capitalism through colonialism, enslavement, and racism'.[76] Expanding the work of anthropologist Donna Haraway,[77] they argued that the 'Plantationocene' might be a more appropriate concept than the 'Anthropocene'. Rather than seeing the steam and factories of western Europe as the active origins of the present globalized world, we should look at how plantation systems converted foreign investment into homogenized tropical landscapes, how they forced people, plants and microbes from all corners of the globe together,[78] and how they were caught up in processes of Indigenous mortality and racialized capital-driven systems of enslavement.[79] Through such a framework, it was the forced labour of plants, landscapes and people that provided the 'model and motor'[80] for the greedy Western factory system that is seen by many as the embodiment of the industrial-based 'Anthropocene' we contend with today.

Some anthropologists and historians have also proposed the

broader term of 'Capitalocene', highlighting how plantation systems were just one part of the origins of a global system that wrapped tropical and extra-tropical landscapes together.[81] This view highlights how from the onset of European colonization in the Atlantic and the Americas tropical lands were mapped, modelled and valued, as open spaces, ready to be put to work. Labour, in the form of slavery and indentured contracts, and nature alike became commodities to be accumulated, appropriated and used. The result was a world in which environments and livelihoods in the tropics – from the silver mines of Bolivia to the coffee plantations of Haiti, from the cattle-modified landscapes of Mexico to the logged hills of the Philippines, and from the Indian farmers desperately planting cotton instead of food to earn enough money to pay heavy taxation to the tea plantations of highland Sri Lanka – were heavily degraded to provide capital and power for populations, factories and governments in the western half of Europe and northern North America. The result was a world in which even human lives across the tropics became reduced to a cheap, often racially based commodity, grist for the mill for national, corporate and public wealth on the other side of the world.[82] This is not to say that oppressed minorities and populations were always powerless, and Chapters 10 and 11 showed us many examples of resistance and shaping of globalized cultures, cuisines and economic systems across the tropics. Overall, however, it is clear that any analysis of anthropocene processes must recognize that the current global role of the tropics in political, economic, social and earth systems in the twenty-first century has not been produced by random, uniform human processes, and certainly not everyone has had an equal seat at the table.

Tropical forests are, of course, not the only environments in which the 'multiple beginnings' of anthropocene processes can be investigated and observed in prehistory and history.[83] Some archaeologists have focused on the origins and spread of pottery in soil deposits around the world as a 'novel' marker of human behaviour and waste,[84] a process now extending back to ~20,000 to 12,000 years ago in East Asia.[85] Alternatively, domesticated animals emerging from the Near

East represent the beginnings of 'distinctive organic remains' that herald a new period of human impacts on the biosphere.[86] Nonetheless, tropical forests are the land-based environments that are perhaps most caught up in biological, climatic and atmospheric earth systems, and even local and regional modification can have widespread repercussions. With their 'keystone' position in planetary function in mind, we can imagine how the disturbance of megafauna through hunting and burning, how the expansion of domesticated animals, emissions-producing crops, and deforestation in the face of expanding and locally developed farming systems, and how the emergence of pre-colonial tropical cities, states and empires within tropical forests might all have had pre-industrial impacts on earth systems, ranging from the local to the global. Working within the anthropocene framework, we can see how these human insertions into earth systems may not represent a clear, global 'Anthropocene' spike, and may not approach the intensity or scale on which such interactions are happening in the twenty-first century. However, Chapters 6 to 9 have undoubtedly demonstrated how almost all tropical forests had had a deep-time history of human manipulation, leaving lasting legacies in terms of species distributions, biodiversity and ecosystem dynamics. We will see in the next chapter how these long histories of Indigenous management show that there are few 'pristine' tropical forests and we must shape our conservation choices and practices accordingly. In turn, the growing, vibrant archaeological, historical and palaeoecological literature in tropical forests forces us to face challenging questions as to what kind of environments we actually want to preserve around the world moving forward – empty and uninhabited, or sustainability managed, used, and perhaps, therefore, better protected?

Tropical forests, and their human, plant, and animal inhabitants, are also certainly not the only environments and communities to have witnessed the impact of colonialism, imperialism and capitalism between the fifteenth and twentieth centuries. In 1541, England, under the rule of Henry VIII, began to extract materials and land from Irish forests and relocate Irish labour as part of colonial rule.[87] Meanwhile, a desire for iron, copper and silver ores had seen significant portions of Central Europe laid waste by the early six-teenth century as a consequence of imperial demands.[88] Beyond Europe, the horrific treatment of Indigenous populations and the atrocities of slavery that occurred as part of growing appropriation of nature and flows of capital were repeated in temperate North Amer-ica,[89] the non-tropical portions of Australia,[90] and southern Africa among other places.[91] Nevertheless, we have seen how the tropics were at the centre of the intensifying global spread and flow of impe-rial power and capital. The mines of the Neotropics quite literally provided the money for European economic and political ventures, tropical plants and landscapes the world over provided the basis for plantation and ranching systems that amassed national and private wealth in Euro-America, and human lives across the tropics were lost, commodified, forcibly relocated, and altered forever by colonial and imperial processes. The scars left behind on our planet, in the form of barren landscapes, political and capitalist economic inequalities, racial conflict and earth systems consequences, can be profoundly seen through the lens of tropical history. They mean that all of us are caught up in the ongoing balancing act between tropical commoditization and conservation, between necessary sought-after infrastructure and economic expansion and planetary health, and between Western demands for national and global action and the difficulties facing often-impoverished tropical nations if they are to act. Only by recog-nizing this can we begin to develop practicable conservation solutions, involving diverse international, governmental and local stakehold-ers, to the crisis now facing some of the most threatened land-based environments on the surface of the Earth.

13. Houses on fire

In January 2019 Swedish student Greta Thunberg declared to world leaders at the World Economic Forum that 'Our House is on Fire'.[1] Her words were to prove disturbingly prophetic as the year wore on, particularly for the tropics. Between June and August, fires swept across the rainforests of the Brazilian Amazon. Although natural fires are rare in these environments, a weakening in environmental protection regulations by the government, long-term trends of deforestation, fragmentation and timber cutting, climate change, and linked regional aridity resulted in the highest levels of active fires and forest clearance seen in a decade.[2] Fires in Brazil, as well as in Peru, Bolivia and Paraguay, had catastrophic consequences for Amazonian biomes and, by August, blankets of smog shot into the high atmosphere, extended across both the Amazon and Paraná River systems, and adorned cities, such as São Paulo, many kilometres away. By January 2020 smoke was arriving in South America from a very different source. Natural wildfires are common in many parts of Australia during the dry season. However, between June 2019 and January 2020 the wildfires of what was to become known as the 'Black Summer', spurred on by human-induced global climatic warming, were so bad that the resulting smog made it all the way to Chile and Argentina.[3] Not only that, but fires even began within Australia's rainforests – environments that scientists had thought would rarely burn. As firefighter Renier van Raders put it when I met him on a recent trip to the Atherton Tablelands, 'In my twenty years in this job, I have never seen these wet forests burn. This year we had two local events where the rainforest canopy was on fire.' These two sets of infernos vividly illustrated the increasingly precarious state of the tropics, and indeed our entire planet, in the twenty-first century – and worrying trends in fire continued in the Amazon Basin in 2020.[4] They also showcased the diverse social,

Warlpiri people burning spinifex to promote future growth. Tanami Desert, Northern Territory, Australia. Aboriginal burning is crucial for the maintenance of biodiversity and for providing protection of ecosystems from the devastating bushfires that are increasing in recent years thanks to human-induced global warming, including within the Wet Tropics Bioregion of Australia.

political, economic and environmental factors that make developing sustainable conservation and policy solutions so challenging.

Scientists looking at changes in forest cover using satellite imagery taken from space between the years 2000 and 2012 have shown that, globally, tropical forests and woodlands are disappearing at a rate of 91,400 square kilometres every year[5] – an area roughly the size of Portugal. It is therefore not hard to see why tropical forests are frequently considered, in academia, policy and the media alike, to be key battlegrounds in the fight for human sustainability. The drivers behind their disappearance can be various. In some cases, it is low-income farmers seeking to plant food to survive or cash crops to scrape a living. In others, it is multinational corporations sponsoring widespread land conversion to maximize productivity and profits.[6] Logging for local, national and international timber needs represents another threat. Although selective, well-managed logging can be sustainable, poorly regulated, corner-cutting approaches, and the building of roads and other forms of infrastructure can lead to forest degradation and fragmentation with impacts that can be ultimately just as bad as wholesale clearance.[7] Beyond deforestation, the diverse plant and animal life of tropical forests faces further challenges. In Africa, nearly 5,000 tonnes of hunted tropical bushmeat are dragged from their leafy homes every year.[8] Here, as well as in Southeast Asia and South America, the 'catch' often includes large-bodied animals, such as primates, which, as we have seen, can have crucial roles in the overall health of forests. Atmospheric CO_2 increases and global warming represent more indirect, though no less significant, human threats for tropical forest life forms, radically altering their distribution and dynamics. This is only going to get worse as raging fires and deforestation in the same forests release yet more CO_2 from the land

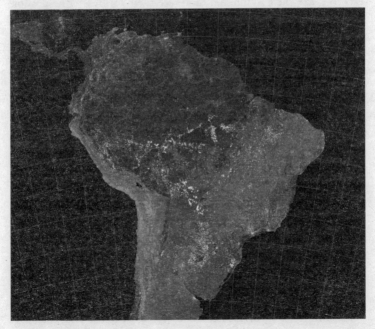

Locations of fires in the Amazon Basin, marked in white, which were detected by MODIS satellite scanning from 15 to 22 August 2019.

into the atmosphere, creating an even hotter and less nurturing 'greenhouse'. In fact, the average carbon emissions produced by tropical deforestation around the world are currently higher than those produced by the entirety of the European Union.[9]

The prominent contribution of tropical deforestation to rising global emissions, as well as the local, regional and global earth systems feedbacks which, as we saw in Chapter 12, are possible should these forests continue to disappear, has meant that the conservation and management of tropical forests has become a major environmental, economic and political priority. This is not just the case for governments in and around the equator but, as demonstrated by the international political fallout during the Amazonian fires of 2019, also those further to the north. And rightly so. Because by 2050 over half the world's human population, and two-thirds of its children, will live in the tropics, looking increasingly intently at tropical forests to make their living and to secure their futures.[10] The possible solutions can seem obvious from Euro-American homes. Cordon off the land like a crime scene, remove all humans to let it return to 'nature', plant more trees, and encourage foreign governments to prohibit logging, hunting and landscape conversion. However, these sometimes patronizing perspectives usually ignore two major issues. First, as we've seen, most tropical forests have not been completely 'natural' for many millennia. Second, the responsibility for the current impact of capitalism on tropical landscapes, as well as the global wealth imbalances at the core of deforestation, hunting and climate change crises, does not lie solely with people in the tropics. Instead, we are all, particularly those of us enjoying comfort in the Western world, liable for the fate of these threatened biomes. While this can be hard to swallow, as we will now see, it is only by acknowledging the above two points that more successful and just conservation, economic, and political policies can be developed.

Given all that we have seen throughout this book it may come as something of a surprise that some conservationists and ecologists have sought to define areas of tropical forest as 'wilderness'. The current rate of global species extinctions is estimated at 1,000 times its natural baseline rate as a result of twenty-first-century human planetary influences.[11] Thanks to the sheer quantity of plant and animal species in tropical forests, a plethora of which are thought to still be waiting to be discovered, conservationists have focused on these ecosystems as bastions in the global fight against species loss. Some have even suggested that by protecting vast areas of 'wilderness' with low human 'footprints' and high levels of biodiversity – i.e. tropical rainforests – we can secure the future of the majority of the Earth's plants and animals.[12] Similar ideas drive the proposal that by adding more trees around the world, notably in dry tropical and sub-tropical areas, we can successfully counteract climate change by returning large areas of land to a 'restored' state.[13] Alternatively, ecologists have emphasized the significance of the condition of the remaining forests – namely whether they remain 'intact' or have been 'degraded' by human actions.[14] Such terminology hints at desires to reach a primeval, undisturbed condition, something clearly visible in many UNESCO designations of natural heritage areas. For example, the Wet Tropics of Australia are said to contain 'relics of the great Gondwanan forest that covered Australia . . . 50–100 million years ago'.[15] Now, the maintenance of intact forests, with their characteristic canopies, stratification and dynamics, can certainly have its benefits. These forests undoubtedly play immensely significant roles in air quality, soil stability, climate and biodiversity protection.[16] In fact, many of the best-protected eco-regions worldwide occur in 'intact' areas of tropical forest within the Neotropics and Southeast Asia.[17] Nevertheless, the fencing off of wholly 'natural' tropical areas is becoming increasingly challenging in the face of human population growth. Moreover, ecologists are recognizing that 'intact' does not have to mean 'pristine', and that many ecological benefits also exist in secondary and 'anthropogenic' forests.[18] As we've seen from the beginning of this book, tropical forests are not static. Instead, they have been constantly on the move thanks to plate tectonics, climate change, and, later, deliberate human interventions. The exclusion of these dynamic histories, as well as their traditional human occupants, can thus misrepresent these forests, often with unintended consequences.

The establishment of protective reserves can still represent a highly effective solution to tropical conservation if done in the right way. However, for many areas of tropical forest, they are simply not an effective, or even viable, option. So, what can be done instead? Can we learn something from the archaeological, historical and palaeoecological insights we have already gained? Bushmeat hunting, for example, is often taken to be inherently unsustainable, leading to 'empty' forests.[19] However, as we saw in Chapter 6's tour of our species' first global expansions, semi-arboreal and arboreal mammals, including two primates (the gray langur and purple-faced leaf monkey) that are now endangered, were consistently hunted by bow-and-arrow-wielding human populations in Sri Lanka's rainforests between 45,000 and 3,000 years ago, without apparent negative impacts.[20] Furthermore, even deliberate early human modifications to tropical forests to maintain clearings, tracks and grassland biomes, dating as far back as 45,000 years, likely stimulated, rather than hindered, populations of prey such as wild boar in Borneo.[21] Certainly, ethnographic and historical observations in Central Africa and South America document that Indigenous hunting methods and cultural restrictions see hunters move between different forest areas and different species to maintain the overall health of their prey communities.[22] Meanwhile, a review of ethnographic research across the Amazon Basin, New Guinea and Australia has revealed the benefits of human burning for the new growth of starch-rich plants as well as for populations of locally hunted birds and ground-dwelling and arboreal mammals.[23] In fact, although prehistoric 'over-hunting' was discussed as a potentially early human earth systems intervention in Chapter 12, the majority of tropical animal extinctions that have been convincingly directly linked to hunting actually post-date European colonialism. In fact, today, the most unsustainable bushmeat hunting occurs where traditional weaponry, local ecological knowledge, and subsistence requirements have been replaced by heavy population pressure, 'trophy' killing by tourists, or a commercial, individualistic drive to use snares and rifle technologies to maximize capture.[24] In the latter case, this is often done to feed a rampant international illegal trade in pets, medicinal ingredients, or valuable items such as ivory.[25] With the right conditions and practices in place, however, hunting in tropical forests, particularly as a local source of food, need not inevitably result in them becoming silent.

We have also seen that there does not necessarily need to be a conflict between farming activities and tropical forests. As shown in Chapter 7, these environments have yielded some of the earliest examples of deliberate cultivation anywhere in the world. Not only that, but Indigenous farming activities in the tropics reveal the possibility of very different avenues to food production than our common Euro-American perceptions of 'agriculture'. Archaeology, palaeoecology and Indigenous knowledge are all now advocating the benefits of such 'usable pasts' for improving food security in the tropics in the twenty-first century. Nigerian archaeobotanist Dr Emuobosa Orijemie of the University of Ibadan, Nigeria, emphasizes how the archaeobotanical record of the last 1,000 years in central Nigeria show that it was 'mixed landscapes of indigenous crops, such as oil palm, groundnuts, yams, cowpea, and pearl millet, and not monoculture' that provided food security for human populations through periods of demographic expansion and climatic variability. Indeed, all around the tropics there are now calls by scientists, local smallholders and Indigenous communities to move away from the precarious world of maximum productivity that has been encouraged by Euro-American expectations of Genetically Modified technologies, monoculture and market demand, and turn, instead, towards the security of diverse traditional grains, legumes, farming systems and extended family-based social networks.[26] These practical legacies of food production are also the reason why, at the very beginning of this book, we saw villages on the banks of the Amazon Basin still sitting astride the fertile human-created Amazonian Dark Earths and combining small fields of manioc and maize with stands of wild Amazon nut (Brazil nut) and palms. Although many imagine agricultural fields and clearance as the prime route towards supposed 'civilization' and development, the millennia of farming practices recorded in the soils of tropical forests around the world suggest very different, more sustainable routes forward for food production, particularly as the tropics and their inhabitants are increasingly threatened by climatic instability.

As a result of the fact that 70% of the world's population will be living in urban settlements by 2050, and that much of this will be focused on the tropics, we also need to come to terms with the fact that cities and their infrastructure will come into increasing conflict with tropical forests.[27] Although high concentrations of people, roads, railways, airports, houses and industry associated with Euro-American ideals of a city seem incompatible with tropical sustainability, Chapter 9 showed us that the tropical archaeological record may once again offer some solutions. In particular, the 'low-density agrarian-based urbanism' seen in Cambodia, North and Central America, the Amazon Basin and Sri Lanka provides models for the low-density distribution of people across a landscape and the intermeshing of food production with urban facilities, administrations and dwellings. Similarly, pre-colonial walled cities in West Africa, such as Benin City, demonstrate how well-organized systems of urban nodes, wild forest, land for cultivation and fallow land formed to protect large populations from enemy incursions but also from resource shortages, a lack of fertile soils, and the loss of crucial forest environments.[28] There is also renewed present-day scientific interest in the soil modifications that many of these past urban landscapes developed and relied upon to maintain their large tropical populations. Although they represent very different settings and socioeconomic contexts to their present counterparts, these archaeological examples are being actively used and pushed by twenty-first-century urban planners as they seek to develop more sustainable models of cities that are more resistant to fluctuations in climate and natural hazards, and have a greater local security in terms of food and livelihood.[29] In particular, they are highlighting how a planned mixture of 'green' and urban space, as opposed to the informal developments currently seen expanding across Latin America and Southeast Asia, can provide a better model for tropical cities. Moreover, they point out that past urban examples highlight the importance of local ownership, consultation and involvement, rather than top-down decisions and economic and political elites, in the ultimate long-term success of tropical urban sprawls.[30]

Accepting the 'deep human history in tropical forests'[31] also allows us to use archaeology and palaeoecology to make some specific suggestions in terms of the species and ecosystems we want to preserve or even try to re-introduce in different parts of the tropics. Take the example of planting more trees. Care clearly needs to be taken as to *which* trees are planted, *where* and *how*. Oil palm has been highlighted, by governments, companies and conservationists alike, as an important route to prosperity for tropical nations and their populations thanks to a large and growing global demand. Where agricultural fields, pastures and grasslands are covered in oil palm trees in a systematic manner, the land can capture more carbon, leading to long-term environmental and economic benefits for local populations,[32] and, despite its bad media press, compared litre-for-litre oil palm threatens fewer species than coconut oil and olive oil.[33] However, in Southeast Asia in particular, remote sensing satellite imagery has shown that much of the land converted to oil palm plantations over the last two decades has actually been native, peat swamp forest, leading to significant carbon emissions through fire, peat drainage and loss of significant biodiversity.[34] Archaeology and history can have a significant role to play in developing more ecologically relevant solutions, highlighting the profitable and sustainable inclusion of economically valuable crops within diverse farming systems, as seen with oil palm in Africa in Chapters 7 and 11. Similarly, palaeoecological research, ranging from the study of ancient plant remains in archaeological sites in Hawai'i[35] to the analysis of ancient plant DNA preserved in lake records at the high-altitude border of Uganda and Rwanda,[36] has helped to bring past landscapes back to life. Records of forests from before the arrival of European colonialism and that existed under different Holocene climatic conditions can better shape 're-wilding' and conservation priorities. Study of ancient animal fossils in the tropics can play a similar role in determining where and how species that have disappeared from the area, but were present in the past, might be locally and productively reintroduced.[37] Nevertheless, ultimately, the question remains – what do we actually want to protect or restore? Are we really always trying to reconstruct a 'natural' landscape that, as we have seen throughout this book, has

Monoculture oil palm plantation in Borneo, Indonesia. In principle oil palm offers a valuable crop for boosting incomes and, when planted on open land, can even improve overall carbon sequestration. Although increasingly regulated, in the last two decades significant conversion of rainforest and peat swamp forests to plantations has occurred, releasing massive amounts of CO_2 into the atmosphere and destroying species-rich habitats.

not truly existed in many parts of the tropics for thousands of years? Or do we rather want to engage with the idea that many tropical forests were being successfully managed for a long time before colonial and imperial forces arrived and subsequently made them the threatened biomes they are today?

This question is becoming increasingly pressing, as the creation of reserves by national governments can damage the rights and livelihoods of Indigenous people whose ancestors have occupied tropical forests for millennia. There is sometimes a narrow conservation assumption that because industrialized, capitalist impacts on forests are devastating, we must lock out *all* human communities. The Twa community, which had lived in the Kahuzi-Biega National Park in the eastern Congo Basin, was forcibly expelled, leading to rampant malnutrition as they were forced to integrate into wage systems[38] and lost their traditional sources of food from the forest. The

Wanniya-laeto of Sri Lanka have been similarly banned from hunting in their traditional forest lands, and now face the gradual disappearance of their culture and even their language. In some cases, the removal of Indigenous groups evidently impacts the forests in a negative manner. In Australia, bureaucratic obstacles hindering Traditional Aboriginal Owners, such as the Jirrbal community, practising burning and forest management on their traditional forest land can have ecological as well as cultural consequences. A reduction in the application of long-held burning practices has meant that dense undergrowth has formed, in the dry sclerophyll forests and rainforests of the region alike.[39] The land now looks 'sick', as Jirrbal elder Desley Mosquito described to me. As the climate dries in the face of human-induced CO_2 emissions, dense forest floors provide the ideal kindling for increasingly frequent forest fires. Significantly, a number of initiatives by the Wet Tropics Management Authority and the Rainforest Aboriginal Peoples are now supporting the application of traditional knowledge within ecologically protected areas.[40] Palaeo-ecological research in the Amazon Basin has demonstrated how drastic changes in forest burning practices between pre-colonial and post-colonial times have had a major bearing on ongoing fire frequency and intensity.[41] A Western scientific arrogance that we should always keep parts of the world fenced off and 'un-degraded' – perhaps as a way to somehow morally offset our own growing destructive impacts – while potentially ecologically productive, can sometimes actually damage the very ecosystems it seeks to protect. Not only that but, like the history of the last five centuries, it has often been undertaken in a way that continues the unjust dismissal and marginalization of the very people who have the best long-term knowledge of local ecological dynamics and thresholds, as well as the most intense experiences of integrating new social, economic and political structures into tropical land management. In short, the people we should be supporting and asking for advice.

We need to accept that colonial and imperial processes, documented in Chapters 10 and 11, have led to a series of environmental, economic and political injustices that tie the hands of people living in the tropics today. At the most fundamental level, five centuries of eco-logical and land-use disruption by Euro-American powers have made tropical communities some of the most vulnerable to climate change and other conservation threats. Challenges of meeting subsistence needs in the face of rampant soil erosion, declining nutrients and invasive species that plague islands such as Haiti, São Tome and Príncipe, Madeira, Mauritius, Madagascar and the Canary Islands can be traced back to profit-driven colonially established plantation agriculture. Similarly, the deforestation of the Philippine uplands and coasts, which has left behind eroding slopes and a vulnerability to flash flooding, is a product of centuries of extractive logging by first Spanish and, later, imperial governments of the United States of America and Japan. Colonial and imperial extractive demands have also impacted the wildlife left behind in certain regions. For ex-ample, in Sri Lanka, massive hunts by British colonial agents in the Horton Plains highlands have meant that these animals, as well as their crucial functions of seed dispersal and forest disturbance, are now completely absent from mountainous regions on the island today. One nineteenth-century British hunter reported that he alone had killed 6,000 elephants.[42] Finally, the massive amount of fossil fuels burned in the name of industrialization, and that have, to date, primarily benefited the Euro-American world, are the primary cause behind the present state of the polar ice-caps and global climatic sys-tems that has seen frightening sea level rise and extreme typhoons and cyclones, respectively, batter the islands of the Pacific, the Indian Ocean and the Caribbean.[43] Given that a significant portion of capital for this process in many cases came from trade in human lives, cheap labour and extractive profits, often from these very same tropical landscapes, we should perhaps be more sickened than we often are by this climate injustice and the current plight of these drowning nations.

Beyond the tropical landscapes themselves, long legacies of coloni-alism and imperialism, by producing very different socioeconomic conditions, have also dictated the flexibility with which local com-munities can now even afford to act in the face of threats. This has included the suppression of local Indigenous traditional knowledge in many regions, through murder, disease and relocation. But it has also involved fundamental emerging wealth imbalances that inevitably shape local priorities. Many tropical nations in Africa are home to some of the fastest growing, but also the poorest, populations anywhere in the world. For example, in Madagascar, more than 92% of the population live on less than US $2 per day. Schemes of protected areas were announced by the government in 2003. However, breakdown in law enforcement and political crisis have hindered their progress, opening the door to illegal logging and mining, as well as bushmeat hunting and slash and burn farming by desperate populations attempting to survive in a drying landscape.[44] In the southern area of the island, twentieth-century French imperial policies that forced mobile pastoralists to adopt more 'modern' sedentary cultivation of cash crops have also increased the contemporary pressure on tropical forests[45] and a lemur population that represents ~20% of the world's primate species. Similarly, stark wealth discrepancies and political instability pose major challenges to coordinated conservation efforts across tropical Africa as well as New Guinea, where small-scale farmers are often the primary drivers behind deforestation. Contrast this to Australia, the richest nation in the world to host tropical forest thanks to its very different colonial history. Here, a far higher level of per capita income, and the economic and social benefits offered by comprehensive access to education, provide a context in which a designated Wet Tropics World Heritage area can be effectively extended and policed across 894,420 hectares in Queensland (an area eight times the size of Greater London[46]), and can profit from the arrival of 5,000,000 tourists (from home and abroad) and an income of AUS $426,000,000 every year.[47]

We should therefore not underestimate the degree to which Western conservation ideals are something of a luxury, born out of the fact that, in many ways, we have already 'made it' in terms of 'development' and industrialization. The centuries of colonial and imperial removal of wealth from tropical nations, with limited return in terms of local investment and infrastructure, have, by contrast, exposed governments in the tropics to tensions between demands for improvements to living conditions at the same time that conservation and climate crises are reaching a tipping point. Brazil is an increasingly wealthy member of the BRIC economic group, with low population growth and relatively high earnings. Nevertheless, its government remains under heavy pressure by its supporters to develop further. The state of Amazonas is Brazil's poorest, but also contains the largest area of tropical forest cover. As Victor and I saw at the start of this book, dense rainforest means that slow boat rides or expensive planes are required to move around within or beyond the state. Political, economic and social incorporation of Amazonas, and its capital, Manaus, into Brazil's federal network has always been challenging, and involved a relatively late history of European contact with Indigenous communities. Poor populations are desperate for access to growing markets or even basic medical supplies, as tragically witnessed during the recent COVID-19 outbreak.[48] Better infrastructure and land conversion, either government-mandated or illegal, are seen as key routes to improving local livelihoods, but will also inevitably represent a major conservation threat to significant areas of tropical forest, necessitating detailed planning.[49] Local situations such as these must be considered when international calls are made for South American nations to stop cutting down trees or Kenya and India are told to stop burning fossil fuels, especially in the context of long histories of Western extraction, exploitation and unequal development efforts.

Multinational corporations, as well as global consumerism, also now have a vice-like grip on tropical nations, thanks to centuries of profits made on the basis of poorly paid labour and conversion of tropical land and an establishment of capitalist-based approaches to agriculture and tropical landscapes. Even relatively wealthy nations, such as Australia, do not escape, and the extent and slow response to the 'Black Summer' can be, at least partially, linked to poor policies on climate change in a nation that relies on coal-mining companies to bolster its economy.[50] Similarly, governments in South America and West and Central Africa have often leased vast areas of tropical forest to dominating corporations seeking oil and gas.[51] In the case of 'cash crops', from coffee in North, Central and South America to tea in Sri Lanka, and from bananas in the Caribbean to soy beans in Brazil,

1. Artist's reconstruction of a Carboniferous forest floor including insects.

2. A photogrammetric model of a preserved footprint from *Dromopus lacertoides* (a lizard-like sauropsid) as it appears on the right-hand side of a slab from Hampstead, Birmingham. The sauropsid group was to later produce the ancestors of all reptiles, including the dinosaurs.

3. Image of an angiosperm leaf recovered from the Cerrejón Formation in northern Colombia. The density of veins in the leaf surface remains clearly visible after 60 million years.

4. Artist's reconstruction of the Neotropical forest of Cerrejón with Titanoboa in the foreground.

5. Artist's reconstruction of Jurassic forest environments dominated by gymnosperm conifers and cycads as well as ferns. A stegosaur and sauropod can be seen in and among this vegetation.

6. Fossil of mammaliaform *Vilevolodon diplomylos*, a Jurassic glider, found in north-eastern China. The split slabs are located at the Beijing Museum of Natural History. The red arrows indicate the margins of the skin membrane that would have enabled gliding between treetops.

7. Gliding Jurassic mammaliaform consuming gymnosperm vegetation using well-adapted herbivorous teeth.

8. *Eohippus*, or 'dawn horse', the earliest known ancestor of the modern horse. Its diminutive stature and teeth adapted for eating forest fruits were to change significantly over the course of the Oligocene, Miocene and Pleistocene.

9. Artist's reconstruction of 'Ardi', *Ardipithecus ramidus*.

10. Photographs of cutmarks on an otter bone from the Late Pleistocene Sri Lankan site of Fa-Hien Lena.

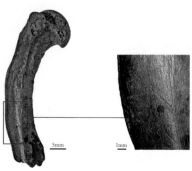

5mm   1mm

11. Preserved microscopic botanical remains from Kuk Swamp shown alongside their modern counterparts. *Left*: a yam starch granule shown above a greater yam plant; *Right*: taro starch mass above a taro plant.

12. The UNESCO-protected Banaue rice terraces in Ifugao on the island of Luzon in the Philippines. These terraces likely have ancient precursors in the region that may have been used for the growth of taro prior to rice.

13. The ancient Classic Maya ceremonial centre of Tikal, Guatemala.

14. Classic Maya water temple next to *cenote* or steep-sided sinkhole fed by groundwater showing the cosmologic, as well as practical, importance of water to these populations.

15. Oblique view of the central temples of the pre-Angkorian urban complex of Sambor Prei Kuk, Cambodia.

16. Aerial overview of the Angkor Wat temple complex in Cambodia.

17. An artist's reconstruction of the bustling streets that connected the Greater Angkor urban network.

18. A visualization of the residential areas inside the Angkor Wat temple enclosure, *c.* 1300 AD.

19. Sixteenth-century engraving of enslaved Indigenous peoples working in the Spanish silver mines of colonial Bolivia.

20. The streets and buildings of Vigan City in the Philippines are today a World Heritage site and show the way the Spanish sought to impose their ideals of urbanism in this part of the tropics.

21. Fishing canoes on the beach near Elmina Castle, Ghana.

22. Forty cast brass plaques created in Benin City (in modern-day Nigeria) in the sixteenth and seventeenth centuries. They are known collectively as the Benin Bronzes. Around 2,500 works of art, which have great political and cultural significance for the people of Nigeria, are scattered through European and North American museums and art galleries.

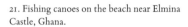

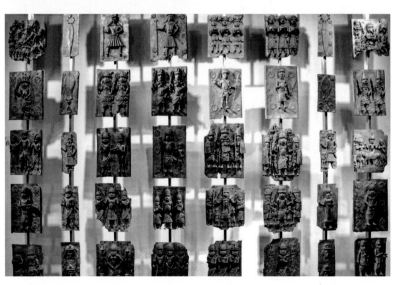

23. An aircraft releasing water over the fires in Queensland's rainforest.

24. The scene left behind following a fire in a tropical forest in Queensland, Australia.

25. Dar es Salaam, Tanzania. Cities such as this in the tropics will, alongside population increase, bring humans into increasing contact with tropical forests around the world.

global demands, including from ourselves, many miles away, have long driven large landowners and small farmers towards cheaper, profit-making monoculture fields. Returning to palm oil, companies relying on this ingredient for products as diverse as lipstick, pizza dough, instant noodles, shampoo, chocolate, cooking oils, packaged bread and biodiesel, catering to growing demand from Asian and Euro-American markets and consumers, have been a major driving force behind Southeast Asian governments having poor control over, or even actively encouraging, unregulated, unsustainable conversion of tropical forests, and their endemic wildlife, to widespread planta-tion systems. Meanwhile, simultaneous rising costs of local land, as well as corporate expansion of their own holdings, can leave the door open to abuse of the land rights of local people,[52] as well as rising prices that leave small-scale farmers and Indigenous populations with little option but to convert, contract, abandon or sell their own plots. More mixed, more sustainable, and often more productive approaches are thus left threatened by aggressive market conditions.[53] The expan-sion of cattle ranching by large landowners across the Amazon Basin to meet regional and global meat markets has represented a similar problem in Neotropical forests.[54]

Euro-American consumerism can even dictate which areas and tropical forest environments are made conservation priorities in the first place. Reserves, whether they are 39,000 square kilometres, like the Tumucumaque National Park in Brazil, or under 2 square kilo-metres, like the Bukit Timah Nature Reserve in Singapore, are expensive – both to establish and maintain. Where tropical govern-ments are able to invest in parks, or in laws prohibiting logging or clearance, they need there to be some form of pay-off.[55] Eco-tourism, travel to protected reserves to explore wildlife, has provided one way in which foreign wealth can be simultaneously used to maintain for-ests with crucial ecosystem roles while also producing additional economic benefits for local communities. The problem is that this means the forests that have received the most attention by protective policies are often those that provoke Western fascination rather than necessarily serve the most pressing local needs.[56] For example, they frequently include the green homes of iconic conservation stars such as the troops of montane gorilla in Uganda's Bwindi Impenetrable National Park, the orang-utans of the Kabili Forest Reserve in northern Borneo, the jaguars of Brazil's Pantanal Reserve, or the cassowaries and tree kangaroos of the Wet Tropics of Queensland. This can be significant, especially as these giants often have a critical 'keystone' position in tropical ecosystems.[57] Nevertheless, it means that governments are less willing to pay for the conservation of less traditionally glamorous forests, including dry tropical forests, montane rainforests, coastal mangroves and peat swamp forests, which often have more unique taxa, are often far more threatened, and have crucial functions in landscape stability and buffering natural hazards.[58] The position of many of these forests on climatic boundary areas, such as the sides of mountains or precipitation gradients between aseasonal and seasonal rainfall regimes, also make them some of the most vulnerable to increased pressures of human land use and climate change.[59]

Tropical nations and their governments are, of course, not completely blameless in the conservation challenges facing the tropics. Political corruption, crime networks, increasing movements towards nationalism and wilful neglect of science[60] can pose real problems for tropical forest conservation. Nor does the world need any repeat of ideologies of condescendingly 'saving' supposedly incapable local tropical populations that were used as justification for many colonial and imperial enterprises and atrocities over the past half a millennium. In fact, local voices such as the Association of Small Island States, the Caribbean Community Climate Change Center, Cooperative Unions of Fairtrade cultivators, Coordination of the Indigenous Organizations of the Brazilian Amazon, and, as mentioned above, Rainforest Aboriginal organizations in Australia are showing themselves perfectly able to express the existential threats that they, and all of us, face in the context of climate change, governmental development strategies and corporate interests, as well as advocating practicable solutions.[61] However, top-down international, or even national, declarations and demands are often blind to the social, economic, political and environmental histories that can limit the power of these groups and their voices. They can also hide the role of Euro-American nations

in continuing to create an imbalanced tropical world through consumer choices and demands and wealthy, exploitative businesses. If
we want to truly protect tropical forests for future generations, it is
local stakeholders and Indigenous populations such as the groups
mentioned above that we must not only listen to, but also practically
support by putting our money where our mouths are. In this way, we
can combine archaeological and palaeoecological insights into how
tropical landscapes have evolved and been managed by humans
through time, with local, traditional ecological and economic knowledge and strategies, to give them, and ourselves, the best chance to
fight back some of the most extreme sustainability challenges to ever
face humanity during its time as a widespread tropical species.

The historical colonial restrictions placed upon significant amounts
of Indigenous and local knowledge that was developed over millennia means that the application of traditional management and
adaptations in tropical forests is not always straightforward and can
sometimes even have negative impacts. Here, archaeology and palaeoecology, as well as the revitalization of Indigenous languages and oral
histories, can enable local stakeholders to engage with ancestral
practices and solutions. For example, in the Caribbean, whereas post-
colonial houses are built of hard materials that are expensive to
replace and dangerous in the face of intensifying cyclones, hurricanes
and earthquakes, the archaeological record of pre-colonial housing
shows something very different. Semi-permanent housing, built on
foundations of resistant materials and structures as well as more light-
weight, easy-to-replace components, was not a sign of 'backwardness',
but an ideal, economic solution to the everyday challenges of living
on these islands.[62] On the Atherton Tablelands of Australia, combined visits of Aboriginal elders, archaeologists and palaeoecologists
to previous sites of occupation such as Urumbal Pocket have enabled
more thorough documentation of the nature and importance of past
human land and resource use through a combination of preserved
oral histories with traces of human food processing, plant and animal
use, and records of controlled burning emerging from the soil.[63]
Sometimes, such inter-disciplinary work and collaboration can even

highlight negative elements of past practices, emphasizing the need for detailed, context-specific reviews and solutions rather than blanket statements or applications. For example, archaeological and environmental investigation of human-made sediment traps at the locations of Konso and Engaruka in Tanzania show that traditional approaches enabled the capture of fertile, fine-grained soils and the avoidance of salinization of fields. However, while these practices improved growing conditions in the valley regions, soil erosion and vegetation loss in the highlands may have resulted in ultimate abandonment and loss of livelihoods.[64] More detailed work of this nature is essential to understand how food security, resilient settlement, and sustainable use of tropical forests and landscapes was achieved, or indeed not, prior to the arrival of domineering colonial, imperial and capitalist forces across much of the tropics – using the past to plan for the future.

Where Protected Reserve initiatives are possible, the involvement of local stakeholders, whether they are recognized Indigenous groups or local farmers, has repeatedly been shown to result in more successful outcomes than when they are ignored. In the Brazilian Amazon, Indigenous classification systems of plants and ecosystem transitions have long been at the centre of attempts to maintain botanical diversity throughout the giant 130,000 square kilometre reserve of the Kayapó and Upper Xingu people. In the Ecuadorean Amazon, Indigenous groups combine with local Non-Government Organizations, and government initiatives to promote eco-tourism to individual villages as a source of additional wealth but also a way to promote tropical forest preservation and conservation.[65] Government and administrative support for Aboriginal ranger groups, such as those of the Jabalbinna, Djabugay and Girringun, have had significant success preserving wildlife in the Wet Tropics of Queensland.[66] Similarly, the allocation and networking of so-called local 'gorilla guardians' in Cameroon has enabled a better understanding of the frequency and distribution of some of the most endangered great apes in Africa, revealing ongoing illicit hunting, and strengthening communication and understanding between local villages, hunters, governments and conservation scientists in order to increase support for protective measures and action.[67] Sometimes, conservation initiatives spring completely from the actions of local interest groups. In the Mamberamo-Foja region of Papua (the Indonesian portion of the New Guinea land area), the Ijabait elders act quite literally as 'guardians' to local tropical forests. They live at strategic sites, such as forest boundaries and rivers, to monitor and control human use of forest and freshwater animals. In each case, the involvement of local groups, whether through cultural and traditional concerns, social contracts, or added economic benefits, has led to the more beneficial imposition of protected areas and a reduction of illegal incursions.[68] As conservationist Professor Douglas Sheil of Wageningen University puts it, 'the inclusion of populations living in tropical forests not only acts as a robust local barrier to incoming external threats, but is also more likely to reconcile economic and subsistence concerns with conservation needs'.

Considering local interest groups is also essential to developing and supporting more sustainable ways of extracting resources and products from tropical landscapes. For example, Reduced Impact Logging (RIL) initiatives use dedicated training for workers, effective patrolling, protection of hillside forests and refined tree cutting to reduce ecosystem and soil disturbance. The long-term ecological and economic benefits of retaining tropical forest cover are clear. However, these programmes also require short-term costs for producers, consumers and governments.[69] In Bolivia, local government laws, enforcement, tax benefits and regulated extractions have been required to encourage the effective use of certified forests.[70] To reduce bushmeat hunting in Africa, governments and NGOs similarly need to consider the provision of protein substitutes given the current importance of wildlife to local diets. On the Yucatán Peninsula, honey production represents the main source of income for as many as 16,000 rural ethnic Maya farmers, and Mexico is the world's fourth largest honey exporter. These Indigenous farmers produce honey traditionally, refusing the use of chemicals, limiting impacts to the forest environment, and taking close care of the health of bees.[71] Nevertheless, the cost of meeting global 'organic' certification standards means that, to reap the benefits, they will require government assistance if they are not to resort to more 'conventional' methods.[72] Similarly, some of the first pioneering growers of organic coffee included Indigenous farmers practising agroforestry in Chiapas and Oaxaca in Mexico. Their approach proved beneficial for soils, ecosystems, and even the carbon cycle,[73] but as organic coffee has been taken up by larger landowners, and official organic status has become harder to sustain in the face of global competition, they have met with difficult choices of whether to risk intensification or revert to cheaper alternatives. Ultimately, 'trade-offs' require governments, companies and NGOs to support local communities in the short term, to cash in ecological and economic benefits of longer-term sustainable approaches to forest use within the tropics.[74]

On a planetary scale, there are also growing movements by Western nations to actively 'pay into' tropical conservation. Most recently, the summit of G7 nations in Biarritz, France, saw a US $22 million aid package agreed upon to help combat and mitigate the Amazon wildfires of 2019. As is often the case, however, the idea of 'aid' proved controversial, as Brazilian President Bolsonaro suggested that he would reject the money as the conditions attached amounted to colonial interference in South American agro-business competition.[75] The Reducing Emissions from Deforestation and Forest Degradation (REDD+) programme represents a more organized, long-term means for wealthier nations, which have often benefited from the exploitation of the tropics, to begin to pay something back to support conservation efforts in and around the equator and, in so doing, simultaneously protect local environments and economies and the global climate. Developed by the United Nations Framework Convention on Climate Change (UNFCCC), the REDD+ programme acknowledges that tropical forests should be valued not only for their extractive potential but primarily by the carbon that they store, protecting our atmosphere from further CO_2 emissions, warming and ultimate climatic disaster.[76] In fact, it places a financial estimate on the benefits of this vital 'carbon sink'. Poorer countries receive results-based payments, subsidized by wealthier nations, for reducing unregulated logging impacts, slowing overall deforestation, preventing fires and restoring tropical forest biomes. Not only that, but REDD+ ensures that Indigenous peoples, small and large businesses, and national governments are all heard during the development and application of these initiatives across the tropics. Although the overall economic and ecological benefits are still being determined, and it has also been seen as a controversial, top-down colonial scheme that can be prone to abuses in its own right,[77] this programme currently supports sixty-four nations across Africa, Asia, the Pacific, Central and South America, and the Caribbean. Furthermore, it does, at least, represent perhaps the first explicit recognition by the international community that the responsibility for tropical forest conservation and sustainability is a global one.[78] If we want to do something about the intensifying loss of these environments, as well as the associated earth systems repercussions, then we will all have to play (and pay) our part.

Which brings us to perhaps the most important way in which we can all limit our own impacts on tropical forests: modifying our consumer habits and choices. Many of us have benefited from globally imbalanced social, economic and political processes that have occurred over the last 500 years, with our own relative comfort relying on the resources, people and environments of the tropics. It still does. Tropical deforestation, and its causes, remain tied into our own choices and histories, whether we want to accept it or not. It will also, sooner or later, impact us. It is time for us all to take action. Whether it is paying a premium on furniture that uses timber from the Forest Stewardship Council (FSC), or studying what our purchasing of 'organic' coffee actually means for Indigenous-led growing initiatives. Whether it is buying Fairtrade chocolate or boycotting clothing companies that use poorly paid, and often child, labour from the tropics. Whether it is in-depth reading about the practical benefits of our next eco-tourism destination for local communities and economies or whether it is taking the time to read biscuit labels to distinguish products that include sustainable palm oil versus mass-produced, uncertified palm oil. Or whether it is reducing our number of car journeys or flights abroad, curbing our emissions, and reducing the climatic impacts on the most vulnerable tropical communities as well as, eventually, ourselves. Back home, be that a tropical city away from forest frontiers or miles and miles away on an icy, dreary morning in north-western Europe, it may seem like these products and facilities just drop into our laps and fly on to our shelves. But their origins, their production, and your empowered ability to purchase and use them all lie within the tropics and the last half millennium of Euro-American economic and political integration with, and often exploitation of, tropical forests and their communities. Our decisions matter.

In the winter of 2018, the supermarket chain Iceland released a Christmas advert that was very different from the cute, comfortable viewing the UK public was used to at this time of year. Teaming up with Greenpeace, the video showed a cartoon orang-utan swinging through the bedroom of a young girl, messing up her possessions as it went. Reaching a bottle of shampoo, 'Rang-tan' suddenly stops and howls, transporting the viewer back to its own, tropical home. Here, fire, destruction and devastation were suddenly strewn across the screen, causing 'Rang-tan' to flee, an ecological refugee. While, as a multinational corporation, people will undoubtedly question Iceland's motives and timing, their move to ban uncertified palm-oil products from their stores is a positive one. Furthermore, the video provides a beautiful artistic vision of the influence all of us can have on tropical homes many thousands of kilometres away from us, merely by picking up the wrong bottle in a supermarket. Do we really want rampant flames and falling trees messing up forests and livelihoods on our behalf? Even if we cannot always see them? In a manner perhaps characteristic of the way the Western world often seeks to bury its head in the sand, the advert was actually banned from television by the monitoring body Clearcast, being considered 'too political' for the Broadcast Code of Advertising and Practice due to the involvement of the conservation organization Greenpeace.[79] That did not stop it receiving millions of views on social media, however. Ultimately, like that video, this chapter, and those preceding it, have sought to repeatedly highlight how all of us are, in some way, implicated in the fate of tropical forests and the success of their conservation. From Indigenous groups to small-scale farmers seeking to survive, from companies extracting profit to governments trying to improve the lot of their citizens, from conservation scientists campaigning for particular species or biomes to each person standing at a supermarket shelf, starting a car, or queuing for a flight, we are all dependent on the tropics for our economic and social situation and all responsible for what happens next. Is it so hard, therefore, to find a common, intense interest in their ongoing existence? They need all of us, now, more than ever. Will we answer?

14. A global responsibility

It's the middle of the night on the shores of Ponta da Castanha on the Téfé River tributary of the Amazon, towards the end of the Brazilian journey that began this book. Lying in my hammock, restless from the heat, mosquitoes, and a hard day of trekking, I felt incredibly small. Just beyond the door of the wooden hut where myself, Victor and our colleagues were sleeping stood environments that have, in some form or other, been on our planet for at least 300 million years. The specific micro-climates, species and structure of the particular Neotropical forests that swayed in the night-time breeze outside extend back to 60 million years ago. That's older than the most significant mountain range of the South American continent, the Andes,[1] and highlights the resilience, but also the ability to change, of these remarkable environments over geological timescales. The building's open holes for windows let the incessant chattering of insects, the swishing of bats, and, towards dawn, the squawking of birds, wash over me. I had never heard such a concentration of wildlife in such close proximity. But these noises quite literally spoke to the role tropical forests have played in the evolution of animal life on this planet – from the early reptilian ancestors of the crocodile our boat driver had seen loitering by our boat that evening to the dinosaurian ancestors of birds. From the gliding mammals of the Jurassic to the hominin ancestors of myself and my companions resting inside. I had never felt more in awe of tropical forests, and their contribution to the story of life on Earth, than I did in that moment.

These environments have often been framed as dangerous, exotic, pristine 'deserts' of pre-industrial human activities.[2] But closing my eyes, I felt incredibly connected to this place. I remembered the pottery we had seen on the forest floor the day before on top of distinctive Amazonian Dark Earth soils, a solid record of the long human history of the Amazon Basin that extends back to between 13,000 and

12,000 years ago.[3] In my mind I retraced my steps through the clusters of palms and Amazon nut (Brazil nut) trees, and the shifting swidden fields of manioc and maize, that stood on the other side of the village and that our host Jucelino had shown us the day before. The vegetation and forest structure around many such villages along the Amazon and its tributaries provide a living archive of the millennia of land management in these environments. When it came time to leave, I took my seat on a western European-operated airline. I thought about the way in which huge distances and home comforts can make those of us living in Europe feel so separate from tropical forests. But also how colonial and capitalist processes that followed European arrival in the Amazon Basin over the last 500 years were visible all around me. How the growth of the Amazon (Brazil) nuts Victor and I were analysing changed as disease, warfare, murder and ongoing marginalization caused Indigenous inhabitants to flee parts of the forest.[4] How the entwinement of Brazil in the transatlantic slave trade[5] helped drive the global availability of coffee in my aeroplane cup and the sugar I was about to put into it. And how the rising pressures to feed smallholders, multinational corporations, ambitious governments, and Western consumers,[6] were driving the deforestation and land conversion taking place before my eyes on the ground passing by below. I might be able to fly away from Brazil physically, but the historical origins of global and regional inequalities, and the importance of increasingly threatened tropical ecosystems to global sustainability, mean that, ultimately, I am still connected to the tropical forests of the Amazon Basin, and indeed the rest of the world. We all are.

Through this book I hope to have convinced you that tropical forests are a part of your history, your present livelihood, and your future security. I hope that the next time you turn on a nature documentary you feel a little closer to the miraculous plant and animal life tropical forests have to offer. I hope that the next time you watch a film with an explorer cutting through a tangle of tropical vines you are less surprised to witness the incredible achievements of pre-industrial human societies rise out of the 'jungle' before them. And I hope that the next time you see a news report about the poor public health of Indigenous populations, racial tensions, multinational corporation mining, deforestation and rampant fires in the tropics you do not look away. Instead, I hope you keep watching. I hope you feel a responsibility. And I hope you begin to take time to research and explore the ways in which you might be able to make a difference. Remarkably, despite our complex global social and communication systems, humans often struggle to empathize with issues unless they are on their doorstep.[7] Yet Chapters 12 and 13 have shown that the environmental, political and economic threats facing populations living in the tropics will, through the connection of tropical forests to earth systems, lead to changes in biodiversity and climate the world over. Not only that, but participation in a colonial and capitalist system that has exploited the tropics for the last half a millennium[8] means that we all have some obligation to help obtain justice in fights against global economic and infrastructural inequality, rampant profit-driven tropical deforestation and land cover change, marginalization of Indigenous rights and knowledge, as well as the ongoing racial violence and systemic discrimination that are still visible across

Amazon nut (Brazil nut) tree growing just outside the village Victor and I stayed in on the Téfé River tributary of the Amazon Basin in 2019.

almost all of Europe and northern North America. Yet 'hope' at this stage won't do us much good. So I will now make one last attempt to show you what is at stake. Let's look at what is happening and what is predicted to happen to (i) tropical forests and their species, (ii) human inhabitants of the tropics, and (iii) the global climatic, economic, social and political situation as a consequence of the threats facing our equatorial ecosystems. Perhaps by seeing what we all stand to lose, we can gain a new urgency in negotiating a more just, resilient future for the tropics.

Studies of twenty-first-century forest loss based on global yearly satellite images at high levels of resolution suggest that deforestation is continuing unabated, despite growing awareness of its planetary impact. The majority of the world's deforestation is happening in the tropics, where total forest loss increased by over 2,000 square kilometres (an area the size of Cheshire[9]) every year in the first decade of this century.[10] Tropical rainforests, such as those of the Amazon Basin, accounted for nearly a third of global forest cover reductions, while tropical dry forests in South America and moist deciduous forests in Africa saw annual increases in total deforestation of 459 and 536 square kilometres a year, respectively. Southeast Asia also saw a particularly large expansion of forest removal over this time period, with Indonesia alone showing an increase in *annual* losses from 10,000 square kilometres to around 20,000 square kilometres (so that an area of forest equivalent to the size of Wales was being removed by 2010[11]). This highlights the pan-tropical nature of rampant deforestation in the twenty-first century.[12] Some predict that, if nothing is done, by the year 2100 Amazonia, the Congo Basin, West Africa, the Caribbean and Wallacea will retain just half their forest cover. Meanwhile, the tropical Andes, Madagascar, the Atlantic forests of Brazil, and New Guinea will have just 26%, 23%, 26% and 18% of their present-day forests still standing.[13]

It is not just complete deforestation that is a concern but, as we saw in Chapter 13, the less obvious degradation of those forests left standing. A recent satellite study found that forest degradation occurred across the same area of forest in the Amazon Basin as complete deforestation did between 1995 and 2017. Since 2017 alone, approximately 1,036,800 square kilometres of Amazonian tropical forest area has been disturbed in some form[14] (an area four times the size of the United Kingdom[15]). One of the main causes of disturbance is wood harvesting and export, and early twenty-first-century extraction rates from selective logging are particularly high for Guyana (13 cubic metres of wood per hectare) and Indonesia (34 cubic metres of wood per hectare).[16] However, the causes behind this damage vary in different parts of the tropics and the use of wood for fuel is by far the largest contributor to forest disturbance in much of tropical Africa.[17] Forests change even further as tropical cities expand. At Dar es Salaam, Tanzania, a capital likely encountered by any of you seeking safari holidays in eastern Africa, researchers found that urban demand for fuel production and high-value timber left a clear impact on the surrounding tropical forests that extended many kilometres away from this capital city.[18] Similarly, the highest rates of forest degradation in a studied region of south-eastern Mexico are, unsurprisingly, currently found in regions with the highest human population density.[19] Even those tropical forests that are left unshaven in the face of human activities are going to have to come to terms with a growing anthropogenic presence.

Perhaps the most emphatic of forest disturbances is fire, and twenty-first-century tropical forest fires are only going to intensify. Studying rates of charcoal accumulation in sediment sequences across the tropical forest environments of the Amazon,[20] as well as the Brazilian Atlantic coast, researchers have concluded that the fire frequency and intensity seen in these areas today have no precedents throughout the 11,700-year span of the Holocene.[21] In fact, based on current 'worst case' scenarios of changing land use and forest fragmentation in the Brazilian Amazon, models predict that the probability of monthly fire occurrence will increase by 73% by 2071–2100, even before expected changes in climate are factored in.[22] This is not just a South American problem, and crucial peat swamp forest ecosystems in Africa and Southeast Asia are increasingly 'in the line of fire'.[23] During even relatively mild climatic conditions in 2002 nearly three-quarters of the area impacted by fire in Borneo was peat swamp forest, and peat fires are now becoming common in every dry season in this part of the world. Over the course of the next decade there could be a total loss of peat swamp forest ecosystems across Southeast Asia as a product of growing fire and permanent human land-use changes.[24] Observations of fire frequency across Australia during the early twenty-first century[25] suggest that weekly bushfire occurrences are also only going to further impact the tropical forests of the Wet Tropics in coming years.

The rate of fire, disturbance and tropical forest structural change will undoubtedly be made worse by human-induced climate change. It is becoming increasingly apparent that a temperature rise of between 1.5 and 2°C from pre-industrial levels will be enough to 'tip' the planet into a completely new climatic and environmental state.[26] Ice collapse and melt has already advanced beyond expected levels, and recent estimates suggest that the entirety of the Arctic ice-cap will melt in summer by 2050.[27] Although growing economies in Latin America, South Asia, Southeast Asia and Africa, in particular, have contributed to growing carbon emissions, tropical nations currently still only contribute around 20% of the global total. Yet they are still taking the brunt of the damage. Average surface temperatures have increased by an average of 0.7°C around the tropics between 1961 and 2018.[28] While this may not seem like a lot, compared to around 1.5°C for a region such as Scotland during the same time period,[29] normally low intra- and inter-annual temperature variation in the tropics means that smaller changes there can have bigger impacts on climate systems, environments and species than larger changes in temperate regions.[30] Atmospheric CO_2 is already at levels only seen 4 million years ago in the Pliocene and, as climate modeller Tim Lenton, who we met in Chapter 1, has pointed out, we are rapidly on our way to levels seen during the 'hothouse' state of the Eocene, when temperatures were 14°C higher than those of the pre-industrial period.[31] One might argue that this could actually be good for tropical forests, given their proliferation at that time, as we explored in Chapter 2, but, given rising human pressures and the fact that climate change will also result in an increasing unpredictability of rainfall in many tropical regions, they will almost certainly be placed in an equal state of disarray to the rest of the planet should these trends continue.

In fact, scientists have shown that a decrease in forest cover of anywhere between 20 and 40%, or a rise of 4°C in temperature, would be enough to begin a permanent transition to savannah environments in most of the central, southern and eastern Amazon.[32] Alongside rises in fire frequency, this would undoubtedly spark an end to some of the largest extents of tropical forest biomes anywhere in the world. Changes in climate and tropical forest habitat are also placing enormous pressures on the rich animal species of these environments. Insects, amphibians and reptiles can all only tolerate small variations in temperature and will face increasing challenges in the decades to come. Birds have been demonstrated to struggle in the face of fragmentation, and isolated patches of Amazonian forest smaller than 1 square kilometre can lose nearly half their bird species within a period of just 15 years.[33] Tigers in India, forest elephants and gorillas in Africa, and orang-utans in Borneo are all considered particularly vulnerable to habitat loss, and their protected areas may soon become too small to sustain viable populations.[34] Over-hunting is also a growing issue, especially given the ongoing profit-driven global trade in tropical animals we saw in Chapter 13 and an increasing number of people seeking adequate protein in the tropics. Using a database of 3,281 records of how hunting has affected mammals around the tropics, researchers have found that around half the existing intact tropical forest areas and 62% of wilderness areas are now, at least in part, already devoid of large mammals, with particularly large declines expected in West and Central Africa and Southeast Asia.[35] Human introduction of invasive species will also continue to reconfigure tropical plant and animal communities, particularly in the face of habitat change. Invasive plants can suppress native species seeking to regrow.[36] Meanwhile, exotic animals often tip the balance of predator–prey relationships. For example, the widespread distribution of feral cats in the Wet Tropics of north-eastern Queensland, which are preying on often-endangered, endemic small mammals, is a cause for considerable concern.[37]

*

It goes without saying, then, that tropical forests are up against it. As, too, are their human inhabitants, according to the latest 'State of the Tropics'[38] report. In the year 2020, almost half the world's total human population lived in the tropics.[39] And by 2050, over half the world's population will call the equatorial regions their home.[40] The future of our species as a whole is undoubtedly reliant on what happens here in the years to come. Many of these people will also increasingly live in cities. By 2050 6.3 billion of the world's human population will live in an urban area.[41] The tropics account for over half this planned increase, and have witnessed growing rural-to-urban migration between the late twentieth and early twenty-first centuries, with over 1.5 billion people currently living in tropical cities.[42] Growing economies and the social, commercial and health opportunities provided by urban networks are driving these trends, particularly in Southeast Asia. Indonesia's urban population, for example, grew from 32 million in 1990 to nearly 150 million by 2018.[43] Nearly three-quarters of the population of Central America, South America and the Caribbean are already living in urban areas.[44] Just under one quarter of the tropical urban population lives in slums,[45] with incredibly poor living conditions that have characterized regions experiencing rapid urbanization since the nineteenth century.[46] These informal settlements are broadly declining, thanks to rising wealth and settlement improvement programmes, and often offer access to diverse income opportunities and cultural connections. Nevertheless, ensuring the sustainability and inclusivity of cities in the tropics, including moves towards 'green space' integration, is a major United Nations concern in the context of environmental conservation and human living standards.[47]

Growing populations will require growing access to food. In Chapters 10 through 13, we followed an intensifying conflict between tropical forests and expanding agricultural, pastoral and extractive land use. The tropics are home to 39% of the planet's agricultural land area[48] and, while this land is used to support food security for its growing population, it also continues to feed the desires of the export market, as it has since at least the seventeenth century. Cropland in tropical countries expanded by 48,000 square kilometres per year between 1999 and 2008, with Nigeria (10,259 square kilometres), Indonesia (5,826 square kilometres), Ethiopia (5,405 square kilometres), Sudan (5,227 square kilometres), and Brazil (4,205 square kilometres) showing the greatest areas of annual expansion.[49] Yet expansion of agricultural production has not necessarily expanded the tropical food base. Islands in the Pacific and Indian Oceans have actually seen a reduction in agricultural land due to pressures to expand tourist and urban infrastructures.[50] Similarly, the rising proportion of agricultural land used for biofuels or cash crops for export may benefit local incomes, but not food security. Increases in meat production in the tropics (a rise of 76% between 2010 and 2017) have also been primarily driven by foreign demand, with Brazil, India and Australia acting as particularly significant meat exporters to China, the US, Japan, Europe and the Middle East.[51] The economic benefits of these changes may not be reflected in local tropical sustainability. These problems are particularly acute because remote sensing and mapping models of soil erosion predict that tropical Africa, South America and Southeast Asia will see increasingly poor, unstable soils over the course of the twenty-first century.[52] The inhabitants of the tropics are going to face increasingly difficult decisions, between exploiting imbalanced economic export markets to improve incomes and developing more productive, sustainable subsistence agricultural systems to meet their own nutritional requirements.

Extreme poverty is defined as 'severe deprivation of basic human needs, including food, safe drinking water, sanitation facilities, health, shelter, education and information'. In positive news, globally, 1 billion fewer people were living under these conditions in 2015 than in 1990.[53] Nevertheless, the majority (85%) of the remaining >100,000,000 of the world's population experiencing extreme poverty live in the tropics. Population growth, conflicts and a complex interplay of political and economic factors mean that in some tropical nations the proportion living under such conditions is actually increasing. Just five nations – Nigeria, the Democratic Republic of the Congo, Ethiopia, India and Bangladesh – house 50% of the human population living in extreme poverty the world over. In terms of 'moderate poverty', where basic needs are met but just barely, and people earn less than $3.20 per person per day, the tropics have also seen much more modest improvement: 1.3 billion people live in moderate poverty in the tropics, double the number for the rest of the world.[54] As we saw in Chapter 13, poverty is a major issue for tropical forest conservation and sustainability, as populations and governments seek to balance demands for food, water and incomes with environmental impacts. The prevalence of poverty across the tropics is widely recognized as a product of colonial processes that continue to shape economies based on agricultural systems with large, poorly paid workforces, infrastructural imbalances, a reliance on foreign investment, and colonially drawn boundaries that can catalyse political and cultural conflicts.[55] However, this has not stopped nations and companies in Europe and northern North America continuing to exploit the situation, while simultaneously releasing more than their fair share of emissions into the atmosphere, which will again disproportionately affect the livelihoods of tropical residents.

Climate change is already beginning to alter and destroy the lives of many people in the tropics. Climate models show that human-induced warming had already increased the likelihood of extremes in precipitation and temperature occurring in the tropics by 2010,[56] and extreme weather events are likely to only become more extreme and more frequent as the twenty-first century continues. Changing precipitation and water run-off modelled under global warming scenarios of 1.5–2.0°C indicate increased flooding in South Asia, Southeast Asia and the western Amazon,[57] with some studies suggesting that fifteen of the twenty most impacted nations will be in tropical or sub-tropical regions.[58] Many of the same regions, as well as other areas in the sub-tropics and tropics, will face increasingly frequent and intense drought periods over the course of the twenty-first century. These changes can have major impacts on human populations. For example, in Malaysia rice makes up 98% of current cereal production. It is estimated that for each 1°C rise in temperature, rice yields will decline by 10%, while wet rice systems may become increasingly unsustainable in the face of growing drought conditions.[59] Meanwhile twenty-first-century flash floods and landslides have resulted in mass destruction and loss of life in cities such as

Vargas in Venezuela.[60] Growing storms, cyclones and hurricanes are also battering the tropics more frequently. In 2017, Hurricane Maria killed over 3,000 people and caused $91.61 billion worth of damage in the north-eastern Caribbean, and this was just one storm in a hyperactive Atlantic hurricane season.[61] Rising sea levels represent a slower but eventually more permanent climate-based change to tropical livelihoods. Together, it is estimated that an increasing intensity and frequency of storms and floods, sea-level rise, and drought and desertification will create around 200 million climate refugees worldwide in the next 50 years,[62] and perhaps as many as 700 million (nearly one-tenth of the world's current population).[63]

Tropical deforestation is also causing a number of public health issues. Tropical forest edges have proven to be a major petri dish for pandemics. It is estimated that roughly 1,700,000 viruses exist in mammals and birds around the world at any one time, though a minute fraction is known and properly described.[64] Tropical forest edges provide the greatest opportunity for humans and their close animal companions to come into contact with these significant epidemiological reservoirs. Deforestation and forest fragmentation, for logging or infrastructural development, create longer, and more ragged, forest edges, where growing numbers of humans and domestic animals become interspersed with populations of wild animals, priming a large human host for viral outbreak.[65] Hunting for bushmeat or medicinal trade further expands opportunities for contact. In fact, deforestation and human exploitation of tropical wildlife have resulted in many of the major disease outbreaks of the last two decades. The 1980s emergence of HIV has been linked to bushmeat hunting in the Democratic Republic of the Congo.[66] Outbreaks of the incredibly deadly Ebola in West Africa, which killed over 11,000 people between 2013 and 2016, have been linked to a single human contact with a bat colony.[67] Bats are thought to be particularly prominent reservoirs for disease, and feed closer to human settlements when their forest homes are disturbed. They are also thought to be the most probable original carrier of the SARS-CoV-2 virus that causes COVID-19,[68] which likely began as a product of human–wildlife interactions in the humid sub-tropics of China.[69] It is predicted that ongoing deforestation and fragmentation in the tropics in the twenty-first century is only going to increase the probability of highly virulent diseases making their way into human populations.

Another challenge facing tropical communities seeking to battle for more sustainable livelihoods is the suppression of Indigenous and traditional knowledge. The tropics are home to a staggering amount of human cultural diversity, including 80% of all living languages.[70] As we have seen, local communities around the tropics have long histories (and indeed prehistories) of recognizing ecological thresholds and dynamics, performing sustainable agroforestry and farming practices, and incorporating new crops, technologies and settlement patterns into their tropical environments. For example, Ifugao communities in the Cordillera Mountains of the Philippines have land right laws, upland cultivation practices, knowledge of soil and water conservation, and forest stand management, collectively known as *muyong*, that are critical for forest maintenance and biodiversity protection.[71] Indigenous languages can be particularly crucial for passing on knowledge, and it has been shown that people speaking the Indigenous Huastec language in western Mexico show more diverse and evenly shared knowledge of tropical plant use than Spanish-speaking Indigenous populations.[72] However, Indigenous populations are currently some of the most marginalized groups within the tropics, facing higher rates of disease, higher child and adult mortality, and significant social disadvantages.[73] Their languages are also disappearing, as young Indigenous people are integrated, often forcibly, into national capitalist economies and education systems. These losses represent a major threat, particularly in light of the fact that Indigenous populations and other local smallholders are crucial to the development of balanced, sustainable approaches to tropical forest use and conservation. This is especially the case given that over one-fifth of the significant portion of the world's carbon stored in tropical forests lies on land belonging to or claimed by Indigenous communities.[74]

You may feel that this concluding chapter has so far resembled the nightmare international newsreel you are fed up with scrolling through on your smartphone, tired of hearing coming over the air-waves, and thoroughly done with watching on your television. But I am asking you to keep reading: your lives are irrevocably intertwined with tropical forests. Even if you struggle to place yourself in the shoes of inhabitants living at the 'ground zeros' of tropical forest sus-tainability, you cannot get away from the fact that before long these problems will have made their way into your back garden. Increas-ingly seasonal, unstable soils and watersheds will reduce the productivity and availability of cash crops (e.g. rubber, oil palm, cocoa), staple crops (e.g. rice) and meat (e.g. sheep, goat, cow) in the tropics, reducing the size of exports making their way to reliant European, North American, and now also Middle Eastern, nations. Furthermore, tropical forest deforestation and degradation will lead to continental and global changes in precipitation, temperature and agricultural growing conditions. Finally, since the Carboniferous period, tropical forests, alongside the oceans (and, later, northern hemisphere forests), have played a major role in capturing carbon and regulating the degree to which global carbon emissions actually translate into rising atmospheric CO_2.[75] It is possible that higher temperatures could bring about faster tropical growth and more rapid fixation of CO_2.[76] However, as tropical deforestation and degrada-tion, including through more frequent fire outbreaks, continue throughout the twenty-first century, particularly in swamp forest areas, the very real prospect of these environments becoming carbon *sources* rather than carbon sinks is a truly terrifying thought.[77] The United Nations imperative to limit increases in global average tem-perature to below 2°C above pre-industrial levels[78] will drift increasingly out of reach. What happens in the tropics in the next century will shape what happens on our entire planet, be in no doubt about that.

The predicted loss of tropical forest plant and animal species also represents a truly global problem, given their enormous contribution to planetary biodiversity.[79] Assuming a complete loss, degradation and alteration of all primary tropical forests by the year 2225, it has been calculated that there will be a global species richness decline of around ~44% for ant species, ~30% for dung beetles and ~20% for trees.[80] For the coming two centuries, extinction rates for different global species groups vary between around 200 and 2,000 extinctions per million species per year, all relating to the loss of habitat in the tropics. As Professor Xingli Giam of the University of Tennessee has pointed out, these kinds of rates are between 2,000 and 20,000 times higher than what might be expected under natural geological and biological conditions.[81] Furthermore, they are 'two or more orders of magnitude higher than extinction rates associated with four of the five previous mass extinction events', including those of the Devonian, Ordovician, Permian, and the Triassic.[82] In fact, they are comparable to the rate predicted for the K/Pg mass extinction that wiped out the dinosaurs.[83] Put together, then, it is now estimated that the deforestation and degradation of tropical forests, not even mentioning the additional contributions that will likely be made by human-induced climate change, would result in a sixth planetary mass extinction event at the end of the next two centuries. Just think about that for a second. The events unfolding around the equator today as we go about our daily lives are predicted to have the same, or greater, impact on the world's biodiversity as geologically known catastrophes that took life on Earth to the very edge of the cliff face.

The public health crises facing the tropics mentioned above have, in the past decade, also now very blatantly blanketed the world. Cases of the Ebola virus in Texas, the United States of America and Glasgow, in the UK, in 2014 showed the potential for this deadly disease to be carried beyond the African tropics,[84] and outbreaks continued in the Democratic Republic of the Congo into 2020. The extra-tropical transmission seen for Ebola so far, however, pales in comparison to that seen as part of the ongoing pandemic of COVID-19. By the end of January 2020, this disease had spread around much of China, which rapidly went into a state of lockdown. Although recent studies suggest it may have made it to Europe as early as the end of December 2019, certainly by March the SARS-CoV-2 virus was transmitting between people in Europe and North America, with Italy and the United States of America becoming new centres of the outbreak. As of the end of January 2021 there were more than 96 million globally documented cases of COVID-19, with more than 2,000,000 deaths.[85]

Although many countries have managed to reduce cases, and a number of vaccines are now available, as I write this book, there is no rapid end in sight. The emotional, psychological, and economic damage caused by death rates, under-staffed hospitals, abandonment of workplaces, the closure of schools, the decline in domestic and international travel, and reduced access to infrastructure and facilities is expected to be dramatic and long-lasting. Nevertheless, while this virus has already changed our lives for ever, further, and perhaps deadlier, pandemics linked to human–wild animal contact in the tropics are likely to only increase as the twenty-first century wears on.[86]

Legacies of colonialism and imperialism met in Chapters 10 and 11 will also still continue to shape the face of global economics, politics and sustainability in the twenty-first century. Gross Domestic Product has increased across the tropics between 1990 and 2018, but so too has its gap with the rest of the world.[87] Significant proportions of the Gross Domestic Product of many tropical nations are made up by pay cheques sent home by diasporic workers.[88] Furthermore, debt has increased across the tropics since 2013, placing tropical nations in a precarious position, particularly when external events such as pandemics require rapid responses.[89] These issues might have traditionally seemed 'far away' to you. However, they are challenges we will all have to deal with. Climate change is expected to have a particularly intense impact on the growing, impoverished inhabitants of the tropics, resulting in increasing numbers of 'climate migrants' that nations across Europe, North America and Australasia will have to contend with. To take one example, at the current pace of sea level rise, the people forced to migrate away from Bangladesh by the end of the century will outnumber all current refugees around the world.[90] Western European and northern North American nations will also have to manage the challenge of reducing global carbon emissions (the majority of which they are responsible for) while respecting and offsetting the needs of tropical nations to grow and attain living conditions that many of their northern hemispheric counterparts have enjoyed for so long after half a millennium of colonialism.

★

The above may suggest that the situation is bleak. But it is necessary to underscore the urgency of the situation at hand. This does not mean that there is no room for positivity. The deep human history of tropical forests revealed by this book highlights the way in which societies have repeatedly adapted to change and challenges in the tropics. Indeed, this ecological flexibility may even define our species.[91] However, we have somewhat lost our way. Global consumer demand, wealth inequalities and political priorities continue to exert pressure on some of the habitats we can least afford to lose. This book has revealed two key avenues out of this dangerous cycle. First, whether it is Pleistocene hunter-gatherers encountering giant sloths in the Amazon or food producers experimenting in highland New Guinea, or whether it is pioneering pre-colonial voyagers to uninhabited Pacific islands or water-tank building urban dwellers in Sri Lanka, successful adaptations to tropical forests are characterized by intimate on-the-ground knowledge of these ecosystems. Only if local smallholders and Indigenous populations are properly consulted and empowered, supported by detailed archaeological, palaeoecological and historical records of past human management, can more beneficial, adaptable ways to live with tropical forests be enacted.[92] Second, we must acknowledge that some of the greatest pressures facing tropical forests and their inhabitants today are a product of long-term colonial and imperial extraction that has enriched nations and social groups in Europe and northern North America and degraded tropical forest environments and local possibilities for economic growth. Moving forward requires that global consumers and governments recognize that they too must bear responsibility for what happens in these environments.

Fortunately, there are some signs that both these avenues are beginning to be recognized and supported. The African Forest Landscape Restoration (AFR 100) Initiative was founded in December 2015 as a country-led effort to restore 100 million hectares of degraded and deforested land across Africa by 2030. Thirty countries within the African Union are signed up to the scheme, and Ethiopia's Innovation and Technology Minister famously declared that the country had planted more than 350 million tree seedlings in just twelve hours as part of the programme in 2019.[93] Importantly, AFR100 was developed with local outcomes as a top priority. It not only plants trees, but also promotes sustainable agroforestry, the inter-cropping of trees with other forms of land use, and erosion control.[94] This means that broader continental, and global, needs for forest cover, and the production of a 'Great Green Wall' to push Sahara desertification back,[95] are balanced with needs to enhance local food security. Farmers working in Burkina Faso and Niger have, for example, through experimental modifications to traditional agroforestry, water retention and soil management practices, converted between 200,000 and 300,000 hectares of land, which produce 80,000 tonnes of food per year.[96] Such initiatives also offer the opportunity to build more equitable futures in the region. For example, women and girls make up >50% of the active agricultural labour force across the tropics. Consultation of them when planning forest restoration projects ensures that educational, infrastructural, ecological and land ownership benefits are shared more widely across those societies living within tropical forest environments.[97]

As we saw in Chapter 13, a growing number of initiatives around the tropics are also seeking to actively combine archaeological, palaeoecological, historical and traditional knowledge to construct more just, effective conservation plans, that integrate smallholders and Indigenous groups in the sustainable management of forests, rather than forcing them out. Where local smallholders in western Ghana have been involved, and supported, in reforestation initiatives, they have been shown to plant a significant number of IUCN Red List threatened species as part of their tree plantations, acknowledging and conserving the most vulnerable portions of their surrounding ecosystems.[98] We have seen how, in Australia, Rainforest Aboriginal People are being given increasing support in applying traditional burning to landscapes, and their ongoing stewardship over endemic plant and animal diversity.[99] Similarly, collaboration with Indigenous Kayapó and Upper Xingu groups produced one of the largest protected tropical forest areas in the world at 130,000 square kilometres in Brazil, as well as effective policies for biodiversity management.[100] Indeed, prior to the rise of Brazil's recent nationalist government, the country had begun to stem the tide of deforestation, recording a dramatic reduction between 2000 and 2012[101] that had significantly offset rises elsewhere around the planet.[102] Further support and realistic assessment of the priorities and capacities of local communities interacting with tropical forests are essential. Colombia and Costa Rica have provided one such route of support, introducing taxes to carbon emissions to fund more sustainable practices by landowners, such as reforestation and mixed agroforestry.[103]

Nonetheless, given the legacies of colonialism and imperialism, it is also essential that extra-tropical governments and consumers recognize the role they have to play in bolstering change. Predictions of global deforestation and fragmentation are greatly reduced should the international REDD+ scheme, which pays tropical nations to reduce emissions from these activities, be successful. Some scenarios and payment plans are expected to lead to the preservation of between 75 and 98% of current forest cover across the tropics,[104] though care must be taken that such schemes do not come to mimic further Western colonial dictation. Many oil palm companies, seeking to retain access to high-value international markets, have begun to voluntarily sign up to sustainability standards of the Roundtable on Sustainable Palm Oil (RSPO) that was set up in 2004. To obtain RSPO certification necessitates consideration of impacts on soil and water sources and air quality, as well as plans to mitigate damage to biodiversity. It has been clearly demonstrated that where this certification system has been applied, livelihoods have improved and negative environmental impacts have been reduced.[105] The power you have as a consumer to shape sustainability has been highlighted by the fact that European countries such as the Netherlands and Belgium, and giant multinational corporations such as Unilever and Carrefour, have, as of 2020, committed to only using palm oil ingredients with 100% RSPO certification in their food and cosmetic products.[106] The international pressure (and support) must be kept up, however, as smallholders often lack the means or incentives to adopt RSPO guidelines,[107] and non-certified markets, particularly in Asia, remain open for business.

We should also perhaps find some optimism in the way in which figureheads such as Greta Thunberg and Vanessa Nakate have led waves of school strikes and protests around the world to demand urgent, and united, political and economic action on environmental justice and climate change.[108] Furthermore, we are potentially at the threshold of a moment in which Euro-American governments are finally forced to directly reckon with centuries of colonialism and imperialism that have left scars of global inequality and legacies of racial prejudice within their own borders. 2020 saw the inequality and discrimination which have plagued European and North American interactions with the tropics since the fifteenth century boil to the surface. The killing of George Floyd by Minneapolis law enforcement on 25 May 2020 saw the Black Lives Matter (BLM) movement soar to national and international prominence. An estimated 15 to 26 million took part in protests across the United States of America to protest against this brutal killing, as well as the deaths of Ahmaud Arbery and Breonna Taylor, in February and March of the same year, respectively. It became one of the largest movements in North American history.[109] National and international support for BLM also expanded well beyond the United States, with protests organized across Australia, Canada, Denmark, France, Germany, Japan, New Zealand and the United Kingdom. We have also seen a rise in demands to deal with the symbols and systems left behind by long-term northern North American and European colonial abuses,[110] as well as the toppling of statues, including that of Edward Colston who we met in Chapter 11, of individuals known for their racist views, participation in slavery, and contribution to the murder of Indigenous populations.[111] We are finding multinational corporations and academic institutions making commitments to the production of more equitable working environments and societies.[112] Admittedly, much of this still remains superficial or 'tokenistic',[113] and a number of politicians and members of our societies remain resistant. But are we perhaps at the cusp of a major social reckoning? One that acknowledges and addresses the centuries of marginalization, persecution and exploitation of certain sectors of society? One that sees European and northern North American nations accept the role they have had in the political, social and economic situation of the nations of the tropics and the challenges facing them today?

★

By now you have seen how tropical forests reach into almost every recess of your life. They have shaped the bouquets you pick up for your loved ones, the scents and displays of your gardens, and, for those of you who celebrate Christmas, the trees at the centre of your living room. They have also played a major role in the evolutionary origins of the ants running along your paving slabs, the birds singing outside your window, many of the stars of your local zoo, and, of course, you, your families, your friends, and the entirety of our species. Tropical forests have also grown into your kitchens, providing the rice in your cupboards, the crucial ingredients for many of your medicines, the chicken in your oven and the eggs in your omelette, the sweetcorn on your barbecue, the coffee in your cup, and the chocolate you treat yourself to when no one is looking. They have also provided the rubber for your bikes and cars, your ability to erase mistakes, your wellington boots, and the insulation surrounding almost anything with an electronic wire in your home and workplace. Tropical forests are all around you. Sadly, so too are the legacies of centuries of imbalanced colonial and capitalist interactions between the western half of Europe and northern North America and the tropical world, moulding the racial tensions, the economic inequalities, the political battlegrounds, and a lot of the social and cultural marginalization in the societies in which you live. These are not distant, exotic environments on the other side of the world. Rather, through an entangled prehistory and history, they have found their way into your homes. No matter where you are.

So, what are you going to do about it? A number of conservation policies are being enacted across the tropics. Trees are being planted to stabilize landscapes and extract carbon from the atmosphere. Protected areas are being developed and enforced to stem the tide of deforestation, degradation and biodiversity loss. And human populations in the tropics are beginning to be given the resources to apply many of the 'usable pasts' we have explored to develop more sustainable, and often Indigenous-led, approaches to farming, agroforestry and economic productivity that integrate forests into new forms of land use. However, those on the frontlines of this key fight for sustainability cannot be left to take a stand alone. They need economic initiatives that redistribute resources from richer nations and richer sectors of tropical societies. They need global consumers to pay for sustainably produced tropical products, consider the footprint of their tropical tourism, and hold poorly regulated multinational corporations to account. They need global voters that see national interests as global interests, that recognize the mistakes and legacies of colonial pasts and the need to fight for fairer societies around the world. They need the inhabitants of Europe and northern North America, alongside China, to cut their significant per capita contributions to global CO_2 emissions to stop further environmental change[114] and to give them a buffer to undertake their own, more sustainable, economic development. Every single one of you could, today, begin to play at least a small part in every single one of these things. And, as this chapter should have shown, it does need to be today. You might act because this book has shown you that you have an affinity to, and a responsibility for, tropical forests. You might act because you actually already felt a moral or empathetic desire to help others and just needed some direction as to how. Or you might act because you have seen that, if you don't, climate change, declining food sources, economic catastrophe, political instability, mass migration and an explosion of pandemic diseases will very soon be knocking at your own door.

Appendix

Timeline of key events

Era	Period		Epoch	Start (Millions of years ago)	Events
Cenozoic	Neogene		Anthropocene? / Holocene	0.0117	■ Industrial revolution ■ European colonialism and the 'Columbian exchange' ■ Massive pre-industrial urban centres (e.g. Greater Angkor) ■ Origins of food production and farming practices
		Quaternary	Pleistocene	2.58	■ *Homo sapiens* on every continent apart from Antarctica by end of period ■ Specialized tropical forest use by populations of our species ■ Evolution of *Homo sapiens* in Africa ■ Dispersal of first hominins outside of Africa to Southeast Asia tropics (*Homo erectus*, *Homo floresiensis*) ■ Glacial cycles become more intense – Milankovitch cycles
			Pliocene	5.33	■ Emergence of the genus *Homo* and stone tool use ■ Hominin evolution continues in Africa (e.g. 'Lucy') ■ Climatic variability, glacial and inter-glacial cycles
			Miocene	23.04	■ First hominins in Africa (e.g. 'Ardi') ■ C_4 grasses expand in different parts of the tropics ■ Drying leads to tropical forest retreat ■ Initial gradual increase in aridity ■ Warm climate with extension of tropical forests and great apes across Eurasia
	Paleogene		Oligocene	33.9	■ Mammals evolve in response to changing environments, horses increasingly begin to graze and get bigger ■ 'Mega-bats' evolve ■ Increasing climatic swings between cooling and drying and warming and wetting
		Tertiary	Eocene	56.0	■ First simian primates in Africa at end of period ■ Perissodactyls (ancestors of horses, tapirs and rhinos) initially very successful, until ungulates expand as climates cool ■ Tropical and sub-tropical forests extend from Europe to South America ■ Rapid warming to middle 'optimum' prior to cooling and forest retreat
			Paleocene	66.0	■ Explosion of mammals in terms of size and diversity, including early small horses ■ First true tropical rainforests (e.g. Cerrejón, Titanoboa), legumes ■ Warmer and wetter climate

Era	Period	Start (Millions of years ago)	Events
Mesozoic	Cretaceous	143.1	■ Period ends with mass extinction in the face of an asteroid ■ Mammals continue to evolve ■ Debated link between dinosaurs and early flowering plant evolution, dispersal and diversification (angiosperms) ■ Angiosperms emerge and expand ■ The remainder of Gondwana breaks up further into Madagascar, India, Australia and Antarctica
	Jurassic	201.4	■ Dinosaurs living within gymnosperm-dominated forests and with some plants perhaps adapting to dinosaur feeding (ginkgo, monkey puzzle trees) ■ First mammals (*Juramaia*) ■ Dinosaurs diversify, clear emergence of three separate lineages (Theropoda, Sauropodomorpha, Ornithischia) ■ Pangaea divides into Laurasia (north) and Gondwana (south), Gondwana in turn begins to break up with Africa (excluding Madagascar), separating from the rest
	Triassic	251.9	■ First dinosaurs make it to the tropics by the end of the period ■ First gymnosperms (conifers, ginkgo, cycads) appear and dominate forests that slowly re-emerge ■ Forests nearly completely absent ■ Arid and unstable tropics
Paleozoic	Permian	298.9	■ Mass extinctions in sea and on land at the end of the period ■ Supercontinent of Pangaea forms ■ Reptiles proliferate and diversify into different lineages ■ 'Rainforest collapse' of Carboniferous forests
	Carboniferous	359.3	■ First reptiles emerge ■ Extensive tropical swamp forests lead to the formation of massive coal fields that later drove the industrial revolution
	Devonian	419.0	■ Period ends with glaciation and extinction ■ First 'arboreal' forms – *Eospermatopteris* ■ First proto-gymnosperms and true trees – *Archaeopteris*, more soil formation and deep rooting systems ■ First amphibians and insects
	Silurian	443.1	■ First fossil body forms of 'higher' vascular plants (xylem, phloem, true roots)
	Ordovician	486.9	■ Ice Age at end of period ■ First direct evidence for green land plants, enhanced weathering
	Cambrian	538.8	■ First non-vascular land plants estimated based on genetic clocks ■ 'Cambrian explosion' and thriving marine ecosystems

Notes and Sources

Preface

1 Domínguez-Rodrigo, M. (2014), 'Is the "Savanna Hypothesis" a dead concept for explaining the emergence of the earliest hominins?', *Current Anthropology*, 55 (1): 59–81.

2 Mellars, P. (2006), 'Why did modern human populations disperse from Africa ca. 60,000 years ago? A new model', *Proceedings of the National Academy of Sciences of the United States of America*, 103: 9381–6.

3 Meggers, B. J. (1971), *Amazonia: Man and Culture in a Counterfeit Paradise*, Illinois: Harlan Davidson.

4 Turnbull, C. (1961), *The Forest People: A Study of the Pygmies of the Congo*, New York: Simon & Schuster.

5 Watson, J. E. M., Evans, T., Venter, O., et al. (2018), 'The exceptional value of intact forest ecosystems', *Nature Ecology & Evolution*, 2: 599–610.

6 Ghazoul, J. (2015), *Forests: A Very Short Introduction*, Oxford: Oxford University Press.

7 Spracklen, D. V., Arnold, S. R., and Taylor, C.M. (2012), 'Observations of increased tropical rainfall preceded by air passage over forests', *Nature*, 489: 282–5.

8 Malhi, Y. (2010), 'The carbon balance of tropical forest regions, 1990–2005', *Current Opinions in Environmental Sustainability*, 2: 237–44.

9 Curry, A. (2016), ' "Green hell" has long been home for humans', *Science*, 354: 268–9.

10 Pimm, S. L., and Raven, P. (2000), 'Extinction by numbers', *Nature*, 403: 843–5.

11 Throughout this book I use Indigenous with a capitalized 'I' to refer to Indigenous peoples, knowledge, communities, history, land management and settlement organization. Where possible I cite the name a given Indigenous community uses for itself. I use indigenous with a lower case 'i' only to refer to plants and animals endemic to a given region.

12 Roberts, P., Buhrich, A., Caetano-Andrade, V.L., et al., (2021), 'Reimagining the relationship between Gondwanan forests and Aboriginal land management in Australia's "Wet Tropics"', *iScience*, 24: 102190. Doi: https://doi.org/10.1016/j.isci.2021.102190./

Chapter 1: Into the light – the beginning of the world as we know it

1 Vandenbrink, J. P., Brown, E. A., Harmer, S. L., and Blackman, B. K. (2014), 'Turning heads: The biology of solar tracking in sunflower', *Plant Science*, 224: 20–26.

2 Appel, H. M., and Cocroft, R. B. (2014), 'Plants respond to leaf vibrations caused by insect herbivore chewing', *Oecologia*, 175: 1257–66.

3 Hughes, S. (1990), 'Antelope activate the acacia's alarm system', *New Scientist*: https://www.newscientist.com/article/mg12717361-200-antelope-activate-the-acacias-alarm-system/?ignored=irrelevant.

4 Yuan Song, Y., Simard, S. W., Carroll, A., et al. (2015), 'Defoliation of interior Douglas-fir elicits carbon transfer and stress signalling to ponderosa pine neighbours through ectomycorrhizal networks', *Nature Scientific Reports*, 5: https://doi.org/10.1038/srep08495.

5 Wohlleben, P. (2016), *The Hidden Life of Trees: What They Feel, How They Communicate: Discoveries From a Secret World*, London: William Collins.

6 The official dates of geological time periods change as new stratified records are found and dating resolution increases. Throughout this book, where dates are provided for an officially ratified geological period (e.g. the Cambrian) I have followed those given in Volumes 1 and 2 of the latest *Geologic Time Scale 2020* (Gradstein, F. M., Ogg, J. G., Schmitz, M. D., and Ogg, G. M. (2020), *The Geologic Time Scale 2020* (1st edition) (Volumes 1 and 2), Amsterdam: Elsevier). This is the standard reference used by many palaeontologists and geologists and is regularly updated as new information is obtained. Where an unofficial division is mentioned (e.g. the 'early' Cambrian), the date refers to that given in the specific scientific literature research under discussion in that section of the book or is otherwise specified in a note. In the case of the 'early' Cambrian I have combined the time period discussed in notes 7 and 8 below with the new dates given for the start of the Cambrian (538.8 million years ago) and the boundary in the middle of the Cambrian between Epoch 2 and the Wuliuan (509.0 million years ago) in the latest *Geologic Time Scale* of Gradstein, et al. (2020).

7 Paterson, J. R., Edgecombe, G. D., and Lee, M. S. Y. (2019), 'Trilobite evolutionary rates constrain the duration of the Cambrian explosion', *Proceedings of the National Academy of Sciences of the United States of America*, 116: 4394–9.

8 Collette, J. H., and Hagadorn, J. W. (2010), 'Three-dimensionally preserved arthropods from Cambrian lagerstätten of Quebec and Wisconsin', *Journal of Palaeontology*, 84: 646–67.

9 Beck, H. E., Zimmermann, N. E., McVicar, T. R., et al. (2018), 'Present and future Köppen-Geiger climate classification maps at 1-km resolution', *Scientific Data*, 5: 180214.

10 Lenton, T. M., Daines, S. J., and Mills, B. J. W. (2018), 'COPSE reloaded: An improved model of biogeochemical cycling over Phanerozoic time', *Earth-Science Reviews*, 178: 1–28.

11 Mills, B. J. W., Krause, A. J., Scotese, C. R., et al., (2019), 'Modelling the long-term carbon cycle, atmospheric CO2, and Earth surface temperature from late Neoproteozoic to present day', *Gondwana Research*, 67: 172–86.

12 Bergman, N. M., Lenton, T. M., and Watson, A. J. (2004), 'COPSE: A new model of biogeochemical cycling over Phanerozoic time', *American Journal of Science*, 304: 397–437.

13 Lenton, Daines and Mills, 'COPSE reloaded: An improved model of biogeochemical cycling over Phanerozoic time'.

14 Bergman, Lenton and Watson, 'COPSE: A new model of biogeochemical cycling over Phanerozoic time'.

15 Ruhfel, B. R., Gitzendanner, M. A., Soltis, P. S., et al. (2014), 'From algae to angiosperms – inferring the phylogeny of green plants (Viridplantae) from 360 plastid genomes', *BMC Evolutionary Biology*, 14: DOI: https://doi.org/10.1186/1471-2148-14-23.

16 Berner, R. A. (2006), 'GEOCARBSULF: A combined model for Phanerozoic atmospheric O2 and CO2', *Geochimica et Cosmochima Acta*, 70: 5653–64.

17 Shimamura, M. (2016), '*Marchantia polymorpha*: Taxonomy, phylogeny and morphology of a model system', *Plant and Cell Physiology*, 57: 230–56.

18 Rubinstein, C. V., Gerrienne, P., de la Puente, G. S., et al. (2010), 'Early Middle Ordovician evidence for land plants in Argentina (eastern Gondwana)', *New Phytologist*, 188: 365–9.

19 Retallack, G. J. (2020), 'Ordovician land plants and fungi from Douglas Dam, Tennessee', *The Palaeobotanist*, 68: https://cpb-us-e1.wpmucdn.com/blogs.uoregon.edu/dist/d/3735/files/2020/09/Retallack-2020-Ordovician-land-plants.pdf.

20 Edwards, D., and Feehan, J. (1980), 'Records of Cooksonia-type sporangia from late Wenlock strata in Ireland', *Nature*, 287: 41–2.

21 Renzaglia, K. S., Nickrent, D. L., Garbary, D. J., et al. (2000), 'Vegetative and reproductive innovations of early land plants: Implications for a unified phylogeny', *Philosophical Transactions of the Royal Society of London B Series: Biological Sciences*, 355: 769–93.

22 Clarke, J. T., Warnock, R. C. M., and Donoghue, P. C. J. (2011), 'Establishing a time-scale for plant evolution', *New Phytologist*, 192: 266–301.

23 ibid.

24 ibid.

25 Puttick, M. N., Morris, J. L., Williams, T. A., et al. (2018), 'The interrelationships of land plants and the nature of the ancestral embryophyte', *Current Biology*, 28: 733–45.

26 Morris, J. L., Puttick, M. N., Clark, J. W., et al. (2018), 'The timescale of early land plant evolution', *Proceedings of the National Academy of Sciences of the United States of America*, 115: E2274–E2283.

27 ibid.

28 Puttick, Morris and Williams, 'The interrelationships of land plants and the nature of the ancestral embryophyte'.

29 Morris, Puttick and Clark, 'The timescale of early land plant evolution'.

30 Sheldrake, M. (2020), *Entangled Life: How Fungi Make Our Worlds, Change Our Minds & Shape Our Futures*, London: Random House.

31 Popkin, G. (2019), ' "Wood wide web" – the underground network of microbes that connects trees – mapped for the first time', *Science*: DOI: 10.1126/science. aay0516.

32 Steidinger, B. S., Crowther, T. W., Liang, J., et al., 'GFBI consortium. 2019. Climatic controls of decomposition drive the global biogeography of forest-tree symbioses', *Nature*, 569: 404–8.

33 Taylor, T. N., Remy, W., Hass, H., Kerp, H. (1995), 'Fossil arbuscular mycorrhizae from the Early Devonian', *Mycologia*, 87: 560–73.

34 Rimington, W. R., Pressel, S., Duckett, J. G., et al. (2018), 'Ancient plants with ancient fungi: liverworts associate with early-diverging arbuscular mycorrhizal fungi', *Proceedings of the Royal Society B Series: Biological Sciences*, 285: https://doi.org/10.1098/rspb.2018.1600.

35 Field, K. J., Pressel, S., Duckett, J. G., et al. (2015), 'Symbiotic options for the conquest of land', *Trends in Ecology and Evolution*, 30: 477–86.

36 NASA (2016), 'Carbon dioxide fertilization greening Earth, study finds': https://www.nasa.gov/feature/goddard/2016/carbon-dioxide-fertilization-greening-earth.

37 Lenton, T. M., Crouch, M., Johnson, M., et al. (2012), 'First plants cooled the Ordovician', *Nature Geoscience*, 5: 86–9.

38 Lenton, T. M., Rockström, J., Gaffney, O., et al. (2019), 'Climate tipping points – too risky to bet against', *Nature*, 575: 592–6.

39 Lenton, Crouch, Johnson, et al., 'First plants cooled the Ordovician'.

40 ibid.

41 Kotyk, M. E., Basinger, J. F., Gensel, P. G., de Freitas, T. A. (2002), 'Morphologically complex plant macrofossils from the Late Silurian of Arctic Canada', *American Journal of Botany*, 89: 1004–13.

42 Petit, R. J., and Hampe, A. (2006), 'Some evolutionary consequences of being a tree', *Annual Review of Ecology Evolution and Systematics*, 37: 187–214.

43 Goldring, W. (1927), 'The oldest known petrified forest', *Science Monthly*, 24: 514–29.

44 Stein, W. E., Mannolini, F., VanAller Hernick, L., et al. (2007), 'Giant cladoxylopsid trees resolve the enigma of the Earth's earliest forest stumps at Gilboa', *Nature*, 446: 904–7.

45 Stein, W. E., Berry, C. M., Hernick, L. V., Mannolini, F. (2012), 'Surprisingly complex community discovered in the mid-Devonian fossil forest at Gilboa', *Nature*, 483: 78–81.

46 Stein, W. E., Berry, C. M., Morris, J. L., et al. (2020), 'Mid-Devonian *Archaeopteris* roots signal revolutionary change in earliest fossil forests', *Current Biology*, 30: 421–31, e2.

47 ibid.

48 Retallack, G. J., and Huang, C. (2011), 'Ecology and evolution of Devonian trees in New York, USA', *Palaeogeography, Palaeoclimatology, Palaeoecology*, 299: 110–28.

49 Stein, Berry, Morris, et al., 'Mid-Devonian *Archaeopteris* roots signal revolutionary change in earliest fossil forests'.

50 Meyer-Berthaud, B., Scheckler, S. E., and Bousquet, J.-L. (2000), 'The development of *Archaeopteris*: new evolutionary characters from the structural analysis of an Early Famennian trunk from southeast Morocco', *American Journal of Botany*, 87: 456–68.

51 Guo, Y., and Wang, D.-M. (2011), 'Anatomical reinvestigation of *Archaeopteris macilenta* from the Upper Devonian (Frasnian) of South China', *Journal of Systematics and Evolution*, 49: 590–97.

52 Morris, J. L., Leake, J.R., Stein, W.E., et al. (2015), 'Investigating Devonian trees as geo-engineers of past climates: linking palaeosols to palaeobotany and experimental geobiology', *Palaeontology*, 58: 787–801.

53 Averill, C., Turner, B.L., and Finzi, A.C. (2014), 'Mycorrhiza-mediated competition between plants and decomposers drives soil carbon storage', *Nature*, 505: 543–5.

54 Morris, Leake, Stein, et al., 'Investigating Devonian trees as geo-engineers of past climates: linking palaeosols to palaeobotany and experimental geobiology'.

55 Algeo, T. J., and Scheckler, S. E. (2010), 'Land plant evolution and weathering rate changes in the Devonian', *Journal of Earth Science*, 21: 75–8.

56 Lenton, Daines and Mills, 'COPSE reloaded: An improved model of biogeochemical cycling over Phanerozoic time'.

57 Isaacson, P. E., Díaz-Martínez, E., Grader, G. W., et al. (2008), 'Late Devonian-earliest Mississippian glaciation in Gondwanaland and its biogeographic consequences', *Palaeogeography, Palaeoclimatology, Palaeoecology*, 268: 126–42.

58 Ghazoul, J., and Sheil, D. (2010), *Tropical Rain Forest Ecology, Diversity, and Conservation*, Oxford: Oxford University Press.

59 Cleal, C. J., Oplustil, S., Thomas, B. A., and Tenchov, Y. (2009), 'Pennsylvanian vegetation and climate in Variscan Euramerica', *Episodes*, 34: 3–12.

60 Ghazoul and Sheil, *Tropical Rain Forest Ecology, Diversity and Conservation*.

61 Thomas, B. A., and Cleal, C. J. (2017), 'Arborescent lycophyte growth in the late Carboniferous coal swamps', *New Phytologist*, 218: 885–90.

62 Wilson, J. P., Montañez, I.P., White, J.D., et al. (2017), 'Dynamic Carboniferous tropical forests: new views of plant function and potential for physiological forcing of climate', *New Phytologist*, 215: 1333–53.

63 Prestianni, C., Rustán, J. J., Balseiro, D., et al. (2015), 'Early seed plants from Western Gondwana: Paleobiological and ecological implications based on Tournaisian (Lower Carboniferous) records from Argentina', *Palaeogeography, Palaeoclimatology, Palaeoecology*, 417: 210–19.

64 Retallack, G. J., and Germanheins, J. (1994), 'Evidence from paleosols for the geological antiquity of rainforest', *Science*, 265: 499–502.

65 Wright, V. P. (2018), 'An early carboniferous humus from South Wales preserved by marine hydromorphic entombment', *Applied Soil Ecology*, 123: 668–71.

66 Lenton, Daines, and Mills, 'COPSE reloaded: An improved model of biogeochemical cycling over Phanerozoic time'.

67 Puttick, Morris and Williams, 'The interrelationships of land plants and the nature of the ancestral embryophyte'.

68 Edwards, D., Kerp, H., Hass, H. (1998), 'Stomata in early land plants: an anatomical and ecophysiological approach', *Journal of Experimental Botany*, 49: 225–78.

69 Duckett, J. G., and Pressel, S. (2018), 'The evolution of the stomatal apparatus: intercellular spaces and sporophyte water relations in bryophytes – two ignored dimensions', *Philosophical Transactions of the Royal Society of London B Series: Biological Sciences*, 373: 20160498.

70 Ruszala, E. M., Beerling, D. J., Franks, P. J., et al. (2011), 'Land plants acquired active stomatal control early in their evolutionary history', *Current Biology*, 21: 1030–35.

71 Garrouste, R., Clément, G., Nel, P., et al. (2012), 'A complete insect from the Late Devonian period', *Nature*, 488: 82–5.

72 Harrison, J. F., Kaiser, A., and VandenBrooks, J. M. (2010), 'Atmospheric oxygen level and the evolution of insect body size', *Proceedings of the Royal Society B Series: Biological Sciences*, 277: 1937–46.

73 Meade, L., Jones, A. S., and Butler, R. J. (2016), 'A revision of tetrapod footprints from the late Carboniferous of the West Midlands, UK', *PeerJ*, 4: e2718 https://doi.org/10.7717/peerj.2718.

74 ibid.

75 ibid.

76 ibid.

Chapter 2: A tropical world

1 Whitmore, T. C. (1998), *An Introduction to Tropical Rainforests* (2nd edition), Oxford: Oxford University Press.

2 Maslin, M. (2005), 'The longevity and resilience of the Amazon rainforest', in Y. Malhi, O. Phillips (eds.), *Tropical Forests & Global Atmospheric Change*, Oxford: Oxford University Press, 167–82.

3 Morley, R. J. (2000), *Origin and Evolution of Tropical Rain Forests*, Chichester: John Wiley and Sons.

4 Tabor, N. J., and Poulsen, C. J. (2008), 'Palaeoclimate across the Late Pennsylvanian–Early Permian tropical palaeolatitudes: A review of climate indicators, their distribution, and relation to palaeophysiographic climate factors', *Palaeogeography, Palaeoclimatology, Palaeoecology*, 268: 293–310.

5 Corlett, R. T., and Primack, R., (2011), *Tropical Rain Forests: An Ecological and Biogeographical Comparison*, London: Wiley-Blackwell.

6 Basset, Y., Cizek, L., Cuénoud, P., et al. (2019), 'Arthropod diversity in a tropical forest', *Science*, 338: 1481–4.

7 Campos-Arceiz, A., and Blake, S. (2011), 'Megagardeners of the forest – the role of elephants in seed dispersal', *Acta Oecologia*, 37: 542–53.

8 DRYFLOR et al. (2015), 'Plant diversity patterns in neotropical dry forests and their conservation implications', *Science*, 353: 1383–7.

9 Walker, R., Lewis, R., Mandimbihasina, A., et al. (2014), 'The conservation of the world's most threatened tortoise: the ploughshare tortoise (Astochelys yniphora) of Madagascar', *Testudo,* 8: 68–75.

10 Cascales-Miñana, B., and Cleal, C. J. (2014), 'The plant fossil record reflects just two great extinction events', *Terra Nova*, 26: 195–200.

11 Barnosky, A. D., Matzke, N., Tomiya, S., et al. (2011), 'Has the Earth's sixth mass extinction already arrived', *Nature*, 471: 51–7.

12 Morley, *Origin and Evolution of Tropical Rain Forests*.

13 Cleal, C. J., Opluštil, S., Thomas, B. A., and Tenchov, Y. (2009), 'Late Moscovian terrestrial biotas and palaeoenvironments of Variscan Euramerica', *Netherlands Journal of Geosciences*, 88: 181–278.

14 Montañez, I. P., Tabor, N. J., Niemeier, D., et al. (2007), 'CO2-forced climate and vegetation instability during Late Paleozoic deglaciation', *Science*, 314: 87–91.

15 Cleal, Opluštil, Thomas and Tenchov, 'Late Moscovian terrestrial biotas and palaeoenvironments of Variscan Euramerica'.

16 Benton, M. J., Tverdokhlebov, V. P., and Surkov, M. V. (2004), 'Ecosystem remodelling among vertebrates at the Permian-Triassic boundary in Russia', *Nature*, 432: 97–100.

17 Cascales-Miñana and Cleal, 'The plant fossil record reflects just two great extinction events'.

18 Linkies, A., Graeber, K., Knight, C., Leubner-Metzger, G. (2010), 'The evolution of seeds', *New Phytologist*, 186: 817–31.

19 Looy, C. V., Brugman, W. A., Dilcher, D. L., and Visscher, H. (1999), 'The delayed resurgence of equatorial forests after the Permian-Triassic ecologic crisis', *Proceedings of the National Academy of Sciences of the United States of America*, 96: 13857–62.

20 Schneebeli-Hermann, E., Hochuli, P. A., Bucher, H., et al. (2012), 'Palynology of the Lower Triassic succession of Tulong, South Tibet – Evidence for early

recovery of gymnosperms', *Palaeogeography, Palaeoclimatology, Palaeoecology*, 339–41: 12–24.

21 Frohlich, M. W., and Chase, M. W. (2007), 'After a dozen years of progress the origin of the angiosperms is still a great mystery', *Nature*, 450: 1184–9.

22 Doyle, J. (2012), 'Molecular and fossil evidence on the origin of angiosperms', *Annual Review of Earth Planetary Science*, 40: 301–26.

23 Morley, R. J. (2011), 'Cretaceous and Tertiary climate change and the past distribution of megathermal rainforests', in M. B. Bush, J. R. Flenley and W. D. Gosling (eds.), *Tropical Rainforest Responses to Climatic Change*, Berlin: Springer-Verlag, 1–34.

24 Silvestro, D., et al. (2021), 'Fossil data support a pre-Cretaceous origin of flowering plants', *Nature Ecology & Evolution*: https://doi.org/10.1038/s41559-020-01387-8.

25 Feild, T. S., Arens, N. C., Doyle, J. A., et al. (2004), 'Dark and disturbed: A new image of early angiosperm ecology', *Paleobiology*, 30: 82–107.

26 Davis, C. C., Webb, C. O., Wurdack, K. J., et al. (2005), 'Explosive radiation supports a mid-Cretaceous origin of modern tropical rain forests', *American Naturalist*, 165: E36–E65.

27 Morley, 'Cretaceous and Tertiary climate change and the past distribution of megathermal rainforests'.

28 Boyce, C. K., and Jung-Eun, L. (2010), 'An exceptional role for flowering plant physiology in the expansion of tropical rainforests and biodiversity', *Proceedings of the Royal Society B Series: Biological Sciences*, 485: 1–7.

29 Gradstein, F. M., Ogg, J. G., Schmitz, M.D., Ogg, G. M. (2020), *The Geologic Time Scale 2020* (1st edition), vols 1 and 2, Amsterdam: Elsevier.

30 Coiro, M., Doyle, J. A., Hilton, J. (2019), 'How deep is the conflict between molecular and fossil evidence on the age of angiosperms?', *New Phytologist*, 223: 83–99.

31 Ghazoul, J. (2016), *Dipterocarp Biology, Ecology, and Conservation*, Oxford: Oxford University Press.

32 Hu, S., Dilcher, D. L., Jarzen, D. M., and Taylor, D. W. (2008), 'Early steps of angiosperm-pollinator coevolution', *Proceedings of the National Academy of Sciences of the United States of America*, 105: 240–45.

33 Duperon-Laudoueneix, M. (1991), 'Importance of fossil woods (conifers and angiosperms) discovered in continental Mesozoic sediments of northern equatorial Africa', *Journal of African Earth Sciences*, 12: 391–6.

34 Wing, S. L., et al. (2009), 'Late Paleocene fossils from the Cerrejón Formation, Colombia, are the earliest record of Neotropical rainforest', *Proceedings of the National Academy of Sciences of the United States of America*, 106: 18627–32.

35 ibid.

36 Head, J. J., et al. (2009), 'Giant boid snake from the Palaeocene neotropics reveals hotter past equatorial temperatures', *Nature*, 457: 715–17.

37 Johnson, K. R., and Ellis, B. (2002), 'A tropical rainforest in Colorado 1.4 Million Years after the Cretaceous–Tertiary boundary', *Science*, 296: 2379–83.

38 Morley, *Origin and Evolution of Tropical Rain Forests*.

39 Morley, R. J. (2003), 'Interplate dispersal routes for megathermal angiosperms', *Perspectives in Plant Ecology, Evolution and Systematics*, 6: 5–20.

40 Morley, 'Cretaceous and Tertiary climate change and the past distribution of megathermal rainforests'.

41 Goldner, A., Herold, N., and Huber, M. (2014), 'Antarctic glaciation caused ocean circulation changes at the Eocene–Oligocene transition', *Nature*, 511: 574–7.

42 Dupont-Nivet, G., Hoorn, C., and Konert, M. (2008), 'Tibetan uplift prior to the Eocene-Oligocene climate transition. Evidence from pollen analysis of the Xining Basin', *Geology*, 36: 987–90.

43 Prasad, V., Strömberg, C. A. E., Alimohammadian, H., Sahni, A. (2005), 'Dinosaur coprolites and the early evolution of grasses and grazers', *Science*, 310: 1177–80.

44 Osborne, C. O. (2008), 'Atmosphere, ecology and evolution: what drove the Miocene expansion of C4 grasslands?', *Journal of Ecology*, 96: 35–45.

45 Metcalfe, S. E., and Nash, D. J. (2012), 'Introduction', in S. E. Metcalfe and D. J. Nash (eds.), *Quaternary Environmental Change in the Tropics*, London: John Wiley & Sons, 1–33.

46 Deplazes, G., Lückge, A., Peterson, L. C., et al. (2013), 'Links between tropical rainfall and North Atlantic climate during the last glacial period', *Nature Geoscience*, 3: 213–17.

47 Hamon, N., Spulchre, P., Donnadieu, Y., et al. (2012), 'Growth of subtropical forests in Miocene Europe: The roles of carbon dioxide and Antarctic ice volume', *Geology*, 40: 567–70.

48 Cerling, T. E., Wang, Y., and Quade, J. (1993), 'Expansion of C4 ecosystems as an indictor of global ecological change in the late Miocene', *Nature*, 361: 344–5.

49 Feakins, S. J., Levin, N. E., Liddy, H. M., et al. (2013), 'Northeast African vegetation change over 12 m.y.', *Geology*, 41: 295–8.

50 Hamon, Spulchre, Donnadieu, et al., 'Growth of subtropical forests in Miocene Europe'.

51 Salzmann, U., Haywood, A. M., Lunt, D. J., et al. (2008), 'A new global biome reconstruction and data-model comparison for the middle Pliocene', *Global Ecology and Biogeography*, 17: 432–47.

52 Martínez-Botí, M. A., Foster, G. L., Chalk, T. B., et al. (2015), 'Plio-Pleistocene climate sensitivity evaluated using high-resolution CO_2 records', *Nature*, 518: 49–54.

53 Bobe, R., and Behrensmeyer, A. K. (2004), 'The expansion of grassland ecosystems in Africa in relation to mammalian evolution and the origin of the genus Homo', *Palaeogeography, Palaeoclimatology, Palaeoecology*, 207: 399–420.

54 Heaney, L. R. (1991), 'A synopsis of climatic and vegetational change in South-east Asia', in *Tropical Forests and Climate. Climatic Change*, 19: 53–61.

55 Dennell R. W., Roebroeks, W. (2005), 'Out of Africa: An Asian perspective on early human dispersal from Africa', *Nature*, 438: 1099–1104.

56 Heaney, 'A synopsis of climatic and vegetational change in Southeast Asia'.

57 Roberts, P. (2019), *Tropical Forests in Prehistory, History, and Modernity*, Oxford: Oxford University Press.

58 Corlett, R. T. (2011), 'Climate change in the tropics: The end of the world as we know it', *Biological Conservation*, 151: 22–5.

59 Hooghiemstra, H., and Van der Hammen, T. (1998), 'Neogene and Quaternary development of the neotropical rain forest: the forest refugia hypothesis, and a literature overview', *Earth-science Reviews*, 44: 147–83.

60 Rabett, R. J. (2012), *Human Adaptation in the Asian Palaeolithic*, Cambridge: Cambridge University Press.

61 Koutavas, A., Lynch-Stieglitz, J., Marchitto, T. M., and Sachs, J. P. (2002), 'El Niño-like pattern in ice age tropical Pacific sea surface temperature', *Science*, 297: 226–30.

62 Pausata, F. S. R., Messori, G., and Zhang, Q. (2016), 'Impacts of dust reduction on the northward expansion of the African monsoon during the Green Sahara period', *Earth and Planetary Science Letters*, 434: 298–307.

Chapter 3: 'Gondwanan' forests and the dinosaurs

1 Turner, A. H., Makovicky, P. J., and Norell, M. A. (2007), 'Feather quill knobs in the dinosaur *Velociraptor*', *Science*, 317: 1721.

2 Brusatte, S. (2018), *The Rise and Fall of the Dinosaurs*, London: Picador.

3 Barrett, P. M., and Rayfield, E. J. (2006), 'Ecological and evolutionary implications of dinosaur feeding behaviour', *Trends in Ecology and Evolution*, 21: 217–24.

4 Hummel, J., Gee, C. T., Südekum, K.-H., et al. (2008), 'In vitro digestibility of fern and gymnosperm foliage: implications for sauropod feeding ecology and diet selection', *Proceedings of the Royal Society B Series: Biological Sciences*, 275: https://doi.org/10.1098/rspb.2007.1728.

5 Colbert, E. H. (1993), 'Feeding strategies and metabolism in elephants and sauropod dinosaurs', *American Journal of Science*, 293A: 1–10.

6 The Paleobiology Database: https://paleobiodb.org/#/.

7 Dunne, E. M., Close, R. A., Button, D. J., et al. (2018), 'Diversity change during the rise of tetrapods and the impact of the "Carboniferous rainforest collapse"', *Proceedings of the Royal Society B Series: Biological Sciences*, 285: https://doi.org/10.1098/rspb.2017.2730.

8 Irmis, R. B., Nesbitt, S. J., Padian, K., et al. (2007), 'A Late Triassic dinosauro-morph assemblage from New Mexico and the rise of the dinosaurs', *Science*, 317: 358–61.

9 Whiteside, J. H., Grogan, D. S., Olsen, P. E., and Kent, D. V. (2011), 'Climati-cally driven biogeographic provinces of Late Triassic tropical Pangea', *Proceedings of the National Academy of Sciences of the United States of America*, 108: 8972–7.

10 Whiteside, J. H., Lindström, S., Irmis, R. B., et al. (2015), 'Extreme ecosystem instability suppressed tropical dinosaur dominance for 30 million years', *Proceedings of the National Academy of Sciences of the United States of America*, 112: 7909–13.

11 Salgado, L., Canudo, J. I., Garrido, A., et al. (2017), 'A new primitive Neornithischian dinosaur from the Jurassic of Patagonia with gut contents', *Nature Scientific Reports*, 7: 42778.

12 Han, F., Forster, C. A., Xu, X., and Clark, J. M. (2017), 'Postcranial anatomy of Yinlong downsi (Dinosauria: Ceratopsia) from the Upper Jurassic Shishu-gou Formation of China and the phylogeny of basal ornithischians', *Journal of Systematic Palaeontology*, 16: 1159–87.

13 van de Schootbrugge, B., Quan, T. M., Lindström, S., et al. (2009), 'Floral changes across the Triassic/Jurassic boundary linked to flood basalt volcan-ism', *Nature Geoscience*, 2: 589–94.

14 Volkheimer, W., Rauhut, O. W. M., Quattrocchio, M. E., and Martínez, M. A. (2008), 'Jurassic paleoclimates in Argentina, a review', *Revista de la Asociación Geológica Argentina*, 63, 549–56.

15 Van Der Meer, D. G., Zeebe, R. E., van Hinsbergen, D. J. J., et al. (2014), 'Plate tectonic controls on atmospheric CO_2 levels since the Triassic', *Proceedings of the National Academy of Sciences of the United States of America*, 111: 4380–85.

16 Yonetani, T., and Gordon, H. B. (2001), 'Simulated changes in the frequency of extremes and regional features of seasonal/annual temperature and precipi-tation when atmospheric CO_2 is doubled', *Journal of Climate*, 14: 1765–79.

17 Upchurch, P., and Barrett, P. M. (2000), 'The evolution of sauropod feeding mechanisms', in H.-D. Sues (ed.), *Evolution of herbivory in terrestrial vertebrates: Perspectives from the fossil record*, Cambridge: Cambridge University Press, 79–122.

18 Hummel, J., and Clauss, M., 'Feeding and digestive physiology', in N. Klein, K. Remes, C. T. Gee, P. M. Sander (eds.), *Biology of the sauropod dinosaurs: Under-standing the life of giants*, Bloomington: Indiana University Press, 11–33.

19 Sander, P. M., Christian, A., Clauss, M., et al. (2011), 'Biology of the sauropod dinosaurs: the evolution of gigantism', *Biological Reviews*, 86: 117–55.

20 Gee, C. T. (2011), 'Dietary options for the Sauropod dinosaurs from an integrated botanical and paleobotanical perspective', in Klein, Remes, Gee and Sander (eds.), *Biology of the sauropod dinosaurs: Understanding the life of giants*, 34–57.

21 ibid.

22 Upchurch, P., and Barrett, P. M. (2005), 'Phylogenetic and taxic perspectives on sauropod diversity', in K. A. Curry Rogers and J. A. Wilson (eds.), *The sauropods. Evolution and paleobiology*, Berkeley: University of California Press, 104–24.

23 Poulsen, J. R., Rosin, C., Meier, A., et al (2018), 'Ecological consequences of forest elephant declines for Afrotropical forests', *Conservation Biology*, 32: 559–67.

24 Mustoe, G. E. (2007), 'Coevolution of cycads and dinosaurs', *The Cycad Newsletter*, 30: 6–9.

25 Leslie, A. (2011), 'Predation and protection in the macroevolutionary history of conifer cones', *Proceedings of the Royal Society B Series: Biological Sciences*, 278: DOI: 10.1098/rspb.2010.2648.

26 Butler, R. J., Barrett, P. M., Kenrick, P., and Penn, M. G. (2009), 'Testing co-evolutionary hypotheses over geological timescales: interactions between Mesozoic non-avian dinosaurs and cycads', *Biological Reviews*, 84: 73–89.

27 Bakker, R. T. (1978), 'Dinosaur feeding behaviour and the origin of flowering plants', *Nature*, 274: 661–3.

28 Weishampel, D. B., and Norman, D. B. (1989), 'Vertebrate herbivory in the Mesozoic; jaws, plants, and evolutionary metrics', *Geological Society of America Special Paper*, 238: 87–101.

29 Barrett, P. M., and Willis, K. J. (2001), 'Did dinosaurs invent flowers? Dinosaur-angiosperm coevolution revisited', *Biological Reviews*, 76: 411–47.

30 ibid.

31 ibid.

32 Weishampel, D. B., Jianu, C.-M. (2000), 'Plant-eaters and ghost lineages: dinosaurian herbivory revisited', in H.-D. Sues (ed.), *Evolution of Herbivory in Terrestrial Vertebrates: Perspectives from the Fossil Record*, Cambridge: Cambridge University Press, 123–43.

33 Erickson, G. M., Krick, B. A., Hamilton, M., et al. (2012), 'Complex dental structure and wear biomechanics in Hadrosaurid dinosaurs', *Science*, 338: 98–101.

34 Molnar, R. E., and Clifford, H. T. (2000), 'Gut contents of a small ankylosaur', *Journal of Vertebrate Palaeontology*, 20: 194–6.

35 Poulsen, Rosin, Meier, et al., 'Ecological consequences of forest elephant declines for Afrotropical forests'.

36 Godefroit, P., Golovneva, L., Shcheptov, S., Garcia, G., Alekseev, P. (2009), 'The last polar dinosaurs: high diversity of latest Cretaceous arctic dinosaurs in Russia', *Naturwussenschaften*, 96: 495–501.

37 Paik, I. S., Kim, H. J., and Huh, M. (2012), 'Dinosaur egg deposits in the Cretaceous Gyeongsang Supergroup, Korea: Diversity and paleobiological implications', *Journal of Asian Earth Sciences*, 56: 135–46.

38 Voeten, D. F. A. E., Cubo, J., de Margerie, E., et al. (2018), 'Wing bone geometry reveals active flight in *Archaeopteryx*', *Nature Communications*, 9: 923: DOI: 10.1038/s41467-018-03296-8.

39 Barrowclough, G. F., Cracraft, J., Klicka, J., and Zink, R. M. (2016), 'How many kinds of birds are there and why does it matter?', *PLOS ONE*, 11: e0166307: DOI: 10.1371/journal.pone.0166307.

40 Dehling, D. M., Peralta, G., Bender, I. M. A., et al. (2020), 'Similar composition of functional roles in Andean seed-dispersal networks, despite high species and interaction turnover', *Ecology, Ecological Society of America*, 101: -03028: DOI: 10.1002/ecy.3028.

41 Gorchov, D. L., Cornejo, F., Ascorra, C., and Jaramillo, M. (1993), 'The role of seed dispersal in the natural regeneration of rain forest after strip-cutting in the Peruvian Amazon', *Vegetatio*, 107: 339–49.

42 David, J. P., Manakadan, R., and Ganesh, T. (2015), 'Frugivory and seed dispersal by birds and mammals in the coastal tropical dry evergreen forests of southern India: A review', *Tropical Ecology*, 56: 41–55.

43 McConkey, K. R., Meehan, H. J., and Drake, D. R. (2004), 'Seed dispersal by Pacific pigeons (*Ducula pacifica*) in Tonga, western Polynesia', *Emu – Austral Ornithology*, 104: 369–76.

44 Bregman, T., Lees, A. C., MacGregor, H. E. A., et al. (2016), 'Using avian functional traits to assess the impact of land-cover change on ecosystem processes linked to resilience in tropical forests', *Proceedings of the Royal Society B Series: Biological Sciences*, 283: 20161289. tttp://dx.doi.org/10.1098/rspb.2016.1289.

Chapter 4: 'Tree-houses' for the first mammals

1 Janis, C. M. (1993), 'Tertiary mammal evolution in the context of changing climates, vegetation, and tectonic events', *Annual Review of Ecology, Evolution, and Systematics*, 14: 467–500.

2 Li, J., Wang, Y., Wang, Y., and Li, C. (2001), 'A new family of primitive mammal from the Mesozoic of western Liaoning, China', *Chinese Science Bulletin*, 46: 782–5; Gore, R. (2020), 'The rise of mammals', *National Geographic*: https://www.nationalgeographic.com/science/prehistoric-world/rise-mammals/.

3 Berg, M. (2016), 'A miniscule model for research', *Lab Animal*, 45: 133: https://doi.org/10.1038/laban.981.

4 Lockyer, C. (1976), 'Body weights of some species of large whales', *Journal du Conseil Permanent International Pour L'exploration de la Mer*, 36: 259–73.

5 Renne, P. R., Sprain, C. J., Richards, et al. (2015), 'State shift in Deccan volcanism at the Cretaceous–Paleogene boundary, possibly induced by impact', *Science*, 350: 76–8.

6 Barnosky, A. D., Matzke, N., Tomiya, S., et al. (2011), 'Has the Earth's sixth mass extinction already arrived?', *Nature*, 471: 51–7.

7 Smith, F. A., Boyer, A. G., Brown, J. H., et al. (2010), 'The evolution of maximum body size of terrestrial mammals', *Science*, 330: 1216–19.

8 Wiejers, J. W. H., Scouten, S., Sluijs, A., et al. (2007), 'Warm arctic continents during the Palaeocene–Eocene thermal maximum', *Earth and Planetary Science Letters*, 261: 230–38.

9 Sarkar, S., Basak, C., Frank, M., et al. (2019), 'Late Eocene onset of the Proto-Antarctic Circumpolar Current', *Scientific Reports*, 9: https://doi.org/10.1038/s41598-019-46253-1.

10 Zachos, J., Pagani, M., Sloan, L., et al. (2001), 'Trends, rhythms, and aberrations in global climate 65 Ma to present', *Science*, 292: 686–93.

11 Retallack, G. (2001), 'Cenozoic expansion of grasslands and climatic cooling', *Journal of Geology*, 109: 407–26.

12 Rogers, C. S., Hone, D. W. E., McNamara, M. E., et al. (2015), 'The Chinese Pompeii? Death and destruction of dinosaurs in the Early Cretaceous of Lujiatun, NE China', *Palaeogeography, Palaeoclimatology, Palaeoecology*, 427: 89–99.

13 Zhang, F., Kearns, S. L., Orr, P. J., et al. (2010), 'Fossilized melanosomes and the colour of Cretaceous dinosaurs and birds', *Nature*, 463: 1075–8.

14 Chen, P.-J., Dong, Z.-M., and Zhen, S.-N. (1998), 'An exceptionally well-preserved theropod dinosaur from the Yixian Formation of China', *Nature*, 391: 147–52.

15 Luo, Z.-X., Yuan, C.-X., Meng, Q.-J., and Ji, Q. (2011), 'A Jurassic eutherian mammal and divergence of marsupials and placentals', *Nature*, 476: 442–5.

16 ibid.

17 Rink, W. J., and Thompson, J. W. (2015), *Encyclopedia of Scientific Dating Methods*, The Netherlands: Springer.

18 Luo, Yuan, Meng and Ji, 'A Jurassic eutherian mammal and divergence of marsupials and placentals'.

19 Maor, R., Dayan, T., Ferguson-Gow, H., and Jones, K. E. (2017), 'Temporal niche expansion in mammals from a nocturnal ancestor after dinosaur extinction', *Nature Ecology & Evolution*, 1: 1889–95.

20 Bhullar, B.-A. S., Manafzadeh, A. R., Miyamae, J. A., et al. (2019), 'Rolling of the jaw is essential for mammalian chewing and tribosphenic molar function', *Nature*, 566: 528–32.

21 Rowe, T. B., Macrini, T. E., and Luo, Z.-X. (2011), 'Fossil evidence on origin of the mammalian brain', *Science*, 332: 955–7.

22 Gill, P. G., Purnell, M. A., Crumpton, N., et al. (2014), 'Dietary specializations and diversity in feeding ecology of the earliest stem mammals', *Nature*, 512: 303–5.

23 dos Reis, M., Inoue, J., Hasegawa, M., et al. (2012), 'Phylogenomic datasets provide both precision and accuracy in estimating the timescale of placental

mammal phylogeny', *Proceedings of the Royal Society B Series: Biological Sciences*, 279: 3491–500.

24 Zheng, X., Bi, S., Wang, X., and Meng, J. (2013), 'A new arboreal haramiyid shows the diversity of crown mammals in the Jurassic period', *Nature*, 500: 199–203.

25 Luo, Z.-X., Ji, Q., Wible, J. R., and Yuan, C.-X. (2003), 'An early Cretaceous tribosphenic mammal and metatherian evolution', *Science*, 302: 1934–40.

26 Ji, Q., Luo, Z.-X., Yuan, C.-X., Wible, J. R., et al. (2002), 'The earliest known eutherian mammal', *Nature*, 416: 816–22.

27 Maor, Dayan, Ferguson-Gow and Jones, 'Temporal niche expansion in mammals from a nocturnal ancestor after dinosaur extinction'.

28 Grossnickle, D. M., Smith, S. M., and Wilson, G. O. (2019), 'Untangling the multiple ecological radiations of early mammals', *Trends in Ecology and Evolution*, 34: 936–49.

29 Luo, Yuan, Meng and Ji, 'A Jurassic eutherian mammal and divergence of marsupials and placentals'.

30 ibid.

31 Shattuck, M. R., and Williams, S. A. (2010), 'Arboreality has allowed for the evolution of increased longevity in mammals', *Proceedings of the National Academy of Sciences of the United States of America*, 107: 4635–9.

32 Meng, Q. J., Ji, Q., Zhang, Y.-G., et al. (2015), 'An arboreal docodont from the Jurassic and mammaliaform ecological diversification', *Science*, 347: 764–8.

33 Ji, Q., Luo, Z.-X., Yuan, C.-X., and Tabrum, A. R. (2006), 'A swimming mammaliaform from the Middle Jurassic and ecomorphological diversification of early mammals', *Science*, 311: 1123–7.

34 Meng, Q -J., Grossnickle, D. M., Liu, D., et al. (2017), 'New gliding mammaliaforms from the Jurassic', *Nature*, 548: 291–6.

35 ibid.

36 Luo, Z.-X., Meng, Q.-J., Grossnickle, D. M., et al. (2017), 'New evidence for mammaliaform ear evolution and feeding adaptation in a Jurassic ecosystem', *Nature*, 548: 326–9.

37 Hu, Y., Meng, J., Wang, Y., and Li, C. (2005), 'Large Mesozoic mammals fed on young dinosaurs', *Nature*, 433: 149–52.

38 Grossnickle, Smith and Wilson, 'Untangling the multiple ecological radiations of early mammals'.

39 Grossnickle, D. M., and Newham, E. (2016), 'Therian mammals experience an ecomorphological radiation during the Late Cretaceous and selective extinction at the K-Pg boundary', *Proceedings of the Royal Society B Series: Biological Sciences*, 283: https://doi.org/10.1098/rspb.2016.0256.

40 Chen, M., Strömberg, C. A. E., and Wilson, G. P. (2019), 'Assembly of modern mammal community structure driven by Late Cretaceous dental evolution, rise of flowering plants, and dinosaur demise', *Proceedings of the National Academy of Sciences of the United States of America*, 116: 9931–40.

41 Sun, G., Ji, Q., Dilcher, D.L., et al. (2002), 'Archaefructaceae, a new basal angiosperm family', *Science*, 296: 899–904.

42 Wilson, G. P., Evans, A. R., Corfe, I. J., et al. (2012), 'Adaptive radiation of multituberculate mammals before the extinction of dinosaurs', *Nature*, 483: 457–60.

43 ibid.

44 Lyson, T. R., Miller, I. M., Bercovici, A. D., et al. (2019), 'Exceptional continental record of biotic recovery after the Cretaceous–Paleogene mass extinction', *Science*, 366: 977–83.

45 ibid.

46 Nichols, D. J., and Johnson, K. R. (2008), *Plants and the K-T Boundary*, Cambridge: Cambridge University Press.

47 Cascales-Miñana, B., and Cleal, C. J. (2014), 'The plant fossil record reflects just two great extinction events', *Terra Nova*, 26: 195–200.

48 Kowalczyk, J. B., Royer, D. L., Miller, I. M., et al. (2018), 'Multiple proxy estimates of atmospheric CO_2 from an early Paleocene rainforest', *Paleoceanography and Paleoclimatology*, 33: 1427–38.

49 Lyson, Miller, Bercovici, et al., 'Exceptional continental record of biotic recovery after the Cretaceous–Paleogene mass extinction'.

50 ibid.

51 Huurdeman, E. P., Frieling, J., Reichgelt, T., et al. (2020), 'Rapid expansion of meso-megathermal rain forests into the southern high latitudes at the onset of the Paleocene–Eocene Thermal Maximum', *Geology*: DOI: 10.1130/G47343.1

52 Janis, C. M. (1989), 'A climatic explanation for patterns of evolutionary diversity in ungulate mammals', *Palaeontology*, 32: 463–81.

53 Prothero, D. R., and Foss, S. E. (eds.) (2007), *The Evolution of Artiodactyls*, Baltimore, Maryland: Johns Hopkins University Press.

54 Gingerich, P. D., ul Haq, M., Zalmout, I. S., et al. (2001), 'Origin of whales from early Artiodactyls: Hands and feet of Eocene Protocetidae from Pakistan', *Science*, 293: 2239–42.

55 Schaal, S., and Ziegler, W. (1993), *Messel: An Insight into the History of Life and of the Earth*, Oxford: Oxford University Press.

56 Collinson, M. E., Manchester, S. R., Wilde, V., and Hayes, P. (2010), 'Fruit and seed floras from exceptionally preserved biotas in the European Paleogene', *Bulletin of Geosciences*, 85: 155–62.

57 Jordano, P. (2000), 'Fruits and frugivory', in M. Fenner (ed.), *The Ecology of Regeneration in Plant Communities*, Wallingford: CAB International, 125–66.

58 Eriksson, O. (2008), 'Evolution of seed size and biotic seed dispersal in angiosperms: paleoecological and neoecological evidence', *International Journal of Plant Sciences*, 169: 863–70.

59 Tiffney, B. H. (1984), 'Seed size, dispersal syndromes, and the rise of the angiosperms: evidence and hypothesis', *Annals of the Missouri Botanical Garden*, 71: 551–76.

60 Kargaranbafghi, F., and Neubauer, F. (2018), 'Tectonic forcing to global cooling and aridification at the Eocene–Oligocene transition in the Iranian plateau', *Global and Planetary Change*, 171: 248–54.

61 Solounias, N., and Semprebon, G. (2002), 'Advances in the reconstruction of ungulate ecomorphology with applications to early fossil equids', *American Museum Novitates*, 3366: 1–49.

62 Semprebon, G. M., Rivals, F., and Janis, C. M. (2019), 'The role of grass vs. exogenous abrasives in the paleodietary patterns of North American ungulates', *Frontiers in Ecology and Evolution*: https://doi.org/10.3389/fevo.2019.00065.

63 ibid.

64 Semprebon, G. M., Rivals, F., Solounias, N., and Hulbert Jr, R. C. (2016), 'Paleodietary reconstruction of fossil horses from the Eocene through Pleistocene of North America', *Palaeogeography, Palaeoclimatology, Palaeoecology*, 442: 110–27.

65 Badlangana, N. L., Adams, J. W., and Manger, P. R. (2009), 'The giraffe (*Giraffa cameloparadlis*) cervical vertebral column: a heuristic example in understanding evolutionary processes?', *Zoological Journal of the Linnean Society*, 155: 736–57.

66 Mitchell, G., and Skinner, J. D. (2003), 'On the origin, evolution and phylogeny of giraffes. *Giraffa camelopardalis*', *Transactions of the Royal Society of South Africa*, 58: 51–73.

67 Dumont, E. R., Dávalaos, L. M., Goldberg, A., et al. (2011), 'Morphological innovation, diversification and invasion of a new adaptive zone', *Proceedings of the Royal Society B Series: Biological Sciences*, 279: https://doi.org/10.1098/rspb.2011.2005.

68 Eriksson, O. (2014), 'Evolution of angiosperm seed disperser mutualisms: the timing of origins and their consequences for coevolutionary interactions between angiosperms and frugivores', *Biological Reviews*, 91: 168–86.

69 Shilton, L. A., Altringham, J. D., Compton, S. G., and Whittaker, R. J. (1999), 'Old World fruit bats can be long-distance seed dispersers through extended retention of viable seeds in the gut', *Proceedings of the Royal Society of London B Series: Biological Sciences*, 266: DOI: https://doi.org/10.1098/rspb.1999.0625.

70 Beard, K. C., Qi, T., Dawson, M. R., et al. (1994), 'A diverse new primate fauna from middle Eocene fissure-fillings in southeastern China', *Nature*, 368: 604–9.

71 Sussman, R. W., Rasmussen, D. T., and Raven, P. H. (2013), 'Rethinking primate origin again', *American Journal of Primatology*, 75: 95–106.

Chapter 5: The leafy cradles of our ancestors

1 Whiten, A., Goodall, J., McGrew, W. C., et al. (1999), 'Cultures in chimpanzees', *Nature*, 399: 682–5.

2 The Chimpanzee Sequencing and Analysis Consortium (2005), 'Initial sequence of the chimpanzee genome and comparison with the human genome', *Nature*, 437: 69–87.

3 Darwin, C. (2004) [1871], *The descent of man, and selection in relation to sex*, London: Penguin.

4 Domínguez-Rodrigo, M. (2014), 'Is the "Savanna Hypothesis" a dead concept for explaining the emergence of the earliest hominins?', *Current Anthropology*, 55(1): 59–81.

5 Dennell, R. W., and Roebroeks, W. (2005), 'Out of Africa: An Asian perspective on early human dispersal from Africa', *Nature*, 438: 1099–1104.

6 Estimated on the basis of the MODIS (Moderate Resolution Imaging Spectroradiometer) Land Cover MCD12Q1 majority land cover type 1, class 2 for 2012 (spatial resolution of 500m). Downloaded from the US Geological Survey Earth Resources Observation System (EROS) Data Center (EDC).

7 Hamon, N., Spulchre, P., Donnadieu, Y., et al. (2012), 'Growth of subtropical forests in Miocene Europe: The roles of carbon dioxide and Antarctic ice volume', *Geology*, 40: 56770.

8 ibid.

9 Nelson, S. (2003), *The extinction of Sivapithecus: faunal and environmental changes in the Siwaliks of Pakistan*, American School of Prehistoric Research Monographs, volume 1, Boston: Brill Academic Publishers.

10 Macchiarelli, R., Bergeret-Medina, A., Marchi, D., and Wood, B. (2020), 'Nature and relationships of *Sahelanthropus tchadensis*', *Journal of Human Evolution*, 149: https://doi.org/10.1016/j.jhevol.2020.102898.

11 White, T., Asfaw, B., Beyene, Y., et al. (2009), '*Ardipithecus ramidus* and the paleobiology of early hominids', *Science*, 326: 75–86.

12 ibid.

13 Haile-Selassie, Y., Suwa, G., and White, T. D. (2004), 'Late Miocene teeth from Middle Awash, Ethiopia, and early hominid dental evolution', *Science*, 303: 1503–5.

14 Prado-Martinez, J., Sudmant, P. H., Kidd, J. M., et al. (2013), 'Great ape genetic diversity and population history', *Nature*, 499: 471–5.

15 White, T. D., Lovejoy, C. O., Asfaw, B., et al. (2015), 'Neither chimpanzee nor human, *Ardipithecus* reveals the surprising ancestry of both', *Proceedings of the National Academy of Sciences of the United States of America*, 112: 4877–84.

16 WoldeGabriel, G., Ambrose, S. H., Barboni, D., et al. (2009), 'The geological, isotopic, botanical, invertebrate, and lower vertebrate surroundings of *Ardipithecus ramidus*', *Science*, 326 (5949): 65–65e5.

17 White, Asfaw, Beyene, et al., '*Ardipithecus ramidus* and the paleobiology of early hominids'.

18 Levin, N. E., Simpson, S. W., Quade, J., et al. (2008), 'Herbivore enamel carbon isotopic composition and the environmental context of *Ardipithecus* at Gona, Ethiopia', in J. Quade and J. G. Wynn (eds.), *The Geology of Early Humans*

in the Horn of Africa, Boulder, Colorado: Geological Society of America Special Paper 446: 215–34.

19 Brunet, M., Guy, F., Pilbeam, D., et al. (2002), 'A new hominid from the Upper Miocene of Chad, Central Africa', *Nature*, 418: 145–51.

20 Pickford, M., Senut, B., Gommery, D., and Treil, J. (2002), 'Bipedalism in *Orrorin tugenensis* revealed by its femora', *Comptes Rendus Palevol* 1 (4): 191–203.

21 Crompton, R. H., Sellers, W. I., and Thorpe, S. K. S. (2010), 'Arboreality, terrestriality and bipedalism', *Philosophical Transactions of the Royal Society B Series*, 365: 3301–14.

22 Elton, S. (2008), 'The environmental context of human evolutionary history in Eurasia and Africa', *Journal of Anatomy*, 212: 377–93.

23 Pusey, A. E., Pintea, L., Wilson, M. L., et al. (2007), 'The contribution of long-term research at Gombe National Park to chimpanzee conservation', *Conservation Biology*, 21: 623–34.

24 Johanson, D. C., and Maitland, A. E. (1981), *Lucy: The Beginning of Humankind*, St Albans: Granada.

25 Latimer, B., and Lovejoy, C. O. (1989), 'The calcaneus of *Australopithecus afarensis* and its implications for the evolution of bipedality', *American Journal of Physical Anthropology*, 78 (3): 369–86.

26 Harcourt-Smith, W. E. H., and Aiello, L. C. (2004), 'Fossils, feet and the evolution of human bipedal locomotion', *Journal of Anatomy*, 204: 403–16.

27 Montgomery, S. (2018), 'Hominin brain evolution: The only way is up', *Current Biology*, 28: R784–R802.

28 Harmand, S., Lewis, J. E., Feibel, C. S., et al. (2015), '3.3-million-year-old stone tools from Lomekwi 3, West Turkana, Kenya', *Nature*, 521: 310–15.

29 Lee-Thorp, J. A., van der Merwe, N. J., and Brain, C. K. (1994), 'Diet of *Anstralopithecus robustus* at Swartkrans from stable carbon isotopic analysis', *Journal of Human Evolution*, 27: 361–72.

30 Sponheimer, M., Alemseged, Z., Cerling, T. E., et al. (2013), 'Isotopic evidence of early hominin diets', *Proceedings of the National Academy of Sciences of the United States of America*, 110 (26): 10513–18.

31 Green, D. J., and Alemseged, Z. (2012), '*Australopithecus afarensis* scapular ontogeny, function, and the role of climbing in human evolution', *Science*, 338: 514–17.

32 Ruff, C. (2009), 'Relative limb strength and locomotion in *Homo habilis*', *American Journal of Physical Anthropology*, 138 (1): 90–100.

33 Sponheimer, M., Passey, B. H., de Ruiter, D. J., et al. (2006), 'Isotopic evidence for dietary variability in the early hominin *Paranthropus robustus*', *Science*, 314: 980–82.

34 Feakins, S. J., Levin, N. E., Liddy, H. M., et al. (2013), 'Northeast African vegetation change over 12 m.y.', *Geology*, 41 (3): 295–8.

35 Bonnefille, R. (2010), 'Cenozoic vegetation, climate changes and hominid evolution in tropical Africa', *Global and Planetary Change*, 72: 390–411.

36 Feakins, Levin, Liddy, et al., 'Northeast African vegetation change over 12 m.y.'.

37 Levin, N. E. (2015), 'Environment and climate of early human evolution', *Annual Review of Earth and Planetary Sciences*, 43: 405–29.

38 Levin, N. E., Brown, F. H., Behrensmeyer, A. K., et al. (2011), 'Paleosol carbonates from the Omo Group: isotopic records of local and regional environmental change in East Africa', *Palaeogeography, Palaeoclimatology, Palaeoecology*, 307: 75–89.

39 Robinson, J. R., Rowan, J., Campisano, J., et al. (2017), 'Late Pliocene environmental change during the transition from Australopithecus to Homo', *Nature Ecology & Evolution*, 1: 0159.

40 White, T. D., WoldeGabriel, G., Asfaw, B., et al. (2006), 'Asa Issie, Aramis and the origin of Australopithecus', *Nature*, 440: 883–9.

41 Saylor, B. Z., Gibert, L., Deino, A., et al. (2019), 'Age and context of mid-Pliocene hominin cranium from Woranso-Mille, Ethiopia', *Nature*, 573: 220–24.

42 Kingston, J. D., and Harrison, T. (2007), 'Isotopic dietary reconstructions of Pliocene herbivores at Laetoli: Implications for early hominin paleoecology', *Palaeogeography, Palaeoclimatology, Palaeoecology*, 243: 272–306.

43 Cerling, T. E., Harris, J. M., Leakey, M. G., et al. (2010), 'Stable carbon and oxygen isotopes in East African mammals: Modern and fossil', in L. Werdelin and W. J. Sanders (eds.), *Cenozoic Mammals of Africa*, London: University of California Press, 941–52.

44 Carbonell, E., Bermúdez de Castro, J. M., Parés, J. M., et al. (2008), 'The first hominin of Europe', *Nature*, 452: 465–9.

45 Lordkipanidze, D., Ponce de León, M. S., Margvelashvili, A., et al. (2013), 'A complete skull from Dmanisi, Georgia, and the evolutionary biology of early *Homo*', *Science*, 342: 326–31.

46 Ashton, N., Lewis, S. G., De Groote, I., et al. (2014), 'Hominin footprints from Early Pleistocene deposits at Happisburgh, UK', *PLOS ONE*, 9 (2): e88329.

47 Scott, G. R., and Gilbert, L. (2009), 'The oldest handaxes in Europe', *Nature*, 461: 82–5.

48 Dennell and Roebroeks, 'Out of Africa: An Asian perspective on early human dispersal from Africa'.

49 Zaim, Y., Ciochon, R. L., Polanski, J. M., et al. (2011), 'New 1.5 million-year-old Homo erectus maxilla from Sangiran (Central Java, Indonesia)', *Journal of Human Evolution*, 61 (4): 363–76.

50 Smith, R. J., and Jungers, W. L. (1997), 'Body mass in comparative primatology', *Journal of Human Evolution*, 32: 523–59.

51 Zhang, Y., and Harrison, T. (2017), '*Gigantopithecus blacki*: a giant ape from the Pleistocene of Asia revisited', *American Journal of Physical Anthropology, Supplement Yearbook of Physical Anthropology*, 162: 153–77.

52 Louys, J., Curnoe, D., and Tong, H. (2007), 'Characteristics of Pleistocene megafauna extinctions in Southeast Asia', *Palaeogeography, Palaeoclimatology, Palaeoecology*, 243: 152–73.

53 Marwick, B. (2009), 'Biogeography of Middle Pleistocene hominins in mainland Southeast Asia: a review of current evidence', *Quaternary International*, 2002: 51–8.

54 Sutikna, T., Tocheri, M. W., Morwood, M. J., et al. (2016), 'Revised stratigraphy and chronology for *Homo floresiensis* at Liang Bua in Indonesia', *Nature*, 532: 366–9.

55 Brumm, A., van den Bergh, G. D., Storey, M., et al. (2016), 'Age and context of the oldest known hominin fossils from Flores', *Nature*, 534: 249–53.

56 Westaway, K. E., Morwood, M. J., Sutikna, T., et al. (2009), '*Homo floresiensis* and the late Pleistocene environments of eastern Indonesia: defining the nature of the relationship', *Quaternary Science Reviews*, 28: 2897–912.

57 Estimated on the basis of the MODIS (Moderate Resolution Imaging Spectroradiometer) Land Cover MCD12Q1 majority land cover type 1, class 2 for 2012 (spatial resolution of 500m). Downloaded from the US Geological Survey Earth Resources Observation System (EROS) Data Center (EDC).

58 Sutikna, T., Tocheri, M. W., Faith, J. T., et al. (2018), 'The spatio-temporal distribution of archaeological and faunal finds at Liang Bua (Flores, Indonesia) in light of the revised chronology for *Homo floresiensis*', *Journal of Human Evolution*, 124: 52–74.

59 Weston, E. M., and Lister, A. M. (2009), 'Insular dwarfism in hippos and a model for brain size reduction in *Homo floresiensis*', *Nature*, 459: 85–8.

60 Rizal, Y., Westaway, K. E., Zaim, Y., et al. (2020), 'Last appearance of *Homo erectus* at Ngandong, Java, 117,000–108,000 years ago', *Nature*, 577: 381–5.

61 Sutikna, Tocheri, Morwood et al., 'Revised stratigraphy and chronology for *Homo floresiensis* at Liang Bua in Indonesia'.

62 Louys, J., and Roberts, P. (2020), 'Environmental drivers of megafauna and hominin extinction in Southeast Asia', *Nature*, 586: 402–6.

63 Potts, R. (2013), 'Hominin evolution in settings of strong environmental variability', *Quaternary Science Reviews*, 73: 1–13.

Chapter 6: On the tropical origins of our species

1 Hublin, J.-J., Ben-Ncer, A., Bailey, S. E., et al. (2017), 'New fossils from Jebel Irhoud, Morocco and the pan-African origin of *Homo sapiens*', *Nature*, 546: 289–92.

2 Stringer, C. (2016), 'The origin and evolution of *Homo sapiens*', *Proceedings of the Royal Society B Series*, 371: https://doi.org/10.1098/rstb.2015.0237.

3 Bailey, R., Head, G., Jenike, M., et al. (1989), 'Hunting and gathering in tropical rain forest: Is it possible?', *American Anthropologist*, 91: 59–82.

4 Curry, A. (2016), ' "Green hell" has long been home for humans', *Science*, 354: 268–9.

5 Bradfield, J., Lombard, M., Reynard, J., and Wurz, S. (2020), 'Further evidence for bow hunting and its implications more than 60,000 years ago: Results of a use-trace analysis of the bone point from Klasies River Main site, South Africa', *Quaternary Science Reviews*, 236: https://doi.org/10.1016/j.quascirev.2020.106295.

6 Tylén, K., Fusaroli, R., Rojo, S., et al. (2020), 'The evolution of early symbolic behaviour in *Homo sapiens*', *Proceedings of the National Academy of Sciences of the United States of America*, 117: 4578–84.

7 Henshilwood, C. S., d'Errico, F., van Niekerk, K. L., et al. (2018), 'An abstract drawing from the 73,000-year-old levels at Blombos Cave, South Africa', *Nature*, 562: 115–18.

8 d'Errico, F., Banks, W. E., Warren, D. L., et al. (2017), 'Identifying early modern human ecological niche expansions and associated cultural dynamics in the South African Middle Stone Age', *Proceedings of the National Academy of Sciences of the United States of America*, 114: 7869–76.

9 Marean, C. W. (2016), 'The transition to foraging for dense and predictable resources and its impact on the evolution of modern humans', *Philosophical Transactions of the Royal Society B Series: Biological Sciences*, 371: https://doi.org/10.1098/rstb.2015.0239.

10 Bird, M., Taylor, D., and Hunt, C. (2005), 'Palaeoenvironments of insular Southeast Asia during the last glacial period: a savanna corridor in Sundaland?' *Quaternary Science Reviews*, 24: 2228–42.

11 Mellars, P. (2006), 'Why did modern human populations disperse from Africa ca. 60,000 years ago? A new model', *Proceedings of the National Academy of Sciences of the United States of America*, 103: 9381–6.

12 Clarkson, C., Jacobs, Z., Marwick, B., et al. (2017), 'Human occupation of northern Australia by 65,000 years ago', *Nature*, 547: 306–10.

13 Ardelean, C. F., Becerra-Valdivia, L., Pedersen, M. W., et al. (2020), 'Evidence of human occupation in Mexico around the Last Glacial Maximum', *Nature*: https://doi.org/10.1038/s41586-020-2509-0.

14 Erlandson, J. M., and Braje, T. J. (2015), 'Coasting out of Africa: The potential of mangrove forests and marine habitats to facilitate human coastal expansion via the Southern Dispersal Route', *Quaternary International*, 382: 31–41.

15 Cann, R. L., Stoneking, M., and Wilson, A. C. (1987), 'Mitochondrial DNA and human evolution', *Nature*, 325: 31–6.

16 Schelebusch, C. M., Skoglund, P., Sjödin, P., et al. (2012), 'Genomic variation in seven Khoe-San groups reveals adaptation and complex African history', *Science*, 338: 374–9.

17 Sankararaman, S., Mallick, S., Dannermann, M., et al. (2014), 'The genomic landscape of Neanderthal ancestry in present-day humans', *Nature*, 507: 354–7.

18 Meyer, M., Kircher, M., Gansauge, M.-T., et al. (2012), 'A high-coverage genome sequence from an archaic Denisovan individual', *Science*, 338: 222–6.

19 Jacobs, G. S., Hudjashov, G., Saag, L., et al. (2019), 'Multiple deeply divergent Denisovan ancestries in Papuans', *Cell*, 177: 1010–21.

20 van de Loosdrecht, M., Bouzouggar, A., Humphrey, L., et al. (2018), 'Pleistocene North African genomes link Near Eastern and sub-Saharan African human populations', *Science*, 360: 548–52.

21 Prendergast, M. E., and Sawchuk, E. (2018), 'Boots on the ground in Africa's ancient DNA "revolution": archaeological perspectives on ethics and best practices', *Antiquity*, 92: 803–15.

22 McDougall, I., Brown, F. H., and Fleagle, J. G. (2005), 'Stratigraphic placement and age of modern humans from Kibish, Ethiopia', *Nature*, 433: 733–6.

23 Clark, J. D., Beyene, Y., WoldeGabriel, G., et al. (2003), 'Stratigraphic, chronological and behavioural contexts of Pleistocene *Homo sapiens* from Middle Awash, Ethiopia', *Nature*, 423: 747–52.

24 Henshilwood, d'Errico, van Niekirk, et al., 'An abstract drawing from the 73,000-year-old levels at Blombos Cave, South Africa'.

25 Potts, R., Behrensmeyer, A. K., Faith, J. T., et al. (2018), 'Environmental dynamics during the onset of the Middle Stone Age in eastern Africa', *Science*: DOI: 10.1126/science.aao2200.

26 Henshilwood, C. S., and Dubreuil, B. (2011), 'The Still Bay and Howiesons Poort, 77–59 ka: Symbolic material culture and the evolution of the mind during the African Middle Stone Age', *Current Anthropology*, 52: 361–400.

27 Barham, L. (2001), 'Central Africa and the emergence of regional identity in the Middle Pleistocene', in L. S. Barham and K. Robson-Brown (eds.), *Human Roots: Africa and Asia in the Middle Pleistocene*, Bristol: Western Academic and Specialist Press, 65–80.

28 Scerri, E., Thomas, M. G., Manica, A., et al. (2018), 'Did our species evolve in subdivided populations across Africa, and why does it matter?', *Trends in Ecology and Evolution*, 33: 582–94.

29 Neubauer, S., Hublin, J.-J., and Gunz, P. (2018), 'The evolution of modern human brain shape', *Science Advances*, 4: DOI: 10.1126/sciadv.aao5961.

30 Harvati, K., Stringer, C., Grün, R., et al. (2011), 'The Later Stone Age calvaria from Iwo Eleru, Nigeria: morphology and chronology', *PLOS One*, 6: e24024: DOI: 10.1371/journal.pone.0024024.

31 Bergström, A., Stringer, C., Hajdinjak, M., et al. (2021), 'Origins of modern human ancestry', *Nature*, 590: 229–37.

32 Scerri, Thomas, Manica, et al., 'Did our species evolve in subdivided populations across Africa, and why does it matter?'

33 Tryon, C. (2019), 'The Middle/Later Stone Age transition and cultural dynamics of late Pleistocene East Africa', *Evolutionary Anthropology*, 28: 267–82.

34 Taylor, N. (2016), 'Across rainforests and woodlands: A systematic reappraisal of the Lupemban Middle Stone Age in Central Africa', in S. C. Jones and B. A.

Stewart (eds.), *Africa from MIS 6-2: Population dynamics and paleoenvironments*, Dordrecht: Springer, 273–99.

35 Barham, 'Central Africa and the emergence of regional identity in the Middle Pleistocene'.

36 Mercader, J. (2002), 'Forest People: The role of African rainforests in human evolution and dispersal', *Evolutionary Anthropology*, 11: 117–24.

37 Taylor, 'Across rainforests and woodlands: A systematic reappraisal of the Lupemban Middle Stone Age in Central Africa'.

38 Perry, G. H., Verdu, P. (2017), 'Genomic perspectives on the history and evolutionary ecology of tropical rainforest occupation by humans', *Quaternary International*, 448: 150–57.

39 ibid.

40 Shipton, C., Roberts, P., Armitage, S., et al. (2018), 'A 78,000-year-old record of tropical adaptation and complex human behaviour in coastal East Africa', *Nature Communications*: https://doi.org/10.1038/s41467-018-04057-3.

41 Blome, M. W., Cohen, A. S., Tryon, C. A., et al. (2012), 'The environmental context for the origins of modern human diversity: A synthesis of regional variability in African climate 150,000–30,000 years ago', *Journal of Human Evolution*, 62: 563–92.

42 Gamble, C. (1993), *Timewalkers: The prehistory of global colonization*, Stroud: Alan Sutton.

43 Wedage, O., Amano, N., Langley, M. C., et al. (2019), 'Specialized rainforest hunting by *Homo sapiens* 45,000 years ago', *Nature Communications*, 10: 739: DOI: 10.1038/s41467-019-08623-1.

44 Roberts, P., Perera, N., Wedage, O., et al. (2015), 'Direct evidence for human reliance on rainforest resources in late Pleistocene Sri Lanka', *Science*, 347: 1246–9.

45 Roberts, P., Perera, N., Wedage, O., et al. (2017), 'Fruits of the forest: Human stable isotope ecology and rainforest adaptations in Late Pleistocene and Holocene (~ 36 to 3 ka) Sri Lanka', *Journal of Human Evolution*, 106: 102–18.

46 Wedage, Amano, Langley, et al., 'Specialized rainforest hunting by *Homo sapiens* 45,000 years ago'.

47 Roberts, P. (2017), 'Forests of plenty: Ethnographic and archaeological rainforests as hotspots of human activity', *Quaternary International*: DOI: 10.1016/j.quaint.2017.03.041.

48 Langley, M. C., Amano, N., Wedage, O., et al. (2020), 'Bows-and-arrows and complex symbolic displays at 48,000 years bp in the South Asian tropics', *Science Advances*, 6: DOI: 10.1126/sciadv.aba3831.

49 Liu, W., Martinón-Torres, M., Cai, Y.-J., et al. (2015), 'The earliest unequivocally modern humans in southern China', *Nature*, 526: 696–700.

50 Westaway, K. E., Louys, J., Awe, R. D., et al. (2017), 'An early modern human presence in Sumatra 73,000–63,000 years ago', *Nature*, 548: 322–5.

51 Barker, G., and Farr, L. (eds.) (2016), *Archaeological Investigations in the Niah Caves, Sarawak. The Archaeology of the Niah Caves, Sarawak*, Volume 2, McDonald Institute for Archaeological Research, Cambridge.

52 O'Connor, S., Ono, R., and Clarkson, C. (2011), 'Pelagic fishing at 42,000 years before the present and the maritime skills of modern humans', *Science*, 334: 1117–21.

53 Roberts, P., Louys, J., Zech, J., et al. (2020), 'Direct evidence for initial coastal colonization and subsequent diversification in the human occupation of Wallacea', *Nature Communications*, 11: https://doi.org/10.1038/s41467-020-15969-4.

54 ibid.

55 Gosden, C. (2010), 'When humans arrived in the New Guinea Highlands', *Science*, 330: 41–2.

56 Summerhayes, G. R., Leavesley, M., Fairbairn, A., et al. (2010), 'Human adaptation and plant use in highland New Guinea 49,000 to 44,000 years ago', *Science*, 330: 78–81.

57 Kershaw, A. P. (1986), 'The last two glacial-interglacial cycles from northeastern Australia: implication for climate change and Aboriginal burning', *Nature*, 322: 47–9.

58 Cosgrove, R., Field, J., and Ferrier, Å. (2007), 'The archaeology of Australia's tropical rainforests', *Palaeogeography, Palaeoclimatology, Palaeoecology*, 251: 150–73.

59 Ardelean, Becerra-Valdivia, Pedersen, et al., 'Evidence of human occupation in Mexico around the Last Glacial Maximum'.

60 Dillehay, T. D., Ocampo, C., Saavedra, J., et al. (2015), 'New archaeological evidence for an early human presence at Monte Verde, Chile', *PLOS ONE*, 10: e0145471.

61 Roosevelt, A. C., da Costa, M. L., Machado, C. L., et al. (1996), 'Paleoindian cave dwellers in the Amazon: The peopling of the Americas', *Science*, 272: 373–84.

62 Rademaker, K., Hodgins, G., Moore, K., et al. (2014), 'Paleoindian settlement of the high-altitude Peruvian Andes', *Science*, 346: 466–9.

63 Nash, D. J., Coulson, S., Staurset, S., et al. (2010), 'Going the distance: Mapping mobility in the Kalahari Desert during the Middle Stone Age through multi-site geochemical provenancing of silcrete artefacts', *Journal of Human Evolution*, 96: 113–33.

64 Ossendorf, G., Groos, A. R., Bromm, T., et al. (2019), 'Middle Stone Age foragers resided in high elevations of the glaciated Bale Mountains, Ethiopia', *Science*, 365: 583–7.

65 Zhang, X. L., Ha, B. B., Wang, S. J., et al. (2018), 'The earliest human occupation of the high-altitude Tibetan Plateau 40 thousand to 30 thousand years ago', *Science*, 362: 1049–51.

66 Pitulko, V. V., Tikhonov, A. N., Pavlova, E. Y., et al. (2016), 'Early human presence in the Arctic: Evidence from 45,000-year-old mammoth remains', *Science*, 351: 260–63.

67 Roberts, P., and Stewart, B. (2018), 'Defining the "generalist-specialist" niche for Pleistocene *Homo sapiens*', *Nature Human Behaviour*: DOI: 10.1038/s41562-018-0394-4.

68 Stringer, C., and Galway-Witham, J. (2018), 'When did modern humans leave Africa?', *Science*, 359: 389–90.

Chapter 7: Farmed forests

1 Boivin, N. L., Zeder, M. A., Fuller, D. Q., et al. (2016), 'Ecological consequences of human niche construction: Examining long-term anthropogenic shaping of global species distributions', *Proceedings of the National Academy of Sciences of the United States of America*, 113: 6388–96.

2 Price, T. D., and Bar-Yosef, O. (2011), 'The origins of agriculture: New data, new ideas', *Current Anthropology*, 52: S163–S174.

3 Rowley-Conwy, P. (2011), 'Westward Ho! The spread of agriculture from Central Europe to the Atlantic', *Current Anthropology*, 52: S431–S451.

4 Denham, T. P. (2018), *Tracing early agriculture in the highlands of New Guinea: Plot, mound and ditch*, Oxford: Routledge.

5 Fuller, D., and Hildebrand, E. (2013), 'Domesticating plants in Africa', in P. Mitchell and P. Lane (eds.), *The Oxford Handbook of African Archaeology*, Oxford: Oxford University Press, 507–25.

6 Iriarte, J., Elliott, S., Maezumi, S. Y., et al. (2020), 'The origins of Amazonian landscapes: Plant cultivation, domestication and the spread of food production in tropical South America', *Quaternary Science Review*, 248: https://doi.org/10.1016/j.quascirev.2020.106582.

7 Roberts, P., Hunt, C., Arroyo-Kalin, M., et al. (2017), 'The deep human prehistory of global tropical forests and its relevance for modern conservation', *Nature Plants*, 3: 17093.

8 Barton, H., and Denham, T. P. (2018), 'Vegecultures and the social-biological transformations of plants and people', *Quaternary International*, 489: 17–25.

9 Allen, J., Gosden, C., and White, J. P. (1989), 'Human Pleistocene Adaptations in the Tropical Island Pacific: Recent Evidence from New Ireland, a Greater Australian Outlier', *Antiquity*, 63: 548–61.

10 Golson, J., Denham, T., Hughes, P., et al. (2017), *Ten Thousand Years of Cultivation at Kuk Swamp in the Highlands of Papua New Guinea* (Terra Australis 46), Canberra: Australian National University Press.

11 Golson, J. (1989), 'The origins and development of New Guinea agriculture', in D. R. Harris and G. C. Hillman (eds.), *Foraging and Farming: The Evolution of Plant Exploitation*, London: Unwin Hyman, 109–36.

12 Haberle, S. G., Lentfer, C., and Denham, T. (2017), 'Palaeoecology', in Golson, Denham, Hughes, et al. (eds.), 145–62.

13 Bulmer, S. (1991), 'Variation and change in stone tools in the highlands of Papua New Guinea: The witness of Wanelek', in A. Pawley (ed.), *Man and a Half: Essays in Pacific Anthropology and Ethnobiology in Honour of Ralph Bulmer*, Auckland: The Polynesian Society, 470–78.

14 Denham, T. P. (2016), 'Revisiting the past: Sue Bulmer's contribution to the archaeology of Papua New Guinea', *Archaeology in Oceania*, 51: 5–10.

15 Denham, T. P., Haberle, S. G., Lentfer, C., et al. (2003), 'Origins of agriculture at Kuk Swamp in the Highlands of New Guinea', *Science*, 301: 189–93.

16 Roberts, P., Gaffney, D., Lee-Thorp, J., and Summerhayes, G. (2017), 'Persistent tropical foraging in the Highlands of Terminal Pleistocene–Holocene New Guinea', *Nature Ecology & Evolution*, 1: 0044: DOI: 10.1038/s41559-016-0044.

17 Gaffney, D., Ford, A., and Summerhayes, S. (2015), 'Crossing the Pleistocene-Holocene transition in the New Guinea Highlands: Evidence from the lithic assemblage of Kiowa rockshelter', *Journal of Anthropological Archaeology*, 39: 223–46.

18 Franklin, L. C. (2013), 'Corn', in A. F. Smith (ed.), *The Oxford Encyclopedia of Food and Drink in America*, 2nd ed., Oxford: Oxford University Press, 551–8.

19 Piperno, D. R., Ranere, A. J., Holst, I., et al. (2009), 'Starch grain and phytolith evidence for early ninth millennium BP maize from the Central Balsas River Valley, Mexico', *Proceedings of the National Academy of Sciences of the United States of America*, 106: 5019–24.

20 Kistler, L., Maezumi, S. Y., de Souza, J. G., et al. (2019), 'Multiproxy evidence highlights a complex evolutionary legacy of maize in South America', *Science*, 362: 1309–13.

21 Meggers, B. J. (1971), *Amazonia: Man and Culture in a Counterfeit Paradise*, Illinois: Harlan Davidson.

22 Olsen, K. M. (2002), 'Population history of *Manihot esculenta* (Euphorbiaceae) inferred from nuclear DNA sequences', *Molecular Ecology*, 11: 901–11.

23 Spooner, D. M., McLean, K., Ramsay, G., et al. (2005), 'A single domestication for potato based on multilocus amplified fragment length polymorphism genotyping', *Proceedings of the National Academy of Sciences of the United States of America*, 102: 14694–9.

24 Clement, C. R., de Cristo-Araújo, M., d'Eeckenbrugge, G. C., et al. (2010), 'Origin and domestication of native Amazonian crops', *Diversity*, 2: 72–106.

25 Scaldaferro, M. A., Barboza, G. E., and Acosta, M. C. (2018), 'Evolutionary history of the chili pepper *Capsicum baccatum* L. (Solanaceae): domestication in South America and natural diversification in the seasonally dry tropical forests', *Biological Journal of the Linnean Society*, 124: 466–78.

26 Zarrillo, S., Gaikwad, N., Lanaud, C., et al. (2018), 'The use and domestication of *Theobroma cacao* during the mid-Holocene in the upper Amazon', *Nature Ecology & Evolution*, 2: 1879–88.

27 Clement, de Cristo-Araújo and d'Eeckenbrugge, 'Origin and domestication of native Amazonian crops'.

28 Iriarte, Elliott, Maezumi, et al., 'The origins of Amazonian landscapes: Plant cultivation, domestication and the spread of food production in tropical South America'.

29 Denham, T., Barton, H., Castillo, C., et al. (2020), 'The domestication syndrome in vegetatively propagated field crops', *Annals of Botany*: https://doi.org/10.1093/aob/mcz212.

30 Thomas, E., Alcazar Caicedo, C., McMichael, C. H., et al. (2015), 'Uncovering spatial patterns in the natural and human history of Brazil nut (*Bertholletia excelsa*) across the Amazon Basin', *Journal of Biogeography*, 42: 1367–82.

31 Clement, C. R., Denevan, W. M., Heckenberger, M. J., et al. (2015), 'The domestication of Amazonia before European Conquest', *Proceedings of the Royal Society B Series: Biological Sciences*, 282: https://doi.org/10.1098/rspb.2015.0813.

32 Clement, C. R., Rodrigues, D. P., Alves-Pereira, A., et al. (2016), 'Crop domestication in the upper Madeira River basin', *Boletim do Museu Paraense Emílio Goeldi. Ciências Humanas*, 11: https://doi.org/10.1590/1981.81222016000100010.

33 Lombardo, U., Iriarte, J., Hilbert, L., et al. (2020), 'Early Holocene crop cultivation and landscape modification in Amazonia', *Nature*, 581: 190–93.

34 Scaldaferro, Barboza and Acosta, 'Evolutionary history of the chili pepper *Capsicum baccatum* L. (Solanaceae): domestication in South America and natural diversification in the seasonally dry tropical forests'.

35 Fuller, D. Q. (2006), 'Agricultural origins and frontiers in South Asia: A working synthesis', *Journal of World Prehistory*, 20: 1–86.

36 Asouti, E., and Fuller, D. Q. (2008), *Trees and Woodlands of South India. Archaeological Perspectives*, Walnut Creek, California: Left Coast Press.

37 Fuller, D. Q., Boivin, N., Hoogervorst, T., and Allaby, R. (2011), 'Across the Indian Ocean: the prehistoric movement of plants and animals', *Antiquity*, 85: 543858.

38 Morrison, K. (2002), 'Historicizing adaptation, adapting to history: forager-traders in South and Southeast Asia', in K. Morrison and L. Junker (eds.), *Forager-Traders in South and Southeast Asia*, Cambridge: Cambridge University Press, 1–20.

39 Yang, Y. D., Liu, L., Chen, X., and Speller, C. F. (2008), 'Wild or domesticated: DNA analysis of ancient water buffalo remains from north China', *Journal of Archaeological Science*, 35: 2778–85.

40 Wang, M.-S., Thakur, M., Peng, M.-S., et al. (2020), '863 genomes reveal the origin and domestication of chicken', *Cell Research*: https://doi.org/10.1038/s41422-020-0349-y.

41 Heckenberger, M., and Neves, E. G. (2009), 'Amazonian archaeology', *Annual Review of Anthropology*, 38: 251–66.

42 Barton, H., and Denham, T. (2011), 'Vegeculture and Social Life in Island Southeast Asia', in G. Barker and M. Janowski (eds.), *Why Cultivate? Anthropological and Archaeological Approaches to Foraging-Farming Transitions in Southeast Asia*, Cambridge: McDonald Institute Monographs, 17–25.

43 Molina, J., Sikora, M., Garud, N., et al. (2011), 'Molecular evidence for a single evolutionary origin of domesticated rice', *Proceedings of the National Academy of Sciences of the United States of America*, 108: 8351–6.

44 Gutaker, R. M., Groen, S. C., Bellis, E. S., et al. (2020), 'Genomic history and ecology of the geographic spread of rice', *Nature Plants*, 6: 492–502.

45 Fuller, D. Q., and Qin, L. (2009), 'Water management and labour in the origins and dispersal of Asian rice', *World Archaeology*, 41: 88–111.

46 Weber, S., Lehman, H., Barela, T., et al. (2010), 'Rice or millets: early farming strategies in prehistoric central Thailand', *Archaeological and Anthropological Sciences*, 2: 79–88.

47 Deng, Z., Hung, H.-C., Carson, M. T., et al. (2020), 'Validating earliest rice farming in the Indonesian Archipelago', *Scientific Reports*, 10: https://doi.org/10.1038/s41598-020-67747-3.

48 Barron, A., Datan, I., Bellwood, P., et al. (2020), 'Sherds as archaeological assemblages: Gua Sireh reconsidered', *Antiquity*, 94: 1325–36.

49 Bellwood, P. (1993), 'Cultural and biological differentiation in Peninsular Malaysia: the last 10,000 years', *Asian Perspectives*, 32: 37–60.

50 Krigbaum, J. (2003), 'Neolithic subsistence patterns in northern Borneo reconstructed with stable carbon isotopes of enamel', *Journal of Anthropological Archaeology*, 22: 292–304.

51 Neumann, K., Bostoen, K., Höhn, A., et al. (2011), 'First farmers in the Central African rainforest: A view from southern Cameroon', *Quaternary International*, 249: 53–62.

52 Garcin, Y., Deschamps, P., Ménot, G., et al. (2018), 'Early anthropogenic impact on Western Central African rainforests 2,600 y ago', *Proceedings of the National Academy of Sciences of the United States of America*, 115: 3261–6.

53 Wotzka, H. P. (2019), 'Ecology and culture of millets in African rainforests: Ancient, historical, and present-day evidence', in B. Eichhorn and A. Höhn (eds.), *Trees, Grasses and Crops. People and Plants in Sub-Saharan Africa and Beyond*, Dr. Rudolf Habelt GmbH, 407–29.

54 Hamilton, A. C., Karamura, D., and Kakudidi, E. (2016), 'History and conservation of wild and cultivated plant diversity in Uganda: Forest species and banana varieties as case studies', *Plant Diversity*, 38 (1): 23–44.

55 Bleasdale, M., Wotzka, H.-P., Eichhorn, B., et al. (2020), 'Isotopic and microbotanical insights into Iron Age agricultural reliance in the Central African rainforest', *Communications Biology*, 3: https://doi.org/10.1038/s42003-020-01324-2.

Chapter 8: Island paradises lost?

1 Russell, J. C., Kueffer, C. (2019), 'Island biodiversity in the Anthropocene', *Annual Review of Environment and Resources*, 44: 31–60.

2 ibid.

3 Blackburn, T. M., Cassey, P., Duncan, R. P., et al. (2004), 'Alien extinction and mammalian introductions on oceanic islands', *Science*, 305: 1955–8.

4 Turvey, S. T., and Cheke, A. S. (2008), 'Dead as a dodo: the fortuitous rise to fame of an extinction icon', *Historical Biology*, 20: 149–63.

5 Martin, P. S., and Steadman, D. W. (1999), 'Prehistoric Extinctions on Islands and Continents', in R. D. E. MacPhee (ed.), *Extinctions in Near Time*, Advances in Vertebrate Paleobiology, volume 2, Boston, MA: Springer.

6 Grayson, D. K. (2001), 'The archaeological record of human impacts on animal populations', *Journal of World Prehistory*, 15: 1–68.

7 Diamond, J. (2005), *Collapse: How Societies Choose to Fail or Survive*, London: Penguin Books.

8 Fordham, D. A., Brook, B. W. (2010), 'Why tropical island endemics are acutely susceptible to global change', *Biodiversity and Conservation*, 19: 329–42.

9 Huebert, J. M., and Allen, M. (2020), 'Anthropogenic forests, arboriculture, and niche construction in the Marquesas Islands (Polynesia)', *Journal of Anthropological Archaeology*, 57: https://doi.org/10.1016/j.jaa.2019.101122.

10 Kirch, P. V. (1982), 'Transported landscapes', *Natural History*, 91: 32–5.

11 Forster, J., Lake, I. R., Watkinson, A. R., and Gill, J. A. (2011), 'Marine biodiversity in the Caribbean UK overseas territories: Perceived threats and constraints to environmental management', *Marine Policy*, 35: 647–57.

12 Napolitano, M. F., DiNapoli, R. J., Stone, J. H., et al. (2019), 'Reevaluating human colonization of the Caribbean using chronometric hygiene and Bayesian modeling', *Science Advances*, 5: https://doi.org/10.1126/sciadv.aar7806.

13 Siegel, P. E., Jones, J. G., Pearsall, D. M., et al. (2015), 'Paleoenvironmental evidence for first human colonization of the eastern Caribbean', *Quaternary Science Reviews*, 129: 275–95.

14 Pagán-Jiménez, J. R., Rodríguez-Ramos, R., Reid, B. A., et al. (2015), 'Early dispersals of maize and other food plants into the southern Caribbean and northeastern South America', *Quaternary Science Reviews*, 123: 231–46.

15 Steadman, D. W., Martin, P. S., MacPhee, R. D. E., et al. (2005), 'Asynchronous extinction of late Quaternary sloths on continents and islands', *Proceedings of the National Academy of Sciences of the United States of America*, 102: 11763–8.

16 Cooke, S. B., Dávalos, L. M., Mychajliw, A. M., et al. (2016), 'Anthropogenic extinction dominates Holocene declines of West Indian mammals', *Annual Review of Ecology, Evolution, and Systematics*, 48: 301–27.

17 Rivera-Collazo, I. C. (2015), 'Por el camino verde: long-term tropical socioecosystem dynamics and the Anthropocene as seen from Puerto Rico', *Holocene*, 25: 1604–11.

18 Fitzpatrick, S. M. (2015), 'The Pre-Columbian Caribbean: Colonization, Population Dispersal, and Island Adaptations', *PaleoAmerica* 1 (4): 305–31.

19 Keegan, W. F. (2006), 'Archaic influences in the origins and development of Taíno societies', *Caribbean Journal of Science*, 42: 1–10.

20 Giovas, C. M., LeFebvre, M. J., and Fitzpatrick, S. M. (2012), 'New records for prehistoric introduction of Neotropical mammals to the West Indies: evidence from Carriacou, Lesser Antilles', *Journal of Biogeography*, 39: 476–87.

21 Fitzpatrick, S. M., and Keegan, W. F. (2007), 'Human impacts and adaptations in the Caribbean Islands: an historical ecology approach', *Earth and Environmental Science Transactions of the Royal Society of Edinburgh*, 98: 29–45.

22 Cooke, Dávalos, Mychajliw, et al., 'Anthropogenic extinction dominates Holocene declines of West Indian mammals'.

23 Newsom, L. A., and Wing, E. S. (2004), *On Land and Sea: Native American Uses of Biological Resources in the West Indies*, Tuscaloosa: University of Alabama Press.

24 Turvey, S. T., Weksler, M., Morris, E. L., and Nokkert, M. (2010), 'Taxonomy, phylogeny, and diversity of the extinct Lesser Antillean rice rats (Sigmodontinae: *Oryzomyini*), with description of a new genus and species', *Zoological Journal of the Linnaean Society*, 160: 748–72.

25 Giovas, C. M., Clark, M., Fitzpatrick, S. M., and Stone, J. (2013), 'Intensifying collection and size increase of the tessellated nerite snail (*Nerita tessellata*) at the Coconut Walk site, Nevis, northern Lesser Antilles, AD 890–1440', *Journal of Archaeological Science*, 40: 4024–38.

26 Bellwood, P. (2005), *First Farmers*, Oxford: Blackwell.

27 Posth, C., Nagele, K., Colleran, H., et al. (2018), 'Language continuity despite population replacement in Remote Oceania', *Nature Ecology & Evolution*, 2: 731–40.

28 Kirch, P. V. (2017), *On the Road of the Winds: An Archaeological History of the Pacific Islands before European Contact*, California: University of California Press.

29 Montenegro, Á., Callaghan, R. T., and Fitzpatrick, S. M. (2016), 'Using seafaring simulations and shortest-hop trajectories to model the prehistoric colonization of Remote Oceania', *Proceedings of the National Academy of Sciences of the United States of America*, 113: 12685–90.

30 Larson, G., Cucchi, T., Fujita, M., et al. (2007), 'Phylogeny and ancient DNA of Sus provides insights into Neolithic expansion in Island Southeast Asia and Oceania', *Proceedings of the National Academy of Sciences of the United States of America*, 104: 4834–9.

31 Nogueira-Filho, S. L. G., Nogueira, S. S. C., and Fragoso, J. M. V. (2009), 'Ecological impacts of feral pigs in the Hawaiian Islands', *Biodiversity and Conservation*, 18: https://doi.org/10.1007/s10531-009-9680-9.

32 Kirch, P. V. (2001), 'Pigs, humans, and tropic competition on small Oceania islands', in A. Anderson and T. Murray (eds.), *Australian Archaeologist: Collected Papers in Honour of Jim Allen*, Canberra: Australian National University Press, 427–39.

33 Kirch, 'Transported landscapes'.

34 Kinaston, R. L., Bedford, S. B., Spriggs, M., et al. (2016), 'Is there a "Lapita diet"? A comparison of Lapita and post-Lapita skeletal samples from four Pacific island archaeological sites', in M. Oxenham and H. Buckley (eds.), *The*

Routledge Handbook of Bioarchaeology in Southeast Asia and the Pacific Islands, London: Routledge, 427–61.

35 Fall, P. L. (2010), 'Pollen evidence for plant introductions in a Polynesian tropical island ecosystem, Kingdom of Tonga', in S. G. Haberle, J. Stevenson and M. Prebble (eds.), *Altered Ecologies: Fire, Climate and Human Influence on Terrestrial Landscapes*, Canberra: Australian National University Press, 253–71.

36 Morrison, A. E., and Hunt, T. L. (2007), 'Human impacts on the nearshore environment: An archaeological case study from Kaua'i, Hawaiian Islands', *Pacific Science*, 61: 325–45.

37 Steadman, D. W. (2006), *Extinction and Biogeography of Tropical Pacific Birds*, Chicago: University of Chicago Press.

38 Hunt, T. L. (2007), 'Rethinking Easter Island's ecological catastrophe', *Journal of Archaeological Science*, 34: 485–502.

39 Tromp, M., Matisoo-Smith, E., Kinaston, R., et al. (2020), 'Exploitation and utilization of tropical rainforests indicated in dental calculus of ancient Oceanic Lapita culture colonists', *Nature Human Behaviour*: https://doi.org/10.1038/s41562-019-0808-y.

40 ibid.

41 Maxwell, J. J., Howarth, J. D., Vandergoes, M. J., et al. (2016), 'The timing and importance of arboriculture and agroforestry in a temperate East Polynesia Society, the Moriori, Rekohu (Chatham Island)', *Quaternary Science Reviews*, 149: 306–25.

42 Lambrides, A. B. J., Weisler, M. I. (2016), 'Pacific Islands ichthyoarchaeology: Implications for the development of prehistoric fishing studies and global sustainability', *Journal of Archaeological Research*, 24: 275–324.

43 Giovas, C. (2006), 'No pig atoll: island biogeography and the extirpation of a Polynesian domesticate', *Asian Perspectives*, 45: 69–95.

44 Matisoo-Smith, E. (2007), 'Animal translocations, genetic variation, and the human settlement of the Pacific', in J. S. Friedlaender (ed.), *Genes, Language and Culture History in the Southwest Pacific*, Oxford: Oxford University Press, 157–70.

45 Ladefoged, T. N., McCoy, M. D., Asner, G. P., et al. (2011), 'Agricultural potential and actualized development in Hawai'i: an airborne LiDAR survey of the leeward Kohala field system (Hawai'i Island)', *Journal of Archaeological Science*, 38: 3605–19.

46 Kirch, P. V. (2010), *How Chiefs Became Kings: Divine Kingship and the Rise of Archaic States in Ancient Hawai'i*, Berkeley, CA: University of California Press.

47 Kirch, P., and Yen, D. (1982), *Tikopia: Prehistory and Ecology of a Polynesian Outlier*, Honolulu: Bernice P. Bishop Museum Bulletin, 238.

48 Ladefoged, T. N., Stevenson, C. M., Haoa, S., et al. (2010), 'Soil nutrient analysis of Rapa Nui gardening', *Archaeology in Oceania*, 45: 80–85.

49 Rainbird, P. (2002), 'A message for our future? The Rapa Nui (Easter Island) ecodisaster and Pacific Island environments', *World Archaeology*, 33: 436–51.

50 Douglass, K., and Zinke, J. (2015), 'Forging ahead by land and by sea: Archae-
    ology and Paleoclimate reconstruction in Madagascar', *African Archaeological
    Review*, 32: 267–99.

51 Hansford, J., Wright, P. C., Rasoamiaramanana, A., et al. (2018), 'Early Holo-
    cene human presence in Madagascar evidenced by exploitation of avian
    megafauna', *Science Advances*, 4: eaat6925.

52 Anderson, A., Clark, G., Haberle, S., et al. (2018), 'New evidence of mega-
    faunal bone damage indicates late colonization of Madagascar', *PLOS ONE*,
    13: e0204368.

53 Douglass, K., Hixon, S., Wright, H. T., et al. (2019), 'A critical review of
    radiocarbon dates clarifies the human settlement of Madagascar', *Quaternary
    Science Reviews*, 221: 105878.

54 Burney, D. A., Burney, L. P., Godfrey, L. R., et al. (2004), 'A chronology for
    late prehistoric Madagascar', *Journal of Human Evolution*, 47: 25–63.

55 Crowley, B. E. (2010), 'A refined chronology of prehistoric Madagascar and the
    demise of the megafauna', *Quaternary Science Reviews*, 29: 2591–603.

56 Martin, P. S. (1984), 'Prehistoric overkill: The global model', in P. S. Martin
    and R. G. Klein (eds.), *Quaternary extinctions: A prehistoric revolution*, Tucson:
    University of Arizona Press, 354–403.

57 Godfrey, L. R., Scroxton, N., Crowley, B. E., et al. (2019), 'A new interpretation
    of Madagascar's megafaunal decline: The "Subsistence Shift Hypothesis"', *Jour-
    nal of Human Evolution*, 130: 126–40.

58 Douglass, K., Antonites, A. R., Quintana Morales, E. M., et al. (2018), 'Multi-
    analytical approach to zooarchaeological assemblages elucidates Late Holocene
    coastal lifeways in southwest Madagascar', *Quaternary International*, 471: 111–31.

59 Burney, Burney, Godfrey, et al., 'A chronology for late prehistoric Madagascar'.

60 Crowley, 'A refined chronology of prehistoric Madagascar and the demise of
    the megafauna'.

61 Li, H., Sinha, A., André, A. A., et al. (2020), 'A multimillennial climatic con-
    text for the megafaunal extinctions in Madagascar and Mascarene Islands',
    Science Advances, 6: eabb2459.

62 Godfrey, Scroxton, Crowley, et al., 'A new interpretation of Madagascar's
    megafaunal decline: The "Subsistence Shift Hypothesis"'.

63 Crowther, A., Lucas, L., Helm, R., et al. (2016), 'Ancient crops provide first
    archaeological signature of the westward Austronesian expansion', *Proceedings
    of the National Academy of Sciences of the United States of America*, 113: 6635–40.

64 Radimilahy, C. M., and Crossland, Z. (2015), 'Situating Madagascar: Indian
    Ocean dynamics and archaeological histories', *Azania: Archaeological Research
    in Africa*, 50: 495–518.

65 Godfrey, Scroxton, Crowley, et al., 'A new interpretation of Madagascar's
    megafaunal decline: The "Subsistence Shift Hypothesis"'.

66 Dewar, R. E., and Richard, A. F. (2012), 'Madagascar: A history of arrivals, what
    happened, and will happen next', *Annual Review of Anthropology*, 41: 495–517.

67 Crowther, Lucas, Helm, et al., 'Ancient crops provide first archaeological sig-
   nature of the westward Austronesian expansion'.

68 Schwitzer, C., Mittermeier, R. A., Johnson, S. E., et al. (2014), 'Averting lemur
   extinctions amid Madagascar's political crisis', *Science*, 343: 842–3.

69 Kaufmann, J. C. (2004), 'Prickly pear cactus and pastoralism in southwest
   Madagascar', *Ethnology*, 43: 345e361.

70 Galván, B., Hernández, C. M., Alberto, V., et al. (1999), 'Poblamiento pre-
   histórico en la costa de Buena Vista del Norte (Tenerife). El conjunto
   arqueológico Fuente-Arena', *Investigaciones Arqueológicas en Canarias*, 6:
   9–257.

71 Morales, J., Rodríguez, A., Alberto, V., et al. (2009), 'The impact of human
   activities on the natural environment of the Canary Islands (Spain) during the
   pre-Hispanic stage (3rd–2nd century BC to 15th century AD): an overview',
   Environmental Archaeology, 14: 27–36.

72 Rando, J. C., and Perera, M. A. (1994), 'Primeros datos de ornitofagia entre los
   aborígenes de Fuerteventura (Islas Canarias)', *Archeofauna*, 3: 13–19.

73 Nogué, S., de Nascimento, L., Fernández-Palacios, J. M., et al. (2013), 'The
   ancient forests of La Gomera, Canary Islands, and their sensitivity to environmental change', *Journal of Ecology*, 101: 368–77.

74 Machado, C. (1995), 'Approche paléoécologique et ethnobotanique du site
   archéologique "El Tendal" (N-E de l'Ile de La Palma, Archipel des Canaries)',
   in CTHS (eds.), *L'Homme Préhistorique et la Mer*, 120 Congrés, Aix-en-Provence:
   CTHS, 179–86.

75 Gangoso, L., Donázar, J. A., Scholz, S., et al. (2006), 'Contradiction in conservation of island ecosystems: Plants, introduced herbivores and avian scavengers
   in the Canary Islands', *Biodiversity and Conservation*, 15: https://doi.org/10.1007/
   s10531-004-7181-4.

76 Crosby, A. W. (1984), 'An ecohistory of the Canary Islands: a precursor of
   European colonization in the New World and Australasia', 8: 214–35.

77 Boivin, N. L., Zeder, M. A., Fuller, D. Q., et al. (2016), 'Ecological consequences of human niche construction: Examining long-term anthropogenic
   shaping of global species distributions', *Proceedings of the National Academy of
   Sciences of the United States of America*, 113: 6388–96.

Chapter 9: Cities in the 'jungle'

1 Diamond, J. (2005), *Collapse: How Societies Choose to Fail or Survive*, London:
   Penguin Books.

2 Meggers, B. J. (1971), *Amazonia: Man and Culture in a Counterfeit Paradise*, Illinois: Harlan Davidson.

3 Webster, D. (2002), *The Fall of the Ancient Maya*, London: Thames & Hudson.

4 Bacus, E. A., and Lucero, L. J. (eds.) (1999), *Complex Polities in the Ancient Tropical World*, Archaeological Papers of the American Anthropological Association 9, Arlington, Virginia; American Anthropological Association.

5 Algaze, G. (2018), 'Entropic cities: The paradox of urbanism in ancient Mesopotamia', *Current Anthropology*, 59: 23–54.

6 Wengrow, D. (2006), *The Archaeology of Early Egypt: Social Transformations in North-East Africa, c. 10,000 to 2,650 BC*, Cambridge: Cambridge University Press.

7 Flad, R. (2018), 'Urbanism as technology in early China', *Archaeological Research in Asia*, 14: 121–34.

8 Childe, V. G. (1950), 'The Urban Revolution', *Town Planning Review*, 21: 3–17.

9 Postgate, J. N. (1992), *Early Mesopotamia: Society and Economy at the Dawn of History*, London and New York: Routledge.

10 Webster, *The Fall of the Ancient Maya*.

11 Martin, S., and Grube, N. (2008), *Chronicle of the Maya kings and queens: deciphering the dynasties of the ancient Maya*, London: Thames & Hudson.

12 Met Office (2020), Greenwich Park climate station, https://www.metoffice.gov.uk/research/climate/maps-and-data/uk-climate-averages/u10hb54gm.

13 Scarborough, V. L. (1993), 'Water management in the southern Maya lowlands: an accretive model for the engineered landscape', *Research in Economic Anthropology*, 7: 17–69.

14 Beach, T., Dunning, N., Luzzadder-Beach, S., et al. (2006), 'Impacts of the ancient Maya on soils and soil erosion in the central Maya Lowlands', *Caterna*, 65: 166–78.

15 Webster, *The Fall of the Ancient Maya*.

16 Fletcher, R. (2012), 'Low-density, agrarian-based urbanism: scale, power and ecology', in M. E. Smith (ed.), *The comparative archaeology of complex societies*, Cambridge: Cambridge University Press, 285–320.

17 Webster, D., Murtha, T., Straight, K. D., et al. (2007), 'The Great Tikal Earthworks Revisited', *Journal of Field Archaeology*, 32: 41–64.

18 Chase, A. F., Chase, D. Z., Weishampel, J. F., et al. (2011), 'Airborne LiDAR, archaeology, and the ancient Maya landscape at Caracol, Belize', *Journal of Archaeological Science*, 38: 387–98.

19 Ross, N. J. (2011), 'Modern tree species composition reflects ancient Maya "forest gardens" in northwest Belize', *Ecological Applications*, 21: 75–84.

20 Ford, A., and Nigh, R. (2015), *The Maya Forest Garden: Eight Millennia of Sustainable Cultivation of the Tropical Woodlands*, London: Routledge.

21 Lucero, L. J. (1999), 'Water control and Maya politics in the southern Maya lowlands', in E. A. Bacus and L. J. Lucero (eds.), *Complex Polities in the Ancient Tropical World* (Archaeological Papers of the American Anthropological Association 9), Arlington; Virginia: American Anthropological Association, 34–49.

22 Medina-Elizalde, M., and Rohling, E. J. (2012), 'Collapse of Classic Maya civilization related to modest reduction in precipitation', *Science*, 335: 956–9.

23 McNeil, C. L., Burney, D. A., and Burney, L. P. (2010), 'Evidence disputing deforestation as the cause for the collapse of the ancient Maya polity of Copan, Honduras', *Proceedings of the National Academy of Sciences of the United States of America*, 107: 1017–22.

24 Thompson, K. M., Hood, A., Cavallaro, D., and Lentz, D. L. (2015), 'Connecting contemporary ecology and ethnobotany to ancient plant use practices of the Maya at Tikal', in D. L. Lentz, N. Dunning and V. Scarborough (eds.), *Tikal: Paleoecology of an Ancient Maya City*, Cambridge: Cambridge University Press, 124–51.

25 Cook, B. I., Anchukaitis, K. J., Kaplan, J. O., et al. (2012), 'Pre-Columbian deforestation as an amplifier of drought in Mesoamerica', *Geophysical Research Letters*, 39: L16706.

26 Douglas, P. M. J., Demarest, A. A., Brenner, M., and Canuto, M. A. (2016), 'Impacts of climate change on the collapse of lowland Maya civilization', *Annual Review of Earth and Planetary Sciences*, 44: 613–45.

27 Hoggarth, J. A., Breitenbach, S. F. M., Culleton, B. J., et al. (2016), 'The political collapse of Chichén Itzá in climatic and cultural context', *Global and Planetary Change*, 138: 25–42.

28 Masson, M. A., Peraza Lope, C. (2014), *Kukulkan's Realm: Urban Life at Ancient Mayapán*, Boulder: University Press of Colorado.

29 Chase, D. Z., and Chase, A. F. (2006), 'Framing the Maya Collapse: Continuity, discontinuity, method, and practice in the Classic to Postclassic southern Maya lowlands', in G. M. Schwartz and J. J. Nichols (eds.), *After Collapse: The Regeneration of Complex Societies*, Tucson: University of Arizona Press, 168–87.

30 Coe, M. D. (1999), *The Maya*, London: Thames & Hudson.

31 Ford and Nigh, *The Maya Forest Garden: Eight Millennia of Sustainable Cultivation of the Tropical Woodlands*.

32 Aimers, J. J. (2007), 'What Maya collapse? Terminal Classic variation in the Maya lowlands', *Journal of Archaeological Research*, 15: 329–77.

33 O'Reilly, D. J. W. (2007), *Early Civilizations of Southeast Asia*, New York: AltaMira Press.

34 Pottier, C. (2012), 'Beyond the temples: Angkor and its territory', in A. Haendel (ed.), *Old Myths and New Approaches: Interpreting Ancient Religious Sites in Southeast Asia*, Clayton, Melbourne, Australia: Monash University Publishing, 12–27.

35 Evans, D., Pottier, C., Fletcher, R., et al. (2007), 'A comprehensive archaeological map of the world's largest preindustrial settlement complex at Angkor, Cambodia', *Proceedings of the National Academy of Sciences of the United States of America*, 104: 14277–82.

36 Evans, D. H., Fletcher, R. J., Pottier, C., et al. (2013), 'Uncovering archaeological landscapes at Angkor using lidar', *Proceedings of the National Academy of Sciences of the United States of America*, 110: 12595–600.

37 Higham, C. (2001), *The Civilization of Angkor*, London: Weidenfeld & Nicolson.

38 Fletcher, R., Penny, D., Evans, D., et al. and Authority for the Protection and Management of Angkor and the Region of Siem Reap (APSARA) Department of Monuments and Archaeology Team (2008), 'The water management network of Angkor, Cambodia', *Antiquity*, 82: 658–70.

39 Buckley, B. M., Anchukaitis, K. J., Penny, D., et al. (2010), 'Climate as a contributing factor in the demise of Angkor, Cambodia', *Proceedings of the National Academy of Sciences of the United States of America*, 107: 6748–52.

40 Lucero, L. J., Fletcher, R., and Coningham, R. (2015), 'From "collapse" to urban diaspora: the transformation of low-density, dispersed agrarian urbanism', *Antiquity*, 89: 1139–54.

41 Bandaranayake, S. (2003), 'The Pre-Modern City in Sri Lanka: The "First" and "Second" Urbanisation', in P. J. J. Sinclair (ed.), *The Development of Urbanism from a Global Perspective*, Uppsala: Uppsala University.

42 Coningham, R. A. E. (1999), *Anuradhapura: The British-Sri Lankan Excavations at Anuradhapura Salgaha Watta 2*, British Archaeological Reports, International Series 824, Oxford: Archaeopress Press.

43 BBC NEWS (2020), 'Hagia Sophia: Former Istanbul museum welcomes Muslim worshippers': https://www.bbc.com/news/world-europe-53506445.

44 Coningham, R. A. E, Gunawardhana, P., Manuel, M., et al. (2007), 'The state of theocracy: Defining an Early Medieval Hinterland in Sri Lanka', *Antiquity*, 81: 699–719.

45 Mail, D. (2017), 'The official boundaries of the city of Liverpool are far too small – and it matters', CityMonitor: https://citymonitor.ai/politics/official-boundaries-city-liverpool-are-far-too-small-and-it-matters-3319.

46 Lucero, Fletcher and Coningham, 'From "collapse" to urban diaspora: the transformation of low-density, dispersed agrarian urbanism'.

47 Gilliland, K., Simpson, I. A., Adderley, W. P., et al. (2013), 'The dry tank development and disuse of water management infrastructure in the Anuradhapura hinterland, Sri Lanka', *Journal of Archaeological Science*, 40: 1012–28.

48 Meggers, B. (1954), 'Environmental limitations of the development of culture', *American Anthropologist*, 56: 801–24.

49 Heckenberger, M. J., Kuikuro, A., Tabata Kuikuro, U. T., et al. (2003), 'Amazonia 1492: Pristine Forest or Cultural Parkland?', *Science*, 301: 1710–14.

50 Heckenberger, M., and Neves, E. G. (2009), 'Amazonian archaeology', *Annual Review of Anthropology*, 38: 251–66.

51 Roosevelt, A. (1999), 'The Development of Prehistoric Complex Societies: Amazonia, a Tropical Forest', in E. A. Bacus, L. J. Lucero and J. Allen (eds.), *Complex Polities in the Ancient Tropical World*, Arlington: American Anthropological Association, 13–34.

52 Hermenegildo, T., O'Connell, T. C., Guapindaia, V. L. C., and Neves, E. G. (2017), 'New evidence for subsistence strategies of late pre-colonial societies of the mouth of the Amazon based on carbon and nitrogen isotopic data', *Quaternary International*, 448: 139–49.

53 Roosevelt, 'The Development of Prehistoric Complex Societies: Amazonia, a Tropical Forest'.

54 Koch, A., Brierley, C., Maslin, M. M., and Lewis, S. L. (2019), 'Earth system impacts of the European arrival and Great Dying in the Americas after 1492', *Quaternary Science Reviews*, 207: 13–36.

55 Rostain, S. (2014), *Islands in the Rainforest: Landscape Management in Pre-Columbian Amazonia*, London: Routledge.

56 de Souza, J. G., Schaan, D. P., Robinson, M., et al. (2018), 'Pre-Columbian earth-builders settled along the entire southern rim of the Amazon', *Nature Communications*, 9: 1125: https://doi.org/10.1038/s41467-018-03510-7.

57 Schmidt, M. J., Py-Daniel, A. R., de Paula Moraes, C., et al. (2014), 'Dark earths and the human built landscape in Amazonia: a widespread pattern of anthrosol formation', *Journal of Archaeological Science*, 42: 152–65.

58 Roosevelt, 'The Development of Prehistoric Complex Societies: Amazonia, a Tropical Forest'.

59 Scarborough, V. L., and Lucero, L. (2010), 'The non-hierarchical development of complexity in the semi-tropics: water and cooperation', *Water History*, 2: 185–205.

60 McIntosh, S. K. (1999), 'Pathways to complexity: an African perspective', in S. K. McIntosh (ed.), *Beyond Chiefdoms: Pathways to Complexity in Africa*, Cambridge: Cambridge University Press, 1–30.

61 Carson, J. F., Whitney, B. S., Mayle, F. E., et al. (2014), 'Environmental impact of geometric earthwork construction in pre-Columbian Amazonia', *Proceedings of the National Academy of Sciences of the United States of America*, 111: 10497–502.

62 Piperno, D. R., McMichael, C., and Bush, M. B. (2017), 'Further evidence for localized, short-term anthropogenic forest alterations across pre-Columbian Amazonia', *Proceedings of the National Academy of Sciences of the United States of America*, 114: E4118–E4119.

63 Lucero, L. J., and Gonzalez Cruz, J. (2020), 'Reconceptualizing urbanism: Insights from Maya cosmology', *Frontiers in Sustainable Cities*, 2: DOI: 10.3389/frsc.2020.00001.

64 Simon, D., and Adam-Bradford, A. (2016), 'Archaeology and contemporary dynamics for more sustainable, resilient cities in the peri-urban interface', *Water Science and Technology*, 72: 57–83.

65 Miller, S. W. (2007), *An Environmental History of Latin America*, Cambridge: Cambridge University Press.

Chapter 10: Europe and the tropics in the 'Age of Exploration'

1 https://apnews.com/article/mexico-mexico-city-columbus-day-60bdc08a7606641a4825d467d97c5f6c.

2 Mineo, L. (2020), 'A day of reckoning', *Harvard Gazette*: https://news.harvard. edu/gazette/story/2020/10/pondering-putting-an-end-to-columbus-day-and-a-look-at-what-could-follow/.

3 Elliott, J. H. (2006), *Empires of the Atlantic World: Britain and Spain in America, 1492–1830*, Yale: Yale University Press.

4 Steinberg, P. E. (1999), 'Lines of division, lines of connection: Stewardship in the World Ocean', *Geographical Review*, 89: 254–64.

5 Jackson, P. (2000), 'The Mongol Empire, 1986–1999', *Journal of Medieval History*, 26: 189–210.

6 O'Rourke, K. H., Williamson, J. G. (2009), 'Did Vasco da Gama matter for European markets?', *Economic History Review*, 62: 655–84.

7 Nichols, D. L., and Rodríguez-Alegría, E. (2017), *The Oxford Handbook of the Aztecs*, Oxford: Oxford University Press.

8 Bridges, E. J. (2016), 'Vijayanagara Empire', *The Encyclopaedia of Empire*: https:// doi.org/10.1002/9781118455074.wbeoe424.

9 Cissé, M., McIntosh, S. K., Dussubieux, L., et al. (2013), 'Excavations at Gao Saney: New evidence for settlement growth, trade, and interaction on the Niger Bend in the first millennium CE', *Journal of African Archaeology*, 11: 9–37.

10 Koch, A., Brierley, C., Maslin, M. M., and Lewis, S. L. (2019), 'Earth system impacts of the European arrival and Great Dying in the Americas after 1492', *Quaternary Science Reviews*, 207: 13–36.

11 Hofman, C., Mol, A., Hoogland, M., and Rojas, R. V. (2014), 'Stage of encounters: migration, mobility and interaction in the pre-colonial and early colonial Caribbean', *World Archaeology*, 46: 590–609.

12 Mitchell, P. (2005), *African Connections: Archaeological Perspectives on Africa and the Wider World*, Lanham, Maryland: AltaMira Press.

13 Junker, L. L. (1999), *Raiding, Trading, and Feasting: The Political Economy of Philippine Chiefdoms*, Honolulu: University of Hawaii Press.

14 Wolf, E. (1982), *Europe and the People Without History*, Berkeley: University of California Press.

15 Crosby, A. W. (1972), *The Columbian Exchange: Biological and Cultural Consequences of 1492*, Westport, CT: Greenwood Press.

16 Cañizares-Esguerra, J. (2006), *Puritan Conquistadors: Iberianizing the Atlantic 1550–1700*, Stanford: Stanford University Press.

17 Cavanagh, E. (2016), 'Corporations and business associations from the commercial revolution to the Age of Discovery: Trade, jurisdiction and the state, 1200–1600', *History Compass*, 14: 493–510.

18 Curry, A. (2016), ' "Green hell" has long been home for humans', *Science*, 354: 268–9.

19 Roberts, P. (2017), ' "Forests of Plenty": Ethnographic and archaeological rainforests as hotspots of human activity', *Quaternary International*: DOI: 10.1016/j. quaint.2017.03.041.

20 Ramsey, J. F. (1973), *Spain: The Rise of the First World Power*, Alabama: University of Alabama Press.

21 Schwarz, G. R. (2008), 'The Iberian caravel: Tracing the development of a ship of discovery', in F. Vieira de Castro and K. Custer (eds.), *Edge of Empire*, Casal de Cambra (Portugal): Caleidoscópio – Ediċao e Artes Gráficas, 23–42.

22 Hemming, J. (2004), *The Conquest of the Incas* (2nd edition), New York: Harvest.

23 Livi Bacci, M. (2003), 'Return to Hispaniola: Reassessing a demographic catastrophe', *Hispanic American Historical Review*, 83: 3–51.

24 Vågene, Å. J., Herbig, A., Campana, M. G., et al. (2018), '*Salmonella enterica* genomes from victims of a major sixteenth-century epidemic in Mexico', *Nature Ecology & Evolution*, 2: 520–28.

25 McNeill, W. (1977), *Plagues and Peoples*, London: Anchor.

26 Koch, Brierley, Maslin and Lewis, 'Earth system impacts of the European arrival and Great Dying in the Americas after 1492'.

27 Nichols, D. L. (2018), *Agricultural practices and environmental impacts of Aztec and Pre-Aztec Central Mexico*, Oxford: Oxford Research Encyclopaedias: DOI: 10.1093/acrefore/9780199389414.013.175.

28 Weaver, M. P. (2019), *The Aztecs, Maya, and their Predecessors: Archaeology of Mesoamerica* (3rd edition), London: Routledge.

29 Daniel, D. A. (1992), 'Tactical factors in the Spanish conquest of the Aztecs', *Anthropological Quarterly*, 65: 187–94.

30 Valdeón, R. A. (2013), 'Doña Marina/La Malinche. A historiographical approach to the interpreter/traitor', *Target*, 25: 157–79.

31 Mann, C. (2011), *1493: Uncovering the New World Columbus Created*. New York: Vintage Books.

32 Haskett, R. S. (1991), ' "Our suffering with the Taxco Tribute": Involuntary mine labour of indigenous society in central New Spain', *Hispanic American Historical Review*, 71: 447–75.

33 Diel, L. B. (2010), 'The spectacle of death in early colonial New Spain in the Manuscrito del aperreamiento', in J. Beusterien and C. Cortez (eds.), *Death and Afterlife in the Early Modern Hispanic World*, Hispanic Issues On Line, 7: 144–63.

34 Borsdorf, A., and Stadel, C. (2015), *The Andes: A Geographical Portrait* (translated by B. Scott and C. Stadel), Dordrecht: Springer.

35 Shimada, I. (ed.) (2015), *The Inka Empire: A Multidisciplinary Approach*, Austin: University of Texas Press.

36 Cieza da León, P. D. (1998), *The Discovery and Conquest of Peru* (translated by A. P. Cook and N. D. Cook – original 1553), Durham, NC: Duke University Press.

37 Hemming, *The Conquest of the Incas*.

38 de Espinosa, A. (1907), *The Guanches of Tenerife* (translated by C. Markham – original 1594), London: The Hakluyt Society.

39 Cook, N. D. (1998), *Born to Die: Disease and New World Conquest 1492–1650*, Cambridge: Cambridge University Press.

40 Cook, N. D. (1993), 'Disease and depopulation of Hispaniola, 1492–1518', *Colonial Latin American Review*, 2: 213–45.

41 Hemming, J. (2009), *Tree of Rivers: The Story of the Amazon*, London: Thames & Hudson.

42 Livi-Bacci, M. (2006), 'The depopulation of Hispanic America after the conquest', *Population and Development Review*, 32: 199–232.

43 Newson, L. A. (2009), *Conquest and Pestilence in the Early Spanish Philippines*, Manoa: University of Hawaii Press.

44 Scott, W. H. (1982), *Cracks in the Parchment Curtain and Other Essays in Philippine History*, Quezon City: New Day Publishers.

45 Newson, *Conquest and Pestilence in the Early Spanish Philippines*.

46 Acabado, S. B. (2010), 'Landscapes and the archaeology of the Ifugao agricultural terraces: Establishing antiquity and social organisation', *Hukay: Journal for Archaeological Research in Asia and the Pacific*, 15: 31–61.

47 Bassani, E., and Fagg, W. (1988), *Africa and the Renaissance, Art in Ivory*, London: Centre for African Art, Prestel-Verlag.

48 Mitchell, P. (2005), *African Connections: Archaeological Perspectives on Africa and the Wider World*, Lanham, Maryland: AltaMira Press.

49 Fage, J. D. (1980), 'Slaves and society in western Africa, c. 1445–c. 1700', *Journal of African History*, 21: 289–310.

50 Nunn, N., and Qian, N. (2010), 'The Columbian Exchange: A history of disease, food, and ideas', *Journal of Economic Perspectives*, 24: 163–88.

51 Koch, Brierley, Maslin and Lewis, 'Earth system impacts of the European arrival and Great Dying in the Americas after 1492'.

52 Adorno, R. (2000), *Guaman Poma: Writing and Resistance in Colonial Peru* (2nd edition), Austin: University of Texas Press.

53 Crosby, *The Columbian Exchange: Biological and Cultural Consequences of 1492*.

54 Earle, R. (2010), ' "If You Eat Their Food . . .": Diets and Bodies in Early Colonial Spanish America', *American Historical Review*, 115: 688–713.

55 Earle, R. (2012), 'The Columbian Exchange', in J. M. Pilcher (ed.), *The Oxford Handbook of Food History*, Oxford: Oxford University Press: DOI: 10.1093/oxfordhb/9780199729937.013.0019.

56 Dumire, W. (2004), *Gardens of New Spain: How Mediterranean Plants and Foods Changed America*, Austin: University of Texas Press.

57 Melville, E. G. K. (1994), *A Plague of Sheep: Environmental Consequences of the Conquest of Mexico*, Cambridge: Cambridge University Press.

58 Bhattacharya, T., and Byrne, R. (2016), 'Late Holocene anthropogenic and climatic influences on the regional vegetation of Mexico's Cuenca Oriental', *Global and Planetary Change*, 138: 56–69.

59 Hooghiemstra, H., Olijhoek, T., Hoogland, M., et al. (2018), 'Columbus' environmental impact in the New World: Land use change in the Yaque River

valley, Dominican Republic', *The Holocene*, 28: https://doi.org/10.1177/0959683618788732.

60 Rifkin, J. (1993), *Beyond Beef: The Rise and Fall of the Cattle Culture*, New York: Plume.

61 Crosby, A. W. (1984), 'An ecohistory of the Canary Islands: A precursor of European colonialization in the New World and Australasia', *Environmental Review*, 8: 214–35.

62 Alexander, R. T., and Álvarez, H. H. (2017), 'Agropastoralism and household ecology in Yucatán after the Spanish Invasion', *Journal of Human Palaeoecology*, 23: 69–79.

63 Norton, M. (2015), 'The chicken or the *Iegue*: Human-animal relationships and the Columbian Exchange', *American Historical Review*, 120: 28–60.

64 Nunn and Qian, 'The Columbian Exchange: A history of disease, food, and ideas'.

65 ibid.

66 Rain, P. (1992), 'Vanilla: Nectar of the Gods', in N. Foster and L. S. Cordell (eds.), *Chilies to Chocolate: Food the Americas Gave the World*, Tucson: University of Arizona Press, 35–45.

67 Nunn and Qian, 'The Columbian Exchange: A history of disease, food, and ideas'.

68 Thornton, E. K., Emery, K. F., Steadman, D. W., et al. (2012), 'Earliest Mexican turkeys (*Melagris gallopavo*) in the Maya region: Implications for pre-Hispanic animal trade and the timing of turkey domestication', *PLOS ONE*: https://doi.org/10.1371/journal.pone.0042630.

69 Amano, N., Bankoff, G., Findley, D. M., et al. (2020), 'Archaeological and historical insights into the ecological impacts of pre-colonial and colonial introductions into the Philippine Archipelago', *The Holocene*: https://doi.org/10.1177/0959683620941152.

70 ibid.

71 McCann, J. C. (2005), *Maize and Grace: Africa's Encounter with a New World Crop, 1500-2000*, Cambridge, MA: Harvard University Press.

72 ibid.

73 Alpern, S. B. (2008), 'Exotic plants of western Africa: Where they came from and when', *History in Africa*, 35: 63–102.

74 Logan, A. (2020), *The Scarcity Slot: Excavating Histories of Food Security in Ghana*, Oakland, CA: University of California Press.

75 Shanahan, T. M., Overpeck, J. T., Anchukaitis, K. J., et al. (2009), 'Atlantic forcing of persistent drought in West Africa', *Science*, 324: 377–80.

76 Logan, A. (2017), 'Will agricultural technofixes feed the world? Short- and long-term tradeoffs of adopting high-yielding crops', in M. Hegmon (ed.), *The Give and Take of Sustainability: Archaeological and Anthropological Perspectives*, Cambridge: Cambridge University Press, 109–24.

77 Dotterweich, M. (2013), 'The history of human-induced soil erosion: Geomorphic legacies, early descriptions and research, and the development of soil conservation – A global synopsis', *Geomorphology*, 201: 1–34.

78 Mann, *1493: Uncovering the New World Columbus Created*.

79 Carney, J., and Rosomoff, R. (2009), *In the Shadow of Slavery: Africa's Botanical Legacy in the Atlantic World*, Berkeley: University of California Press.

80 Carney, J. (2001), *Black Rice: The African Origins of Rice Cultivation in the Americas*, Cambridge, MA: Harvard University Press.

81 Crosby, A. F. (2004), *Ecological Imperialism: The Biological Expansion of Europe, 900–1900* (2nd edition), Austin: University of Texas Press.

82 Brown, K. W. (2012), *A History of Mining in Latin America: From the Colonial Era to the Present*, Albuquerque: University of New Mexico Press.

83 de Araujo Shellard, A. H. (2016), 'History of the colonization of Minas Gerais: An environmental approach', in E. Vaz, C. Joanaz de Melo and L. Costa Pinto (eds.), *Environmental History in the Making*, Environmental History, volume 6, Cham: Springer, 243–57.

84 Hagan, N., Robins, N., Hsu-Kim, H., et al. (2011), 'Estimating historical atmospheric mercury concentrations from silver mining and their legacies in present-day surface soil in Potosí, Bolivia', *Atmospheric Environment*, 45: 7619–26.

85 Miller, S. W. (2007), *An Environmental History of Latin America*, Cambridge: Cambridge University Press.

86 Barretto-Tesoro, G., and Hernandez, V. (2017), 'Power and Resilience: Flooding and Occupation in a Late Nineteenth Century Philippine Town', in C. Beaule (ed.), *Frontiers of Colonialism*, Gainesville, Florida: University Press of Florida, 149–78.

87 Miller, *An Environmental History of Latin America*.

88 Barretto-Tesoro and Hernandez, 'Power and Resilience: Flooding and Occupation in a Late Nineteenth Century Philippine Town'.

89 Miller, *An Environmental History of Latin America*.

90 Cushner, N. P. (1971), *Spain in the Philippines: From Conquest to Revolution*, Quezon City, Philippines: Ateneo de Manila University.

91 Gerona, D. M. (2001), 'The colonial accommodation and reconstitution of native elite in early Provincial Philippines, 1600–1795', in M. D. E. Pérez-Grueso, J. M. Fradera and L. A. Alvarez (eds.), *Imperios y Naciones en el Pacífico*, Madrid: Consejo Superior de Investigaciones Científicas, 265–76.

92 Moore, J. W. (2009), 'Madeira, sugar, and the conquest of nature in the "first" sixteenth century. Part I: From "Island of timber" to sugar revolution, 1420–1506', *Review (Fernand Braudel Center)*, 32: 345–90.

93 Moore, J. W. (2000), 'Sugar and the expansion of the early modern world-economy: Commodity frontiers, ecological transformation, and industrialisation', *Review (Fernand Braudel Center)*, 23: 409–33.

94 Dannenfeldt, K. H. (1985), 'Europe discovers civet cats and civet', *Journal of the History of Biology*, 18: 403–31.

95 Polónia, A., and Pacheco, J. M. (2017), 'Environmental impacts of colonial dynamics 1400–1800: The first global age and the Anthropocene', in G. Austin (ed.), *Economic Development and Environmental History in the Anthropocene: Perspectives on Asia and Africa*, London: Bloomsbury, 23–49.

96 Teixeira, D. M., and Papabero, N. (2010), 'O tráfico de primatas brasileiros nos séculos XVI e XVII', in L. M. Pessôa, W. C. Tavares and S. Salvatore (eds.), *Mamíferos de restingas e manguezais do Brasil*, Rio de Janeiro: Sociedade Brasileira de Mastozoologia & Museu Nacional da UFRJ, 253–82.

97 Polónia and Pacheco, 'Environmental impacts of colonial dynamics 1400–1800: The first global age and the Anthropocene'.

98 Beaule, C. D., and Douglas, J. G. (eds.) (2020), *The Global Spanish Empire: Five Hundred Years of Place Making and Pluralism*, Arizona: University of Arizona Press.

99 Reed, R. R. (1978), *Colonial Manila: The Context of Hispanic Urbanism and Process of Morphogenesis*, Berkeley: University of California Press.

100 Acabado, S. (2017), 'The archaeology of pericolonialism: Responses of the "unconquered" to Spanish conquest and colonialism in Ifugao, Philippines', *International Journal of Historical Archaeology*, 21: 1–26.

101 Bankoff, G. (2013), ' "Deep Forestry": Shapers of the Philippine forests', *Environmental History*, 18: 523–56.

102 Associated Press, 'Columbus statue removed in Mexico City, defaced elsewhere'.

103 Lowrey, M. M. (2020), 'Why more places are abandoning Columbus Day in favour of Indigenous Peoples' Day', *The Conversation*: https://theconversation. com/why-more-places-are-abandoning-columbus-day-in-favor-of-indigenous-peoples-day-124481.

104 Vink, M. (2003), ' "The world's oldest trade": Dutch slavery and slave trade in the Indian Ocean in the seventeenth century', *Journal of World History*, 14: 131–77.

105 Wing, J. T. (2015), *Roots of Empire: Forests and State Power in Early Modern Spain, c. 1500–1750*, Leiden: Brill's Series in the History of the Environment, Volume 4.

106 Miller, *An Environmental History of Latin America*.

107 Newson, *Conquest and Pestilence in the Early Spanish Philippines*.

108 Bjork, K. (1998), 'The link that kept the Philippines Spanish: Mexican merchant interests and the Manila trade, 1571–1815', *Journal of World History*, 9: 25–50.

Chapter 11: Globalization of the tropics

1 Church, J. A., White, N. J., and Hunter, J. R. (2006), 'Sea-level rise at tropical Pacific and Indian Ocean islands', *Global and Planetary Change*, 53: 155–86.

2 Alroy, J. (2017), 'Effects of habitat disturbance on tropical forest biodiversity', *Proceedings of the National Academy of Sciences of the United States of America*, 114: 6056–61.

3 Corlett, R. T., and Primack, R. (2011), *Tropical Rain Forests: An Ecological and Biogeographical Comparison*, London: Wiley-Blackwell.

4 Diamond, J. (1997), *Guns, Germs, and Steel: The Fates of Human Societies*, New York City: W.W. Norton.

5 Soule, E. B. (2018), 'From Africa to the Ocean Sea: Atlantic slavery in the origins of the Spanish Empire', *Atlantic Studies*, 15: 16–39.

6 von Humboldt, A. (1814–29), 'Personal Narrative', Volume 4, 143–4, cited in Wulf, A. (2015), *The Invention of Nature: The Adventures of Alexander von Humboldt the Lost Hero of Science*, London: John Murray, 57–8.

7 Morgan, P. D. (2009), 'Africa and the Atlantic, c. 1450–1820', in J. P. Greene and P. D. Morgan (eds.), *Atlantic History: A Critical Appraisal*, Oxford: Oxford University Press, 223–48.

8 Martinez, J. S. (2012), *The Slave Trade and the Origins of International Human Rights Law*, Oxford: Oxford University Press.

9 Barquera, R., Lamnidis, T. C., Lankapalli, A. K., et al. (2020), 'Origin and health status of first-generation Africans from early colonial Mexico City', *Current Biology*, 30: 2078–91.e11: https://www.sciencedirect.com/science/article/abs/pii/S0960982220304826.

10 Schroeder, H., Ávila-Arcos, M. C., Malaspinas, A.-S., et al. (2015), 'Genome-wide ancestry of 17th-century enslaved Africans from the Caribbean', *Proceedings of the National Academy of Sciences of the United States of America*, 112: 3669–73.

11 Schroeder, H., O'Connell, T. C., Evans, J. A., et al. (2009), 'Trans-Atlantic slavery: Isotopic evidence for forced migration to Barbados', *American Journal of Physical Anthropology*, 139: 547–57.

12 Klein, H. S., and Luna, F. V. (2009), *Slavery in Brazil*, Cambridge: Cambridge University Press.

13 Mintz, S. (1986), *Sweetness and Power: The Place of Sugar in Modern History*, London: Penguin.

14 Calderia, A. M. (2011), 'Learning the Ropes in the Tropics: Slavery and the plantation system on the island of São Tomé', *African Economic History*, 39: 35–71.

15 Borucki, A., Eltis, D., and Wheat, D. (2015), 'Atlantic history and the slave trade to Spanish America', *American Historical Review*, 120: 433–61.

16 Calderia, 'Learning the Ropes in the Tropics: Slavery and the plantation system on the island of São Tomé'.

17 Mann, C. (2011), *1493: Uncovering the New World Columbus Created*, New York: Vintage Books.

18 Vink, M. (2003), ' "The world's oldest trade": Dutch slavery and slave trade in the Indian Ocean in the seventeenth century', *Journal of World History*, 14: 131–77.

19 Shuler, K. A. (2011), 'Life and death on a Barbadian sugar plantation: historic and bioarchaeological views of infection and mortality at Newton Plantation', *International Journal of Osteoarchaeology*, 21: 66–81.

20 Bethell, L. (1987), *Colonial Brazil*, Cambridge: Cambridge University Press.

21 Emmer, P. C. (translated by C. Emery) (2006), *The Dutch Slave Trade, 1500–1850*, New York: Berghahn Books.

22 Meniketti, M. (2006), 'Sugar mills, technology, and environmental change: A case study of colonial agro-industrial development in the Caribbean', *Journal of the Society for Industrial Archaeology*, 32: 53–80.

23 Fick, C. (2000), 'Emancipation in Haiti: From plantation labour to peasant proprietorship', *Slavery and Abolition*, 21: 11–40.

24 Morgan, K. (2007), *Slavery and the British Empire: From Africa to America*, Oxford: Oxford University Press.

25 Allen, R. B. (2010), 'Satisfying the "want for labouring people": European slave trading in the Indian Ocean, 1500–1850', *Journal of World History*, 21: 45–73.

26 Russell-Wood, A. J. R. (1977), 'Technology and society: The impact of gold mining on the institution of slavery in Portuguese America', *The Journal of Economic History*, 37: 59–83.

27 Bradley, K., and Cartledge, P. (2011), *The Cambridge World History of Slavery*, Cambridge: Cambridge University Press.

28 Kelley, S. M. (2019), 'New world slave traders and the problem of trade in goods: Brazil, Barbados, Cuba, and North America in comparative perspective', *English Historical Review*, 134: 302–33.

29 Richardson, D., Schwarz, S., and Tibbles, A. (2009), *Liverpool and Transatlantic Slavery*, Liverpool: Liverpool University Press.

30 Barquera, Lamnidis, Lankapalli, et al., 'Origin and health status of first-generation Africans from early colonial Mexico City'.

31 Williamson, K. (2019), 'Most slave shipwrecks have been overlooked until now', *National Geographic*: https://www.nationalgeographic.com/culture/2019/08/most-slave-shipwrecks-overlooked-until-now/.

32 Edwards, P., and Rewt, P. (1994), *The Letters of Ignatius Sancho*, Edinburgh: Edinburgh University Press (Letter 14, 56).

33 Inikori, J. (2002), *Africans and the Industrial Revolution in England: A Study in International Trade and Economic Development*, Cambridge: Cambridge University Press.

34 MacEachern, S. (2018), *Searching for Boko Haram: A History of Violence in Central Africa*, Oxford: Oxford University Press.

35 Bellagamba, A., Greene, S. E., and Klein, M. A. (eds.) (2013), *African Voices on Slavery and the Slave Trade*, Cambridge: Cambridge University Press.

36 Hicks, D. (2020), *The Brutish Museums: The Benin Bronzes, Colonial Violence and Cultural Restitution*, London: Pluto Press.

37 Landers, J. (2007), 'Slavery in the Spanish Caribbean and the Failure of Abolition', *Review (Fernand Braudel Center)*, 31: 343–71.

38 Thornton, J. (1998), *Africa and Africans in the Making of the Atlantic World, 1400–1800* (2nd edition), New York: Cambridge University Press.

39 ibid.

40 DeCorse, C. (2001), 'Introduction', in C. DeCorse (ed.), *West Africa during the Atlantic Slave Trade: Archaeological Perspectives*, Leicester: Leicester University Press, 1–13.

41 Behrendt, S. D., Latham, A. J. H., and Northrup, D. (2010), *The Diary of Antera Duke: An Eighteenth-Century African Slave Trader*, Oxford: Oxford University Press.

42 Mitchell, P. (2005), *African Connections: Archaeological Perspectives on Africa and the Wider World*, Lanham, Maryland: AltaMira Press.

43 Williams, H. V. (2010), 'Queen Nzinga (Njinga Mbande)', in L. M. Alexander and W. C. Rucker (eds.), *Encyclopaedia of African American History*, Santa Barbara, California: ABC-CLIO, 82–4.

44 Price, R. (ed.) (1996), *Maroon societies: Rebel slave communities in the Americas*, Baltimore, Maryland: Johns Hopkins University Press.

45 Richardson, D. (2001), 'Shipboard revolutions, African authority, and the Atlantic Slave Trade', *The William and Mary Quarterly*, 58 (New Perspectives on the Transatlantic Slave Trade): 69–92.

46 Diouf, S. A. (2016), *Slavery's exiles: the story of the American maroons*, New York: New York University Press.

47 Schwartz, S. B. (2017), 'Rethinking Palmares: Slave resistance in colonial Brazil', in D. A. Pargas and F. Rosu (eds.), *Critical Readings on Global Slavery*, Leiden: Brill, 1294–1325.

48 Odewale, A. (2019), 'An archaeology of struggle: Material remnants of a double consciousness in the American South and Danish Caribbean Communities', *Transforming Anthropology*, 27: 114–32.

49 ibid.

50 Odewale, A., Foster II, T., and Toress, J. M. (2017), 'In service to a Danish King: Comparing material culture of Royal Enslaved Afro-Caribbeans and Danish soldiers at the Christianised National Historic site', *Journal of African Diaspora Archaeology and Heritage*, 6: 1–39.

51 ibid.

52 Otele, O. (2020), *African Europeans: An Untold History*, London: Hurst Press.

53 Nunn, N., and Qian, N. (2010), 'The Columbian Exchange: A History of Disease, Food, and Ideas', *Journal of Economic Perspectives*, 24: 163–88.

54 Hall, C. (2020), 'The slavery business and the making of "race" in Britain and the Caribbean', *Current Anthropology*, 16: DOI: 10.1086/709845.

55 Meniketti, 'Sugar mills, technology, and environmental change: A case study of colonial agro-industrial development in the Caribbean'.

56 Hersh, J., and Voth, H.-J. (2009), 'Sweet Diversity: Colonial goods and the rise of European living standards after 1492', available at SSRN: http://dx.doi.org/10.2139/ssrn.1443730.

57 Brunache, P. (2019), 'Mainstreaming African diasporic foodways when academia is not enough', *Transforming Anthropology*, 27: 149–63.

58 Brown, C. L. (2012), *Moral Capital: Foundations of British Abolitionism*, Carolina: University of North Carolina Press.

59 Roscoe, P. (2020), 'How the shadow of slavery still hangs over global finance', *The Conversation*: https://theconversation.com/how-the-shadow-of-slavery-still-hangs-over-global-finance-144826?utm_medium=email&utm_campaign=Latest%20from%20The%20Conversation%20for%20August%2024%202020%20-%201711316529&utm_content=Latest%20from%20The%20Conversation%20for%20August%2024%202020%20-%201711316529+CID_58678a94df8ae31f0740b74c6db57bf7&utm_source=campaign_monitor_uk&utm_term=How%20the%20shadow%20of%20slavery%20still%20hangs%20over%20global%20finance.

60 Moore, J. W. (2000), 'Sugar and the expansion of the early modern world-economy: Commodity frontiers, ecological transformation, and industrialization', *Review (Fernand Braudel Center)*, 23: 409–33.

61 Barlow, V. (1993), *The Nature of the Islands: Plants and Animals of the Eastern Caribbean*, Dunedin: Chris Doyle Publishing.

62 Meniketti, 'Sugar mills, technology, and environmental change: A case study of colonial agro-industrial development in the Caribbean'.

63 Dunnavant, J. (2019), 'A Historical Ecology of Slavery in the Danish West Indies', talk given for UCI Media: https://www.youtube.com/watch?v=Q8oR_CPxkyQ&feature=youtu.be.

64 Smith, F. H. (2009), *Caribbean Rum: A Social and Economic History*, Gainesville: University of Florida Press.

65 Baptist, E. E. (2016), *The Half Has Never Been Told: Slavery and the Making of American Capitalism*, London: Hachette.

66 Riello, G. (2013), *Cotton: The Fabric that Made the Modern World*, Cambridge: Cambridge University Press.

67 ibid.

68 Menon, M., and Uzramma (2017), *A Frayed History: The Journey of Cotton in India*, Oxford: Oxford University Press.

69 Beckett, S. (2016), *Empire of Cotton: A Global History*, London: Penguin Books.

70 Tarlo, E. (1996), *Clothing Matters: Dress and Identity in India*, Chicago: University of Chicago Press.

71 Behal, R. P. (2006), 'Power structure, discipline, and labour in Assam tea plantations under colonial rule', *International Review of Social History*, 51: 143–72.

72 Bandarage, A. (1983), *Colonialism in Sri Lanka: The Political Economy of the Kandyan Highlands, 1833–1886*, Berlin: Mouton Publishers.

73 Chatterjee, P. (1995), ' "Secure this excellent class of labour": Gender and race in labor recruitment for British Indian tea plantations', *Bulletin of Concerned Asian Scholars*, 27: 43–56.

74 Topik, S. (1998), 'Coffee', in S. Topik and A. Wells (eds.), *The Second Conquest of Latin America: Coffee, Henequen and Oil during the Export Boom, 1850–1930*, Austin: University of Texas Press.

75 Pomeranz, K., and Topik, S. (2018), *The World that Trade Created: Society, Culture, and the World Economy, 1400 to the Present* (4th edition), New York: Routledge.

76 Topik, S. (1999), 'Where is the coffee? Coffee and Brazilian identity', *Luso-Brazilian Review*, 36: 87–92.

77 Hemming, J. (2009), *Tree of Rivers: The Story of the Amazon*, London: Thames & Hudson.

78 Grandin, G. (2010), *Fordlandia: The Rise and Fall of Henry Ford's Forgotten Jungle City*, London: Picador.

79 Hemming, J. (2004), *Amazon Frontier: Defeat of the Brazilian Indians*, London: Pan Macmillan.

80 Caetano Andrade, V. L., Flores, B. M., Levis, C., et al. (2019), 'Growth rings of Amazon nut trees (*Bertholletia excelsa*) as living record of historical human disturbance in Central Amazonia', *PLOS ONE*, 14(4): e0214128: DOI: 10.1371/journal.pone.0214128.

81 Tully, J. (2011), *The Devil's Milk: A Social History of Rubber*, New York: Monthly Review Press.

82 Dean, W. (1987), *Brazil and the Struggle for Rubber: A Study in Environmental History*, New York, NY: Cambridge University Press.

83 Ross, C. (2017), 'Developing the rain forest: Rubber, environment and economy in Southeast Asia', in G. Austin (ed.), *Economic Development and Environmental History in the Anthropocene: Perspectives on Asia and Africa*, London: Bloomsbury, 199–218.

84 Louis, W. R. (1964), 'Roger Casement and the Congo', *Journal of African History*, 5: 99–120.

85 Loadman, J. (2005), *Tears of the Tree*, Oxford: Oxford University Press.

86 Tully, *The Devil's Milk*.

87 Robins, J. E. (2018), 'Smallholders and machines in the West African palm oil industry, 1850–1950', *African Economic History*, 46: 69–103.

88 Mann, K. (2009), 'Owners, Slaves, and the Struggle for Labour in the Commercial Transition at Lagos', in R. Law (ed.), *From Slave Trade to 'Legitimate' Commerce: The Commercial Transition in Nineteenth-Century West Africa*, Cambridge: Cambridge University Press, 144–71.

89 Watkins, C. (2011), 'Dendezeiro: African oil palm agroecologies in Bahia, Brazil, and implications for development', *Journal of Latin American Geography*, 10: 9–33.

90 McNeill, W. H. (1999), 'How the potato changed the world's history', *Social Research*, 66: 67–83.

91 ibid.

92 Fitzgerald, P., and Lambkin, B. (2008), *Migration in Irish History 1607–2007*, London: Palgrave Macmillan.

93 Morrison, K. D., and Hauser, M. W. (2015), 'Risky business: Rice and inter-colonial dependencies in the Indian and Atlantic Oceans', *Atlantic Studies*, 12: 371–92.

94 ibid.

95 ibid.

96 Moberg, M., and Striffler, S. (eds.) (2003), *Banana Wars: Power, Production, and History in the Americas*, Durham and London: Duke University Press.

97 ibid.

98 Chapman, P. (2007), *Bananas: How the United Fruit Company Shaped the World*, Edinburgh: Canongate.

99 Moberg and Striffler (eds.), *Banana Wars: Power, Production and History in the Americas*.

100 Anderson, J. L. (2004), 'Nature's currency: The Atlantic mahogany trade and the commodification of nature in the eighteenth century', *Early American Studies*, 2: 47–80.

101 ibid.

102 ibid.

103 ibid.

104 Bankoff, G. (2013), ' "Deep Forestry": Shapers of the Philippine forests', *Environmental History*, 18: 523–56.

105 Pearson, M., and Lennon, J. (2010), *Pastoral Australia. Fortunes, Failures & Hard Yakka: A Historical Overview*, Victoria, Australia: CSIRO.

106 Frost, W. (1997), 'Farmers, government, and the environment: The settlement of Australia's "Wet Frontier", 1870–1920', *Australian Economic History Review*, 37: 19–38.

107 Ferrier, Å. (2015), *Journeys into the Rainforest: Archaeology of Culture Change and Continuity on the Evelyn Tableland, North Queensland*, Canberra: Australian National University.

108 Brown, L. (2010), 'Monuments to freedom, monuments to nation: The politics of emancipation and remembrance in the eastern Caribbean', *Slavery and Abolition*, 23: 93–116.

109 Cushman, G. T. (2013), *Guano and the Opening of the Pacific World: A Global Ecological History*, Cambridge: Cambridge University Press.

110 Allen, R. B. (2014), 'Slaves, convicts, abolitionism and the global origins of the post-emancipation indentured labor system', *Slavery and Abolition*, 35: 328–48.

111 Bass, D. (2013), *Everyday Ethnicity in Sri Lanka: Up-country Tamil Identity Politics*, London: Routledge.

112 Firth, S. (1976), 'The transformation of the labour trade in German New Guinea, 1899–1914', *The Journal of Pacific History*, 11: 51–65.

113 Ramasamy, P. (1992), 'Labour control and labour resistance in the plantations of colonial Malaya', *The Journal of Peasant Studies*, 19: 87–105.

114 Holloway, T. H. (2004), 'Immigrants on the Land: Coffee and Society in São Paulo, 1886–1934', North Carolina: University of North Carolina Press.

115 Scharlin, C., and Villanueva, L. V. (2000), *Philip Vera Cruz: A Personal History of Filipino Immigrants and the Farmworkers Movement* (3rd edition), Washington: University of Washington Press.

116 Reyes, M. (2008), 'Migration and Filipino children left behind: A literature review', Quezon City, Philippines: Miriam College/UNICEF, retrieved from http://www.unicef.org/philippines/Synthesis_StudyJuly12008.pdf.

117 Hobson, J. A. (2005) [1902], *Imperialism: A Study*, New York: Cosimo.

118 Wallerstein, I. (2007), 'The ecology and the economy: What is rational?', in A. Hornborg, J. R. McNeill and J. Martinez-Alier (eds.), *Rethinking Environmental History: World-System History and Global Environmental Change*, Plymouth: AltaMira Press, 379–90.

119 von Humboldt, A. (1814–29), 'Personal Narrative', Volume 4, 143–4, cited in Wulf, A. (2015), *The Invention of Nature: The Adventures of Alexander von Humboldt, the Lost Hero of Science*, London: John Murray, 57–8.

120 McKittrick, K. (2013), 'Plantation futures', *Small Axe: A Caribbean Journal of Criticism*, 17: 1–15.

121 Moulton, A. A., and Popke, J. (2017), 'Greenhouse governmentality: Protected agriculture and the changing biopolitical management of agrarian life in Jamaica', *Environment and Planning D: Society and Space*, 35: 714–732l.

122 Moulton, A. A., and Machado, M. R. (2019), 'Bouncing forward after Irma and Maria: Acknowledging colonialism, problematizing resilience and thinking climate justice', *Journal of Extreme Events*, 6: 1940003: DOI: 10.1142/S2345737619400037.

123 Byerlee, D. (2014), 'The fall and rise again of plantations in tropical Asia: History repeated?', *Land*, 3: 574–97.

124 Banerjee, A. (2005), 'History, institutions, and economic performance: The legacy of colonial land tenure systems in India', *American Economic Review*, 95: 1190–213.

125 Rudel, T. K. (2005), *Tropical Forests: Regional Paths of Destruction and Regeneration in the Late Twentieth Century*, New York: Columbia University Press.

126 ibid.

127 Pao, H.-T., and Tsai, C.-M. (2010), 'CO_2 emissions, energy consumption and economic growth in BRIC countries', *Energy Policy*, 38: 7850–60.

128 Houghton, J. T., Ding, Y., Griggs, D. J., et al. (IPCC) (2001), *Climate Change 2001: The Scientific Basis. Contribution of Working Group I to the Third Assessment Report of the Intergovernmental Panel on Climate Change*, Cambridge: Cambridge University.

129 Myers, N. (1988), 'Tropical forests and their species: going, going ... ?', in E. O. Wilson (ed.), *Biodiversity*, Washington: National Academy Press, 28–35.

130 Movuh, M. C. Y. (2012), 'The colonial heritage and post-colonial influence, entanglements and implications of the concept of community forestry by the example of Cameroon', *Forest Policy and Economics*, 15: 70–77.

131 Barber, S. (2020), 'Death by racism', *The Lancet, Infectious Diseases*, 20: DOI: https://doi.org/10.1016/S1473-3099(20)30567-3.

132 Levang, P., Dounias, E., and Sitorus, S. (2012), 'Out of the forest, out of poverty?', *Forests, Trees and Livelihoods*, 15: 211–35.

133 Morris, D. A. (1984), *The Origins of the Civil Rights Movement: Black Communities Organizing for Change*, New York: The Free Press.

134 Erfani-Ghettani, R. (2018), 'Racism, the Press and Black Deaths in Police Custody in the United Kingdom', in M. Bhatia, S. Poynting, and W. Tufail (eds.), *Media, Crime and Racism*, Palgrave Studies in Crime, Media and Culture, Palgrave Macmillan, 255–75.

135 Wilson, E. O. (ed.), (1988), *Biodiversity*, Washington, D.C.: National Academy Press.

Chapter 12: A tropical 'Anthropocene'?

1 Walker, J. D., Geissman, J. W., Bowring, S. A., and Babcock, L. E. (2013), 'The Geological Society of America Geologic Time Scale', *Geological Society of America Bulletin*, 125: 259–72.

2 Malhi, Y. (2017), 'The concept of the Anthropocene', *Annual Review of Environment and Resources,* 42: 77–104.

3 Ronda, M. (2013), 'Mourning and Melancholia in the Anthropocene', *Post 45*: http://post45.org/2013/06/mourning-and-melancholia-in-the-anthropocene/.

4 Zalasiewicz, J., Waters, C. N., Summerhayes, C. P., et al. (2017), 'The Working Group on the Anthropocene: Summary of evidence and interim recommendations', *Anthropocene*, 19: 55–60.

5 Crutzen, P. J. (2002), 'Geology of Mankind', *Nature*, 415: 23.

6 Ruddiman, W. F. (2003), 'The Anthropogenic Greenhouse Era Began Thousands of Years Ago', *Climatic Change*, 61, 261–93.

7 Ruddiman, W. F., Ellis, E. C., Kaplan, J. O., and Fuller, D. Q. (2015), 'Defining the epoch we live in', *Science*, 348: 38–9.

8 Ellis, E., Maslin, M., Boivin, N., and Bauer, A. (2016), 'Involve social scientists in defining the Anthropocene', *Nature*, 540: 192–3.

9 Roberts, P., Boivin, N., and Kaplan, J. (2018), 'Finding the anthropocene in tropical forests', *Anthropocene*, 23: 5–16.

10 Malhi, Y., Gardner, T. A., Goldsmith, G. R., et al. (2014), 'Tropical forests in the Anthropocene', *Annual Review of Environment and Resources*, 39: 125–59.

11 ibid.

12 Pimm, S. L., and Raven, P. (2000), 'Extinction by numbers', *Nature*, 403: 843–5.

13 Malhi, Y. (2012), 'The productivity, metabolism and carbon cycle of tropical forest vegetation', *Journal of Ecology*, 100: 65–75.

14 Brandon, K. (2014), *Ecosystem Services from Tropical Forests: Review of Current Science*, Washington D.C.: Center for Global Development Working Paper 380.

15 Wilson, J. B., Peet, R. K., Dengler, J., and Pärtel, M. (2012), 'Plant species richness: the world records', *Journal of Vegetation Science*, 23: 796–802.

16 Brugmann, H. (2020), 'Tree diversity reduced to the bare essentials', *Science*, 368: 128–9.

17 DRYFLOR, et al. (2015), 'Plant diversity patterns in neotropical dry forests and their conservation implications', *Science*, 353: 1383–7.

18 Watts, J. (2020), 'Tolkien was right: giant trees have towering role in protecting forests', *Guardian Online*: https://www.theguardian.com/environment/2020/apr/09/tolkien-was-right-giant-trees-have-towering-role-in-protecting-forests.

19 Larsen, M. C. (2017), 'Contemporary human uses of tropical forested watersheds and riparian corridors: Ecosystem services and hazard mitigation, with examples from Panama, Puerto Rico, and Venezuela', *Quaternary International*, 448: 190–200.

20 Samarakoon, M. B., Tanaka, N., and Iimura, K. (2013), 'Improvement of Effectiveness of Existing *Casuarina Equisetifolia* Forests in Mitigating Tsunami Damage', *Journal of Environmental Management*, 114: 105–14.

21 Powers, J. S., Montgomery, R. A., Adair, E. C., et al. (2009), 'Decomposition in tropical forests: A pan-tropical study of the effects of litter type, litter placement and mesofaunal exclusion across a precipitation gradient', *Journal of Ecology*, 97: 801–11.

22 Rüger, N., Condit, R., Dent, D. H., et al. (2020), 'Demographic trade-offs predict tropical forest dynamics', *Science*, 368: 165–8.

23 Lutz, J. A., et al. (2018), 'Global importance of large-diameter trees', *Global Ecology and Biogeography*, 27: 849–64.

24 Malhi, Gardner, Goldsmith, et al., 'Tropical forests in the Anthropocene'.

25 Sheil, D. (2014), 'How plants water our planet: advances and imperatives', *Trends in Plant Science*, 19: 209–11.

26 Met Office (2020), Tynemouth, UK Climate Averages: https://www.metoffice.gov.uk/research/climate/maps-and-data/uk-climate-averages/gcybzz9xh.

27 Lawrence, D., and Vandecar, K. (2015), 'Effects of tropical deforestation on climate and agriculture', *Nature Climate Change*, 5: 27–36.

28 Nogherotto, R., Coppola, E., Giorgi, F., and Mariotti, L. (2013), 'Impact of Congo Basin deforestation on the African monsoon', *Atmospheric Science Letters*, 14: 45–51.

29 Medvigy, D., Walko, R. L., Otte, M. J., and Avissar, R. (2013), 'Simulated changes in northwest U.S. climate in response to Amazon deforestation', *Journal of Climatology*, 26: 9115–36.

30 Werth, D., and Avissar, R. (2005), 'The local and global effects of African deforestation', *Geophysical Research Letters*, 32: L12704.

31 Prevedello, J. A., Winck, G. R., Weber, M. M., et al. (2019), 'Impacts of forestation and deforestation on local temperature across the globe', *PLOS ONE*: https://doi.org/10.1371/journal.pone.0213368.

32 Malhi, Y. (2010), 'The carbon balance of tropical forest regions, 1990–2005', *Current Opinions in Environmental Sustainability*, 2: 237–44.

33 Alkama, R., and Cescatti, A. (2016), 'Biophysical climate impacts of recent changes in global forest cover', *Science*, 351: 600–604.

34 Lawrence and Vandecar, 'Effects of tropical deforestation on climate and agriculture'.

35 ibid.

36 Malhi, Y. (2019), 'Does the Amazon provide 20% of our oxygen?': http://www.yadvindermalhi.org/blog/does-the-amazon-provide-20-of-our-oxygen.

37 Betts, R., Sanderson, M., and Woodward, S. (2008), 'Effects of large-scale Amazon forest degradation on climate and air quality through fluxes of carbon dioxide, water, energy, mineral dust and isoprene', *Philosophical Transactions of the Royal Society of London B Series: Biological Sciences*, 363: https://doi.org/10.1098/rstb.2007.0027.

38 Hoffmann, W. A., Schroeder, W., and Jackson, R. B. (2003), 'Regional feedbacks among fire, climate, and tropical deforestation, *Journal of Geophysical Research Atmosphere*, 108: https://doi.org/10.1029/2003JD003494.

39 Lamb, K. (2019), 'Indonesian forest fires spark blame game as smoke closes hundreds of Malaysia schools', *Guardian Online*: https://www.theguardian.com/world/2019/sep/12/indonesia-forest-fires-spark-blame-game-as-smoke-closes-hundreds-of-malaysia-schools.

40 Barrett, K. S. C., Jaward, F. M., and Stuart, A. L. (2019), 'Forest filter effect for polybrominated diphenyl ethers in a tropical watershed', *Journal of Environmental Management*, 248: 109279.

41 Lawrence and Vandecar, 'Effects of tropical deforestation on climate and agriculture'.

42 Roberts, P., and Petraglia, M. (2015), 'Pleistocene rainforests: barriers or attractive environments for early human foragers?', *World Archaeology*, 47: 718–39.

43 Doughty, C. E., Wolf, A., Morueta-Holme, N., et al. (2015), 'Megafauna extinction, tree species range reduction, and carbon storage in Amazonian forests', *Ecography*, 39: 194–203.

44 Davis, M., Faurby, S., and Svenning, J.-C. (2018), 'Mammal diversity will take millions of years to recover from the current biodiversity crisis', *Proceedings of the National Academy of Sciences of the United States of America*, 115: 11262–7.

45 Doughty, Wolf, Morueta-Holme et al., 'Megafauna extinction, tree species range reduction, and carbon storage in Amazonian forests'.

46 Fairbairn, A. S., Hope, G. S., and Summerhayes, G. R. (2006), 'Pleistocene occupation of New Guinea's highland and subalpine environments', *World Archaeology*, 38: 371–86.

47 Kershaw, A. P. (1986), 'Climatic change and Aboriginal burning in north-east Australia during the last two glacial/interglacial cycles', *Nature*, 322: 47–9.

48 Haberle, S. G., Rule, S., Roberts, P., et al. (2010), 'Paleofire in the wet tropics of northeast Queensland, Australia', *PAGES*, 18: 78–80.

49 de Fátima Rossetti, D., de Toledo, P. M., Moraes-Santos, H. M., and de Araújo Santos Jr, A. E. (2004), 'Reconstructing habitats in central Amazonia using megafauna, sedimentology, radiocarbon, and isotope analysis', *Quaternary Research*, 61: 289–300.

50 Ruddiman, 'The Anthropogenic Greenhouse Era Began Thousands of Years Ago'.

51 Fuller, D. Q., van Etten, J., Manning, K., et al. (2011), 'The contribution of rice agriculture and livestock pastoralism to prehistoric methane levels', *The Holocene*, 21: 743–59.

52 Athens, J. S., and Ward, J. V. (2004), 'Holocene vegetation, savanna origins and human settlement of Guam', in V. Attenbrow and R. Fullagar (eds.), *A Pacific Odyssey: Archaeology and Anthropology in the Western Pacific. Papers in Honour of Jim Specht. Records of the Australian Museum, Supplement* 29, Sydney: Australian Museum, 15–30.

53 Levis, C., Flores, B. M., Moreira, P. A., et al. (2018), 'How people domesticated Amazonian forests', *Frontiers in Ecology and Evolution*, 5: 10.3389/fevo.2017.00171.

54 Maezumi, S. Y., Alves, D., Robinson, M., et al. (2018), 'The legacy of 4,500 years of polyculture agroforestry in the eastern Amazon', *Nature Plants*, 4: 540–47.

55 Koch, A., Brierley, C., Maslin, M. M., and Lewis, S. L. (2019), 'Earth system impacts of the European arrival and Great Dying in the Americas after 1492', *Quaternary Science Reviews*, 207: 13–36.

56 Cook, B. I., Anchukaitis, K. J., Kaplan, J. O., et al. (2012), 'Pre-Columbian deforestation as an amplifier of drought in Mesoamerica', *Geophysical Research Letters*, 39: L16706.

57 Chepstow-Lusty, A. J., Bennett, K. D., Fjelså, J., et al. (1998), 'Tracing 4000 years of environmental history in the Cuzco area, Peru, from the pollen record', *Mountain Research and Development*, 18: 159–72.

58 Koch, Brierley, Maslin and Lewis, 'Earth system impacts of the European arrival and Great Dying in the Americas after 1492'.

59 Levis, Flores, Moreira, et al., 'How people domesticated Amazonian forests'.

60 Arroyo-Kalin, M. (2010), 'The Amazonian Formative: crop domestication and anthropogenic soils', *Diversity*, 2: 473–504.

61 Penny, D., Hall, T., Evans, D., and Polkinghorne, M. (2018), 'Geoarchaeological evidence from Angkor, Cambodia, reveals a gradual decline rather than a catastrophic 15th-century collapse', *Proceedings of the National Academy of Sciences of the United States of America*, 116: 4871–6.

62 Klein Goldewijk, K., Beusen, A., Doelman, J., et al. (2017), 'Anthropogenic land use estimates for the Holocene – HYDE 3.2', *Earth System Science Data*, 9, 927–53.

63 Koch, Brierley, Maslin and Lewis, 'Earth system impacts of the European arrival and Great Dying in the Americas after 1492'.

64 Lewis, S. L., and Maslin, M. A. (2015), 'Defining the anthropocene', *Nature*, 519: 171–80.

65 ibid.

66 ibid.

67 Anonymous (1603), 'Descripcion de la Villa y Minas de Potosí – Ano de 1603', in: *Relaciones Geograficas de Indias,* ed. Ministerio de Fomento, Vol. II, 113–36, Madrid: Ministerio de Fomento, quoted in J. W. Moore (2017), 'The Capitalocene, Part I: on the nature and origins of our ecological crisis', *Journal of Peasant Studies*, 44: 594–630, 24.

68 Miller, S. W. (2007), *An Environmental History of Latin America*, Cambridge: Cambridge University Press.

69 Moore, J. W. (2017), 'The Capitalocene, Part I: on the nature and origins of our ecological crisis', *Journal of Peasant Studies*, 44: 594–630, 24.

70 Wing, J. T. (2015), *Roots of Empire*, Leiden: Brill.

71 Moore, 'The Capitalocene, Part I: on the nature and origins of our ecological crisis'.

72 ibid.

73 Wolfe, A. P., Hobbs, W. O., Birks, H. H., et al. (2013), 'Stratigraphic expressions of the Holocene–Anthropocene transitions revealed in sediments from remote lakes', *Earth Science Reviews*, 116: 17–34.

74 Steffen, W., Grinevald, J., Crutzen, P., and McNeill, J. (2011), 'The Anthropocene: conceptual and historical perspectives', *Philosophical Transactions of the Royal Society A: Mathematical, Physical and Engineering Sciences*, 369: https://doi.org/10.1098/rsta.2010.0327.

75 Davis, J., Moulton, A. A., Van Sant, L., and Williams, B. (2018), 'Anthropocene, Capitalocene, . . . Plantationocene?: A manifesto for ecological justice in an age of global crises', *Geography Compass*, 2019: e12438.

76 ibid.

77 Haraway, D. (2015), 'Anthropocene, Capitalocene, Plantationocene, Chthulucene: Making Kin', *Environmental Humanities*, 6: 159–65.

78 ibid.

79 Yusoff, K. (2018), *A Billion Black Anthropocenes or None*, Minneapolis: University of Minnesota Press.

80 Davis, Moulton, Van Sant and Williams, 'Anthropocene, Capitalocene, . . . Plantationocene?: A manifesto for ecological justice in an age of global crises'.

81 Moore, 'The Capitalocene, Part I: on the nature and origins of our ecological crisis'.

82 ibid.

83 Edgeworth, M., de B. Richter, D., Waters, C., et al. (2015), 'Diachronous beginnings of the Anthropocene: The lower bounding surface of anthropogenic deposits', *Anthropocene Review*, 2: 33–58.

84 ibid.

85 Craig, O. E., Saul, H., Lucquin, A., et al. (2013), 'Earliest evidence for the use of pottery', *Nature*, 496: 351–4.

86 Edgeworth, Richter, Waters, et al., 'Diachronous beginnings of the Anthropocene: The lower bounding surface of anthropogenic deposits'.

87 Deckard, S. (2016), 'World-ecology and Ireland: The Neoliberal ecological regime', *Journal of World-Systems Research*, 22: 145–76.

88 Moore, J. W. (2007), 'Silver, ecology, and the origins of the modern world, 1450–1640', in A. Hornborg, J. R. McNeill and J. Martinez-Alier (eds.), *Rethinking Environmental History: World-System History and Global Environmental Change*, Plymouth: AltaMira Press, 123–42.

89 Woolford, A., Benvenuto, J., and Hinton, A. L. (eds.) (2014), *Colonial Genocide in Indigenous North America*, Durham, NC: Duke University Press.

90 Hinkson, M., and Vincent, E. (eds.), 'Shifting Indigenous Australian Realities: Dispersal, Damage, and Resurgence', Special issue, *Oceania*, 88.

91 Dooling, W. (2007), *Slavery, Emancipation and Colonial Rule in South Africa*, Ohio: Ohio University Press.

Chapter 13: Houses on fire

1 Thunberg, G. (2019), '"Our house is on fire": Greta Thunberg, 16, urges leaders to act on climate', *Guardian Online*: https://www.theguardian.com/environment/2019/jan/25/our-house-is-on-fire-greta-thunberg16-urges-leaders-to-act-on-climate.

2 Barlow, J., Berenguer, E., Carmenta, R., and França, F. (2019), 'Clarifying Amazonia's burning crisis', *Global Change Biology*, 26: 319–21.

3 World Meteorological Organization (2019), 'Australia suffers devastating fires after hottest, driest year on record': https://public.wmo.int/en/media/news/australia-suffers-devastating-fires-after-hottest-driest-year-record.

4 Reuters in Brasîlia (2020), 'Brazil's Amazon rainforest suffers worst fires in a decade': https://www.theguardian.com/environment/2020/oct/01/brazil-amazon-rainforest-worst-fires-in-decade.

5 Hansen, M. C., Potapov, P. V., Moore, R., et al. (2013), 'High-resolution global maps of 21st-century forest cover change', *Science*, 342: 850–53.

6 Corlett, R. T., and Primack, R. (2011), *Tropical Rain Forests: An Ecological and Biogeographical Comparison*, London: Wiley-Blackwell.

7 ibid.

8 Fa, J. E., Peres, C. A., and Meeuwig, J. (2002), 'Bushmeat exploitation in tropical forests: an intercontinental comparison', *Conservation Biology*, 16: 232–7.

9 Gibbs, D., Harris, N., and Seymour, F. (2018), 'By the numbers: The value of tropical forests in the climate change equation', World Resources Institute: https://www.wri.org/blog/2018/10/numbers-value-tropical-forests-climate-change-equation.

10 The State of the Tropics Project (2020), *State of the Tropics 2020 Report*, Queensland, Australia: James Cook University Press.

11 Pimm, S. L., Jenkins, C. N., Abell, R., et al. (2014), 'The biodiversity of species and their rates of extinction, distribution, and protection', *Science*, 344: 1246752.

12 Pimm, S. L., Jenkins, C. N., and Li, B. V. (2018), 'How to protect half of Earth to ensure it protects sufficient biodiversity', *Science Advances*, 4: DOI: 10.1126/sciadv.aat2616.

13 Bastin, J.-F., Fiengold, Y., Garcia, C., et al. (2019), 'The global tree restoration potential', *Science*, 365: 76–9.

14 Watson, J. E. M., Evans, T., Venter, O., et al. (2018), 'The exceptional value of intact forest ecosystems', *Nature Ecology & Evolution*, 2: 599–610.

15 UNESCO, 'Wet Tropics of Queensland': https://whc.unesco.org/en/list/486/.

16 Watson, Evans, Venter, et al., 'The exceptional value of intact forest ecosystems'.

17 Corlett and Primack, *Tropical Rain Forests: An Ecological and Biogeographical Comparison*.

18 Ghazoul, J., and Chazdon, R. (2017), 'Degradation and recovery in changing forest landscapes: A multiscale conceptual framework', *Annual Review of Environment and Resources*, 42: 161–88.

19 Harrison, R. D. (2011), 'Emptying the forest: Hunting and the extirpation of wildlife from tropical nature reserves', *BioScience*, 61: 919–24.

20 Roberts, P. (2018), 'Late Pleistocene tropical rainforest forager sustainability and resilience', in N. Sanz (ed.), *Exploring Frameworks for Tropical Forest Conservation: Managing Production and Consumption for Sustainability*, Mexico City: UNESCO, 116–35.

21 Barker, G., and Farr, L. (eds.) (2016), *Archaeological Investigations in the Niah Caves, Sarawak, The Archaeology of the Niah Caves, Sarawak*, Volume 2, McDonald Institute for Archaeological Research, Cambridge.

22 Van Vliet, N., Milner-Gulland, E. J., Bousquet, F., et al. (2009), 'Effect of small-scale heterogeneity of prey and hunter distributions on the sustainability of bushmeat hunting', *Conservation Biology*, 24: 1327–37.

23 Roberts, P. (2019), *Tropical Forests in Prehistory, History, and Modernity*, Oxford: Oxford University Press.

24 Dounias, E. (2016), 'From subsistence to commercial hunting: Technical shift in cynergetic practices among southern Cameroon forest dwellers during the 20th century', *Ecology and Society*, 21: http://dx.doi.org/10.5751/ES-07946-210123.

25 Corlett and Primack, *Tropical Rain Forests: An Ecological and Biogeographical Comparison*.

26 Logan, A. L., Stump, D., Goldstein, S. T., et al. (2019), 'Usable pasts forum: Critically engaging food security', *African Archaeological Review*, 36: 419–38.

27 United Nations (2011), *Population distribution, urbanization, internal migration and development: an international perspective*, Department of Economic and Social Affairs, Population Division, Publication no. ESA/P/WP/223.

28 Connah, G. (2000), 'African city walls: a neglected source?', in D. M. Anderson and R. Rathbone (eds.), *Africa's Urban Past*, Oxford: James Currey, 36–51.

29 Simon, D., and Adam-Bradford, A. (2014), 'Archaeology and contemporary dynamics for more sustainable, resilient cities in the peri-urban interface', in B. Maheshwari, V. P. Singh and B. Thoradeniya (eds.), *Balanced urban development: Options and strategies for liveable cities*, Cham: Springer, 57–84.

30 ibid.

31 Roberts, P., Hunt, C., Arroyo-Kalin, M., et al. (2017), 'The deep human prehistory of global tropical forests and its relevance for modern conservation', *Nature Plants*, 3: 17093.

32 Carlos Quezada, J., Etter, A., Ghazoul, J., et al. (2019), 'Carbon neutral expansion of oil palm plantations in the Neotropics', *Science Advances*, 5: DOI: 10.1126/sciadv.aaw4418.

33 Meijaard, E., Abrams, J. F., Juffe-Bignoli, D., et al. (2020), 'Coconut oil, conservation and the conscientious consumer', *Current Biology*, 30: 757–8.

34 Pin Koh, L., Miettinen, J., Chin Liew, S., and Ghazoul, J. (2011), 'Remotely sensed evidence of tropical peatland conversion to oil palm', *Proceedings of the National Academy of Sciences*, 108: 5127–32.

35 Burney, D. A., and Burney, L. P. (2007), 'Paleoecology and "inter-situ" restoration on Kaua'i, Hawai'i', *Frontiers in Ecology and the Environment*, 5: 483–90.

36 Boesenkool, S., McGlynn, G., Epp, L. S., et al. (2013), 'Use of ancient sedimentary DNA as a novel conservation tool for high-altitude tropical biodiversity', *Conservation Biology*, 28: 446–55.

37 Louys, J., Corlett, R. T., Price, G. J., et al. (2014), 'Rewilding the tropics, and other conservation translocations strategies in the tropical Asia-Pacific region', *Ecology and Evolution*, 4 (22): 4380–98.

38 Ocheje, P. D. (2007), ' "In the public interest": forced evictions, land rights and human development in Africa', *Journal of African Law*, 51: 173–214.

39 Stanton, P., Stanton, D., Stott, M., and Parsons, M. (2014), 'Fire exclusion and the changing landscape of Queensland's Wet Tropics Bioregion 1. The extent and pattern of transition', *Australian Forestry*, 77: 51–7; and Stanton, P., Parsons,

M., Stanton, D., and Stott, M. (2014), 'Fire exclusion and the changing land-scape of Queensland's Wet Tropics Bioregion 2. The dynamics of transition forests and implications for management', *Australian Forestry*, 77: 58–68.

40 Roberts, P., Buhrich, A., Caetano-Andrade, V. L., et al. (2021), 'Reimagining the relationship between Gondwanan forests and Aboriginal land management in Australia's "Wet Tropics"', *iScience*, 24: 102190, DOI: https://doi.org/10.1016/j.isci.2021.102190.

41 Maezumi, S. Y., Robinson, M., de Souza, J., et al. (2018), 'New insights from Pre-Columbian land use and fire management in Amazonian Dark Earth forests', *Frontiers in Ecology and Evolution*: https://doi.org/10.3389/fevo.2018.00111.

42 Breuste, J., and Dissanyake, L. (2013), 'Socioeconomic and environmental change of Sri Lanka's Central Highlands', in A. Borsdorf (ed.), *Forschen im Gebirge: Christoph Stadel zum 75. Gebursttag*, Christoph Stadel-Festschrift, Vienna: Verlag der Österreichischen Akademie der Wissenschaften.

43 AOSIS (2017), *Rising Tides, Rising Capacity: Supporting a Sustainable Future for Small Island Developing States*, New York, NY: Association of Small Island States, United Nations Development Programme.

44 Schwitzer, C., Mittermeier, R. A., Johnson, S. E., et al. (2014), 'Averting lemur extinctions amid Madagascar's political crisis', *Science*, 343: 842–3.

45 Douglass, K., and Cooper, J. (2020), 'Archaeology, environmental justice, and climate change on islands of the Caribbean and southwestern Indian Ocean', *Proceedings of the National Academy of Sciences of the United States of America*, 117: 8254–62.

46 Gleeson, J. (2019), *Housing in our world cities: London, New York, Paris and Tokyo*, London: Greater London Authority: http://data.london.gov.uk.

47 Wet Tropics Management Authority (2020), World Heritage Area – facts and figures: https://www.wettropics.gov.au/world-heritage-area-facts-and-figures.html.

48 Ferrante, L., and Fearnside, P. M. (2020), 'Protect Indigenous peoples from COVID-19', *Science*, 368: 251.

49 Viana, V. (2010), *Sustainable Development in Practice: Lessons Learned from Amazonas*, London: International Institute for Environment and Development.

50 Edwards, G. A. S. (2019), 'Coal and climate change', *WIREs Climate Change*, 10: e607, https://doi.org/10.1002/wcc.607.

51 Finer, M., and Orta-Martínez, M. (2010), 'A second hydrocarbon boom threatens the Peruvian Amazon: trends, projections, and policy implications', *Environmental Research Letters*, 5: 014012.

52 Meijaard, E., and Sheil, D. (2019), 'The moral minefield of ethical oil palm and sustainable development', *Frontiers in Forests and Global Change* 2: DOI: 10.3389/ffgc.2019.00022.

53 Rist, L., Feintrenie, L., and Levang, P. (2010), 'The livelihood impacts of oil palm: smallholders in Indonesia', *Biodiversity and Conservation*, 19: 1009–24.

54 Müller, H., Rufin, P., Griffiths, P., et al. (2016), 'Beyond deforestation: Differences in long-term regrowth dynamics across land use regimes in southern Amazonia', *Remote Sensing of Environment*, 186: 652–62.

55 Corlett and Primack, *Tropical Rain Forests: An Ecological and Biogeographical Comparison*.

56 Ashton, P. S. (2010), 'Conservation of Borneo biodiversity: do small lowland parks have a role, or are big inland sanctuaries sufficient? Brunei as an example', *Biodiversity and Conservation*, 19: 343–56.

57 Voysey, B. C., McDonald, K. E., Rogers, M. E., et al. (1999), 'Gorillas and seed dispersal in the Lope Reserve, Gabon I: Gorilla acquisition by trees', *Journal of Tropical Ecology*, 15: 23–38.

58 DRYFLOR, et al. (2015), 'Plant diversity patterns in neotropical dry forests and their conservation implications', *Science*, 353: 1383–7.

59 ibid.

60 Levis, C., Flores, B. M., Mazzochini, G. G., et al. (2020), 'Help restore Brazil's governance of globally important ecosystem services', *Nature Ecology & Evolution*, 4: 172–3.

61 Roberts, Buhrich, Caetano-Andrade, et al., 'Reimagining the relationship between Gondwanan forests and Aboriginal land management in Australia's "Wet Tropics"'.

62 Cooper, J. (2012), 'Fail to prepare then prepare to fail: Re-thinking threat, vulnerability and mitigation in the pre-Columbian Caribbean', in J. Cooper and P. Sheets (eds.), *Surviving Sudden Environmental Change: Answers from Archaeology*, Boulder, CO: University Press of Colorado, 91–114.

63 Ferrier, Å. (2015), *Journeys into the Rainforest: Archaeology of Culture Change and Continuity on the Evelyn Tableland, North Queensland*, Canberra: Australian National University.

64 Logan, Stump, Goldstein, et al., 'Usable pasts forum: Critically engaging food security'.

65 Corlett and Primack, *Tropical Rain Forests: An Ecological and Biogeographical Comparison*.

66 Roberts, Buhrich, Caetano-Andrade, et al., 'Primordial Gondwanaland or human forests?: Reimagining the Australian "Wet Tropics" and their conservation'.

67 Nicholas, A., Warren, Y., Bila, S., et al. (2010), 'Successes in community-based monitoring of cross river gorillas (*Gorilla gorilla diehli*) in Cameroon', *African Primates*, 7: 55–60.

68 Sheil, D., Boissière, M., and Beaudoin, G. (2015), 'Unseen sentinels: local monitoring and control in conservation's blind spots', *Ecology and Society*, 20: http://dx.doi.org/10.5751/ES-07625-200239.

69 Putz, F. E., Zuidema, P. A., Synnott, T., et al. (2012), 'Sustaining conservation values in selectively logged tropical forests: the attained and the attainable', *Conservation Letters*, 5: 296–303.

70 Ebeling, J., and Yasué, M. (2009), 'The effectiveness of market-based conservation in the tropics: Forest certification in Ecuador and Bolivia', *Journal of Environmental Management*, 90: 1145–53.

71 Güemes-Ricalde, F. J., Villanueva-G, R., Echazarreta-González, C., et al. (2005), 'Production costs of conventional and organic honey in the Yucatán Peninsula of Mexico', *Journal of Apicultural Research*, 45: 106–11.

72 ibid.

73 Soto-Pinto, L., Anzueto, M., Mendoza, J., et al. (2010), 'Carbon sequestration through agroforestry in indigenous communities of Chiapas, Mexico', *Agroforestry Systems*, 78: https://doi.org/10.1007/s10457-009-9247-5.

74 Ghazoul, J., and Sheil, D. (2010), *Tropical Rain Forest Ecology, Diversity, and Conservation*, Oxford: Oxford University Press.

75 Aguilera, J. (2019), 'Bolsonaro says he won't accept $20 million to fight Amazon fires unless Macron apologises', *TIME*: https://time.com/5662395/bolsonaro-reject-g7-pledge-amazon-fires/.

76 UN-REDD Programme Collaborative Online Workspace (2020), UN-REDD Programme: http://www.unredd.net/regions-and-countries/regions-and-countries-overview.html.

77 Saeed, A.-R., McDermott, C., and Boyd, E. (2018), 'Examining equity in Ghana's national REDD+ process', *Forest Policy and Economics*, 90: 48–58.

78 Ghazoul and Sheil, *Tropical Rain Forest Ecology, Diversity, and Conservation*.

79 Butler, S., and Sweeney, M. (2018), 'Iceland's Christmas TV advert rejected for being political', *Guardian Online*: https://www.theguardian.com/media/2018/nov/09/iceland-christmas-tv-ad-banned-political-greenpeace-orangutan.

Chapter 14: A global responsibility

1 Evenstar, L. A., Stuart, F. M., Hartley, A. J., and Tattitch, B. (2015), 'Slow Cenozoic uplift of the western Andean Cordillera indicated by cosmogenic ^{3}He in alluvial boulders from the Pacific Planation Surface', *Geophysical Research Letters*, 42: 8448–55.

2 Dentan, R. K. (1991), 'Potential Food Sources for Foragers in Malaysian Rainforest: Sago, Yams, and Lots of Little Things', *Bijdragen tot de Taal, Land-en Volkenkunde*, 147: 420–44.

3 Morcote-Ríos, G., Aceituno, F. J., Iriarte, J., et al. (2020), 'Colonisation and early peopling of the Colombian Amazon during the Late Pleistocene and the Early Holocene: New evidence from La Serranía La Lindosa', *Quaternary International*: https://doi.org/10.1016/j.quaint.2020.04.026.

4 Perfecto, I., Vandermeer, J. H., and Wright, A. L. (2009), *Nature's Matrix: Linking Agriculture, Conservation and Food Sovereignty*, Sterling, VA: Earthscan.

5 Hawthorne, W. (2010), *From Africa to Brazil: Culture, Identity, and an Atlantic Slave Trade, 1600–1830*, Cambridge: Cambridge University Press.

6   Sonter, L. J., Herrera, D., Barrett, D. J., et al. (2017), 'Mining drives extensive deforestation in the Brazilian Amazon', *Nature Communications*, 8: https://doi.org/10.1038/s41467-017-00557-w.

7   O'Neill, S., and Nicholson-Cole, S. (2009), ' "Fear won't do it": Promoting positive engagement with climate change through visual and iconic representations', *Science Communication*, 30: 355–79.

8   Sayyid, S. (2017), 'Post-racial paradoxes: rethinking European racism and anti-racism', *Patterns of Prejudice*, 51: 9–25.

9   Boothby, J., and Hull, A. P. (1997), 'A census of ponds in Cheshire, North West England', *Aquatic Conservation: Marine and Freshwater Ecosystems*, 7: 75–9.

10 Hansen, M. C., Potapov, P. V., Moore, R., et al. (2013), 'High-resolution global maps of 21st-century forest cover change', *Science*, 342: 850–53.

11 Future Generations Commissioner for Wales (2020), 'The Future Generations Report 2020': https://www.futuregenerations.wales/wp-content/uploads/2020/06/Chap-3-Resilient.pdf.

12 Hansen, Potapov, Moore, et al., 'High-resolution global maps of 21st-century forest cover change'.

13 Strassburg, B. B. N., Rodrigues, A. S. L., Gusti, M., et al. (2012), 'Impacts of incentives to reduce emissions from deforestation on global species extinctions', *Nature Climate Change*, 2: 350–55.

14 Bullock, E. L., Woodcock, C. E., Souza Jr, C., and Olofsson, P. (2020), 'Satellite-based estimates reveal widespread forest degradation in the Amazon', *Global Change Biology*, 26: 2956–69.

15 Office for National Statistics (2009), 'Area', in I. Macrory (ed.), *Annual Abstract of Statistics*, Basingstoke: Palgrave Macmillan, 3–5.

16 Pearson, T. R. H., Brown, S., and Casarim, F. M. (2014), 'Carbon emissions from tropical forest degradation caused by logging', *Environmental Research Letters*, 9: https://doi.org/10.1088/1748-9326/9/3/034017.

17 Hosonuma, N., Herold, M., De Sy, V., et al. (2012), 'An assessment of deforestation and forest degradation drivers in developing countries', *Environmental Research Letters*, 7: https://doi.org/10.1088/1748-9326/7/4/044009.

18 Ahrends, A., Burgess, N. D., Milledge, S. A. H., et al. (2010), 'Predictable waves of sequential forest degradation and biodiversity loss spreading from an African city', *Proceedings of the National Academy of Sciences of the United States of America*, 107: 14556–61.

19 Romero-Sanchez, M. E., and Ponce-Hernandez, R. (2017), 'Assessing and monitoring forest degradation in a deciduous tropical forest in Mexico via remote sensing indicators, *Forests*, 8: https://doi.org/10.3390/f8090302.

20 Cordeiro, R. C., Turcq, B., Moreira, L. S., et al. (2014), 'Palaeofires in Amazon: Interplay between land use change and palaeoclimatic events', *Palaeogeography, Palaeoclimatology, Palaeoecology*, 415: 137–51.

21 Marlon, J. R. (2020), 'What the past can say about the present and future of fire', *Quaternary Research*, 96: 66–87.

22 Fonseca, M. G., Alves, L. M., Aguiar, A. P. D., et al. (2019), 'Effects of climate and land-use change scenarios on fire probability during the 21st century in the Brazilian Amazon', *Global Change Biology*, 25: 2931–46.

23 Page, S. E., and Hooijer, A. (2016), 'In the line of fire: the peatlands of Southeast Asia', *Proceedings of the Royal Society B Series: Biological Sciences*, 371: https://doi.org/10.1098/rstb.2015.0176.

24 Fonseca, Alves, Aguiar, et al., 'Effects of climate and land-use change scenarios on fire probability during the 21st century in the Brazilian Amazon'.

25 Dutta, R., Das, A., and Aryal, J. (2016), 'Big data integration shows Australian bush-fire frequency is increasing significantly', *Royal Society Open Science*, 3: https://doi.org/10.1098/rsos.150241.

26 Lenton, T. M., Rockström, J., Gaffney, O., et al. (2019), 'Climate tipping points – too risky to bet against', *Nature*, 575: 592–6.

27 SIMIP Community (2020), 'Arctic sea ice in CMIP6', *Geophysical Research Letters*, 47: https://doi.org/10.1029/2019GL086749.

28 State of the Tropics (2020), *State of the Tropics 2020 Report*, Townsville, Australia: James Cook University Press.

29 European Environment Agency (2020), 'Global and European temperatures': https://www.eea.europa.eu/data-and-maps/indicators/global-and-european-temperature-10/assessment.

30 King, A. D., and Harrington, L. J. (2018), 'The inequality of climate change from 1.5 to 2°C of global warming', *Geophysical Research Letters*, 45: 5030–33.

31 Lenton, Rickström, Gaffney, et al., 'Climate tipping points – too risky to bet against'.

32 Lovejoy, T. E., and Nobre, C. (2018), 'Amazon tipping point', *Science Advances*, 4: DOI: 10.1126/sciadv.aat2340.

33 Ferraz, G., Russell, G. J., Stouffer, P. C., et al. (2003), 'Rates of species loss from Amazonian forest fragments', *Proceedings of the National Academy of Sciences of the United States of America*, 100: 14069–73.

34 Thatte, P., Joshi, A., Vaidyanathan, S., et al. (2018), 'Maintaining tiger connectivity and minimizing extinction into the next century: Insights from landscape genetics and spatially-explicit simulations', *Biological Conservation*, 218: 181–91.

35 Benitez-López, A., Santini, L., Schipper, A. M., et al. (2019), 'Intact but empty forests? Patterns of hunting-induced mammal defaunation in the tropics', *PLOS Biology*: https://doi.org/10.1371/journal.pbio.3000247.

36 Ghazoul, J., and Sheil, D. (2010), *Tropical Rain Forest Ecology, Diversity, and Conservation*, Oxford: Oxford University Press.

37 Rowland, J., Hoskin, C. J., and Burnett, S. (2019), 'Distribution and diet of feral cats (*Felis catus*) in the Wet Tropics of north-eastern Australia, with a focus on the upland rainforest', *Wildlife Research*, 47: 649–59.

38 *State of the Tropics 2020 Report*.

39 United Nations Department of Economic and Social Affairs (2019), *2019 revision of world population prospects*, New York: UN DESA.

40 *State of the Tropics 2020 Report.*

41 United Nations (2014), *2014 revision of world urbanization prospects*, New York: UN.

42 *State of the Tropics 2020 Report.*

43 ibid.

44 ibid.

45 ibid.

46 WHO and United Nations Settlements Programme (2010), *Hidden Cities: Unmasking and overcoming health inequalities in urban settings*, Kobe: World Health Organization Centre for Health Development.

47 *State of the Tropics 2020 Report.*

48 ibid.

49 Phelan, B., Bertzky, M., Butchart, S. H. M., et al. (2013), 'Crop expansion and conservation priorities in tropical countries', *PLOS ONE*, 8: https://doi.org/10.1371/journal.pone.0051759.

50 Dookhun, A. (2018), *Final country report LDN Target Setting Programme – Republic of Seychelles*, Bonn, Germany: UNCCD.

51 *State of the Tropics 2020 Report.*

52 Borrelli, P., Robinson, D. A., Fleischer, L. R., et al. (2017), 'An assessment of the global impact of 21st century land use change on soil erosion', *Nature Communications*, 8: https://doi.org/10.1038/s41467-017-02142-7.

53 World Bank (2018), *Poverty and shared prosperity 2018: Piecing together the poverty puzzle*, Washington, DC: World Bank.

54 *State of the Tropics 2020 Report.*

55 ibid.

56 Diffenbaugh, N. S., Singh, D., Mankin, J. S., et al. (2017), 'Quantifying the influence of global warming on unprecedented extreme climate events', *Proceedings of the National Academy of Sciences of the United States of America*, 114: 4881–6.

57 Eccles, R., Zhang, H., and Hamilton, D. (2019), 'A review of the effects of climate change on riverine flooding in subtropical and tropical regions', *Journal of Water and Climate Change*, 10: 687–707.

58 Alfieri, L., Bisselink, B., Dottori, F., et al. (2017), 'Global projections of river flood risk in a warmer world', *Earth's Future*, 5: 171–82.

59 Zhao, Y., Wang, C., and Wang, S. (2005), 'Impacts of present and future climate variability on agriculture and forestry in the humid and sub-humid tropics', *Climatic Change*, 70: 73–116.

60 Larsen, M. C. (2017), 'Contemporary human uses of tropical forested watersheds and riparian corridors: Ecosystem services and hazard mitigation, with examples from Panama, Puerto Rico, and Venezuela', *Quaternary International*, 448: 190–200.

61 Cange, C. W., and McGaw-Césaire, J. (2020), 'Long-term public health responses in high-impact weather events: Hurricane Maria and Puerto Rico as a case study', *Disaster Medicine and Public Health Preparedness*, 14: 18–22.

62 Burkett, M. (2011), 'In search of refuge: Pacific islands, climate-induced migration, and the legal frontier', *AsiaPacific Issues*, 98: 1–8.

63 *State of the Tropics 2020 Report*.

64 Carroll, D., Daszak, P., Wolfe, N. D., et al. (2018), 'The global virome project', *Science*, 359: 872–4.

65 Dobson, A. P., Pimm, S. L., Hannah, L., et al. (2020), 'Ecology and economics for pandemic prevention', *Science*, 369: 379–81.

66 Van Heuverswyn, F., and Peeters, M. (2007), 'The origins of HIV and implications for the global epidemic', *Current Infectious Disease Reports*, 9: 338–46.

67 Olivero, J., Fa, J. E., Real, R., et al. (2017), 'Recent loss of closed forests is associated with Ebola virus disease outbreaks', *Nature Scientific Reports*, 7: https://doi.org/10.1038/s41598-017-14727-9.

68 Dobson, Pimm, Hannah, et al., 'Ecology and economics for pandemic prevention'.

69 Zhou, P., Yang, X.-L., Wang, X.-G., et al. (2020), 'A pneumonia outbreak associated with a new coronavirus of probable bat origin', *Nature*, 579: 270–73.

70 Eberhard, D. M., Simons, G. F., and Fenning, C. D. (eds.) (2020), *Ethnologue: Languages of the World* (23rd edition), Dallas, Texas: SIL International.

71 Camacho, L. D., Gevaña, D. T., Carandang, A. P., and Camacho, S. C. (2015), 'Indigenous knowledge and practices for the sustainable management of Ifugao forests in Cordillera, Philippines', *Journal of Biodiversity Science, Ecosystem Services & Management*, 12: 5–13.

72 Benz, B. F., Cevallos, E. J., Santana, M. F., et al. (2000), 'Losing knowledge about plant use in the sierra de manantlan biosphere reserve, Mexico', *Economic Botany*, 54: 183–91.

73 Stephens, C., Porter, J., Nettleton, C., and Willis, R. (2006), 'Disappearing, displaced, and undervalued: a call to action for Indigenous health worldwide', *The Lancet*, 367: 17–23.

74 Frechette, A., Ginsburg, C., and Walker, W. (2018), *A Global Baseline of Carbon Storage in Collective Lands*, Washington, DC: Rights and Resources Initiative.

75 Mitchard, E. T. A. (2018), 'The tropical forest carbon cycle and climate change', *Nature*, 559: 527–34.

76 Sitch, S., Friedlingstein, P., Gruber, N., et al. (2015), 'Recent trends and drivers of regional sources and sinks of carbon dioxide', *Biogeosciences*, 12: 653–79.

77 Mitchard, 'The tropical forest carbon cycle and climate change'.

78 Schleussner, C.-F., Rogelj, J., Schaeffer, M., et al. (2016), 'Science and policy characteristics of the Paris Agreement temperature goal', *Nature Climate Change*, 6: 827–35.

79 Alroy, J. (2017), 'Effects of habitat disturbance on tropical forest biodiversity', *Proceedings of the National Academy of Sciences of the United States of America*, 114: 6056–61.

80 Giam, X. (2017), 'Global biodiversity loss from tropical deforestation', *Proceedings of the National Academy of Sciences of the United States of America*, 114: 5775–7.

81 Barnosky, A. D., Matzke, N., Tomiya, S., et al. (2011), 'Has the Earth's sixth mass extinction already arrived?', *Nature*, 471: 51–7.

82 Giam, 'Global biodiversity loss from tropical deforestation'.

83 Barnosky, Matzke, Tomiya, et al., 'Has the Earth's sixth mass extinction already arrived?'.

84 Coltart, C. E. M., Lindsey, B., Ghinai, I., et al. (2017), 'The Ebola outbreak, 2013–2016: old lessons for new epidemics', *Philosophical Transactions of the Royal Society B Series: Biological Sciences*, 372: https://doi.org/10.1098/rstb.2016.0297.

85 World Health Organization (2020), 'Coronavirus disease (COVID-19): Situation Report – 192': https://www.who.int/docs/default-source/coronaviruse/situation-reports/20200730-covid-19-sitrep-192.pdf?sfvrsn=5e52901f_4.

86 Dobson, Pimm, Hannah, et al., 'Ecology and economics for pandemic prevention'.

87 *State of the Tropics 2020 Report*.

88 World Bank (2019), 'Migration and remittances – Recent developments and outlook (Migration and Development Brief)', Washington, DC: World Bank.

89 *State of the Tropics 2020 Report*.

90 Biermann, F., and Boas, I. (2007), 'Preparing for a Warmer World: Towards a Global Governance System to Protect Climate Refugees', *Global Governance Working Paper*, 33, Amsterdam: Global Governance Project.

91 Roberts, P., and Stewart, B. (2018), 'Defining the "generalist-specialist" niche for Pleistocene *Homo sapiens*', *Nature Human Behaviour*: DOI: 10.1038/s41562-018-0394-4.

92 Barbour, W., and Schlesinger, C. (2012), 'Who's the boss? Post-colonialism, ecological research and conservation management on Australian Indigenous lands', *Ecological Management & Restoration*, 13: 36–41.

93 Reality Check team BBC News (2019), 'Deforestation: Did Ethiopia plant 350 million trees in a day?': https://www.bbc.co.uk/news/world-africa-49266983.

94 *State of the Tropics 2020 Report*.

95 Goffner, D., Sinare, H., and Gordon, L. J. (2019), 'The Great Green Wall for the Sahara and the Sahel Initiative as an opportunity to enhance resilience in Sahelian landscapes and livelihoods', *Regional Environmental Change*, 19: 1417–28.

96 Reij, C., Tappan, G., and Smale, M. (2009), 'Re-greening the Sahel: farmer-led innovation in Burkina Faso and Niger. Agroenvironmental Transformation in the Sahel: Another kind of "Green Revolution"', IFPRI Discussion Paper, Washington DC: International Food Policy Research Institute.

97 Mbile, P. N., Atangana, A., and Mbenda, R. (2019), 'Women and landscape restoration: a preliminary assessment of women-led restoration activities in Cameroon', *Environment, Development and Sustainability*, 21: 2891–911.

98 Osei, R., Zerbe, S., Beckmann, V., and Boaitey, A. (2018), 'Socio-economic determinants of smallholder plantation sizes in Ghana and options to encourage reforestation', *Southern Forests: a Journal of Forest Science*, 81: 49–56.

99 Roberts, P., Buhrich, A., Caetano-Andrade, V. L., et al. (2021), 'Reimagining the relationship between Gondwanan forests and Aboriginal land management in Australia's "Wet Tropics"', *iScience*, 24: 102190. DOI:https://doi.org/10.1016.j.isci.2021.102190.

100 Corlett, R. T., and Primack, R. (2011), *Tropical Rain Forests: An Ecological and Biogeographical Comparison*, London: Wiley-Blackwell.

101 Fonseca, Alves, Aguiar, et al., 'Effects of climate and land-use change scenarios on fire probability during the 21st century in the Brazilian Amazon'.

102 Hansen, Potapov, Moore, et al., 'High-resolution global maps of 21st-century forest cover change'.

103 Barbier, E. B., Lozano, R., Rodríguez, C. M., and Troëng, S. (2020), 'Adopt a carbon tax to protect tropical forests', *Nature*, 578: 213–16.

104 Strassburg, Rodrigues, Gusti, et al., 'Impacts of incentives to reduce emissions from deforestation on global species extinctions'.

105 Morgans, C. L., Meijaard, E., Santika, T., et al. (2018), 'Evaluating the effectiveness of palm oil certification in delivering multiple sustainability objectives', *Environmental Research Letters*, 13: 064032.

106 Vogelgesang, F., Kumar, U., and Sundram, K. (2018), 'Building a sustainable future together: Malaysian palm oil and European consumption', *Journal of Oil Palm, Environment and Health*, 9: 1–49.

107 Hasanah, N., Komarudin, H., Dray, A., and Ghazoul, J. (2019), 'Beyond Oil Palm: Perceptions of local communities of environmental change', *Frontiers in Forests and Global Change*, 2: DOI: 10.3389/ffgc.2019.00041.

108 Warren, M. (2019), 'Thousands of scientists are backing the kids striking for climate change', *Nature*, 567: 291–2.

109 Buchanan, L., Quoctrung, B., and Patel, J. K. (2020), 'Black Lives Matter may be the largest movement in U.S. History', *The New York Times*: https://www.nytimes.com/interactive/2020/07/03/us/george-floyd-protests-crowd-size.html.

110 Pingue, F. (2020), 'NFL'S Washington team to retire Redskins name and logo', Reuters: https://uk.reuters.com/article/uk-football-nfl-washington/nfls-washington-team-to-retire-redskins-name-and-logo-idUKKCN24E1OJ.

111 Elmi, O. (2020), 'Edward Colston: "Why the statue had to fall"': https://www.bbc.co.uk/news/uk-england-bristol-52965803.

112 Hurst, B. (2020), 'Brands backing Black Lives Matter: it might be a marketing ploy, but it also shows leadership', *The Conversation*: https://theconversation.com/brands-backing-black-lives-matter-it-might-be-a-marketing-ploy-but-it-also-shows-leadership-139874.

113 Batty, D. (2020), 'Universities criticised for "tokenistic" support for Black Lives Matter', *Guardian Online*: https://www.theguardian.com/education/2020/jul/06/universities-criticised-for-tokenistic-support-for-black-lives-matter.

114 Marshall, M. (2020), 'Planting trees doesn't always help with climate change': https://www.bbc.com/future/article/20200521-planting-trees-doesnt-always-help-with-climate-change.

Acknowledgements

This book is indebted to the many generations of ecologists, biologists, earth scientists, botanists, zoologists and conservationists who have made their careers revealing the remarkable secrets of our 'jungles'. These pages are built on their labours of love for environments that they have seen increasingly threatened as the decades have ticked by. Since the first chapter, I have argued that tropical forests have been generally marginalized in studies of human history. Nevertheless, I also hope to have highlighted the work of archaeologists, palaeoanthropologists, anthropologists, historians and palaeoecologists who have persistently bucked this trend – endeavouring to re-centre these habitats in discussions of the evolution of our species, the emergence of food production, pre-industrial urban settlements, and the impacts of European colonialism on landscapes and people. This book is a tribute to them. Beyond researchers, this book should also have demonstrated that advancements in 'science' are far from the only sources of knowledge about tropical forests. Indigenous voices have long emphasized the ecological, cultural and economic importance of these environments, not just as stewards of their own lands, but also for the entire world. Governments and scientists often simply haven't listened closely enough. Or even listened at all. I am incredibly grateful to the Indigenous peoples I have had the great fortune to meet and work with, including, specifically, Uruwaruge Heenbanda and Uruwaruge Wanniya-laeto of the Wanniya-laeto community in Sri Lanka and Desley Mosquito, Barry Hunter and Gerry Turpin of the Jirrbal, Djabugay and Mbabaram Aboriginal groups, respectively, in Australia. Indigenous groups, and their traditional knowledge, must be acknowledged and celebrated instead of ignored or exploited. They offer some of the best hopes for sustainable human use and management of these environments around the world today.

In an era when science is, quite rightly, facing increased scrutiny over its ethics, work environments and treatment of researchers, I have been immensely privileged to be able to call upon some of the most supportive mentors during my explorations of tropical forests. My two PhD supervisors at the University of Oxford set me off on my first encounters with these dynamic environments, and I have never looked back since. Mike Petraglia introduced me to Sri Lanka and its remarkable prehistory. Julia Lee-Thorp gave me the methodological tools to explore it further. Looking back, I am sure I was an incredibly irritating student. But hopefully all the red pen and the constant knocking on their doors was worth it. I will also always be immensely grateful for my first adviser at the same institution, Peter Mitchell. Without whom, I can categorically say that I would be neither an academic nor an archaeologist. I would also, of course, have never even made it into my first tropical forest without the support and kindness of my local collaborators in Sri Lanka – Dr Siran Deraniyagala who put the island on the map of Pleistocene archaeology, Dr Nimal Perera who cemented it there, and Dr Oshan Wedage who is one of the truest, most loyal friends anyone could ever ask for and an incredibly talented researcher. Thanks to them, the future of the discipline in this part of South Asia is in very good hands indeed. I also owe an immense debt of gratitude to my current employer, Nicky Boivin, who has been the most encouraging, understanding and inspirational of mentors as I have gone on to expand my research around the tropical world. Thank you for the opportunity and resources to pursue my passion as a career. I will never forget how incredibly lucky I am, and this book would not exist without your support.

While horror stories of bad collaborations can often ring around the conference circuit and coffee breaks, in the context of tropical forest research, I have been constantly left stunned at the kind encouragement, input and assistance provided by my colleagues – from across the world and across disciplines. This book was no different and the quotes and images included are a permanent testament to the generosity of my colleagues, as well as the diversity of voices, from PhD students to professors, from historians to earth scientists, that are invigorating our understanding of the importance of tropical forests to life on Earth. A huge thank you to each and every one of the people who suggested references, answered my annoying questions, and read through different parts of this work: Victor Lery Caetano Andrade (Preface), Tim Lenton, Silvia Pressel and Luke Meade (Chapter 1), Carlos Jaramillo (Chapter 2), Emma Dunne, Leonardo Salgado and Paul Barrett (Chapter 3), Zhe-Xi Luo, Tyler Lyson and Gina Semprebon (Chapter 4), Yohannes Haile-Selassie, Tim White, Sarah Feakins and Kira Westaway (Chapter 5), Eleanor Scerri, Oshan Wedage and Sue O'Connor (Chapter 6), Tim Denham, Dolores Piperno, José Iriarte and Umberto Lombardo (Chapter 7), Scott Fitzpatrick, Monica Tromp and Kristina Douglass (Chapter 8), Lisa Lucero, Damian Evans and Eduardo Neves (Chapter 9), Alexander Koch, Noel Amano, Amanda Logan and Grace Barretto-Tesoro (Chapter 10), Alicia Odewale, Ayushi Nayak, John Hemming, John Tully, Kathy Morrison, Åsa Ferrier, Meena Menon, Justin Dunnavant and Alex Moulton (Chapter 11), Nicky Boivin, Yadvinder Malhi, Nadja Rüger, Yoshi Maezumi and Janae Davis (Chapter 12), and Renier van Raders, Emuobosa Orijemie, Desley Mosquito and Douglas Sheil (Chapter 13). The views expressed are my own, and they should in no way be blamed for any mistakes made. However, they have also helped make this book what it is.

While the other colleagues that have helped me in writing this book, and pursuing this line of inquiry, are too many to name individually, I would like to particularly thank my local collaborators, beyond Sri Lanka, who have welcomed me into their countries and research environments, and taught me more than I can ever repay. These include Grace Barretto-Tesoro and Francis Gealogo in the Philippines, Letícia Morgana Muller, Hilton Pereira, Eduardo Neves, Eduardo Tamanaha, Carolina Levis and Charles Clement in Brazil, Oscár Solis in Mexico, Emuobosa Orijemie in Nigeria, Åsa Ferrier, Alice Buhrich, Desley Mosquito, Barry Hunter, Gerry Turpin, Richard Cosgrove, Simon Haberle and Janelle Stevenson in Australia, Ravi Korisettar in India, and Mahirta Mahirta in Indonesia. It is their work and their efforts, alongside the Indigenous groups and stakeholders that many of them work with (or are a part of), that are cementing tropical forests as landscapes of cultural and natural heritage of both national and global significance. I would also like to thank the often unnamed administrative and support staff who make all our research a possibility in the first place. For me, that is Anja Schatz, Anja Hannewald, Ellen Richter, Dorit Wammetsberger, Graeme Richardson, Daniela Gütsch, Beate Kerpen, Anna Pallaske, Michelle O'Reilly, Hans Sell, Christian Nagel, Gerd Kusserow, Thomas Baumman, Thomas Melzer and Thomas Brückner at the Max Planck Institute for the Science of Human History. It is Stefanie Schirmer, Ulrike Thüring, Dovydas Jurkenas, Mary Lucas, Erin Scott, Elsa Perruchini, Sara Marzo, Bianca Fiedler and Jana Ilgner in our laboratories. It is Anneke van Heteren (Zoologische Staatssammlung München), Jacques Cuisin, Violaine Nicolas-Colin, Géraldine Véron, Joséphine Lesur and Christine Lefèvre (Muséum National d'Histoire Naturelle), Malcolm McCallum and Tom Gillingwater (University of Edinburgh) for their support in the sampling and repatriation of museum collections. It is also all the visitors of museums, journalists who feature our findings, peer reviewers, journal editors, members of national governments and administrative sectors that process fieldwork permits, teachers, and the taxpayers, voters and politicians who support our work around the world.

It is perhaps not said enough that, while leading a Research Group comes with a lot of responsibility and stress, it also comes with immense benefits. Chief among those is the ability to supervise gifted students and postdoctoral researchers from a variety of backgrounds. It has been an immense privilege to interact with the talented and driven individuals that have already, in some way, passed through my lab: Victor Lery Caetano Andrade, Ayushi Nayak, Maddy Bleasdale, Rebecca Jenner Hamilton, Bob Patalano, Max Findley, Jillian Swift, Anneke Janzen, Alicia Ventresca-Miller, Noel Amano, Asier García-Escárazaga, Eleftheria Orfanou, Oshan Wedage, Michael Ziegler, Phoebe Heddell-Stevens, Patxi Perez Ramallo, Neha Dhavale, Celeste Samec, Verónica Zuccarelli, Xueye Wang, Courtney Culley, Óscar Ricardo Solís Torres, Letícia Morgana Muller, Giulia Riccomi and Clara Boulanger. I have learnt an immense deal from all of you and it has been a privilege to, in some small way, be involved in all your exciting research which I am sure will be setting agendas in archaeology and palaeoecology, including within the tropics, for many years to come. As well as many of the individuals listed above, I am also grateful to a number of people who have, at various points in my career, forced me to put work to one side and (usually) talk about something else for a change. Thank you especially to Tom Boulton, Jonny Rollings, Max Mewes, Christoph Brückner, Christoph Klose, Gerd Gleixner, Florian Ott, Rob Spengler, Andrea Kay, Julien Louys, Gilbert Price, Katerina Douka, Tom Higham, Brian Stewart, Sam Challis, Luíseach Nic Eoin, Rachel King, Mark McGranaghan, Ed Peveler, Abi Tomkins, Olly Beeley, Adam Besant, Vaughan Edmonds, Dan McArthur, Andy Baldock, Alison Crowther, Cosimo Posth, Mathew Stewart, Huw Groucutt and Eleanor Scerri for the good company and for the beer.

This book has been an immense pleasure to write, though it has not been all plain sailing. I am incredibly grateful to my editors, especially Connor Brown in the UK and Eric Henney in the United States of America, for their enthusiasm and patience in crafting this into a final product that was hopefully readable enough for those of you that kindly picked it up! I am also indebted to copy editor Annie Lee for her detailed, patient reading of the manuscript. Thank you also to the arts team at Viking, Penguin Random House in the UK for turning some of my scribbles and ramblings into effective artwork that brings the story of tropical forests to life. The same goes to those fantastic creators of the 'palaeoart' used in different parts of the book: Jeff Gage for his fantastic drawing of Titanoboa, Jay Matternes for his *Ardipithecus ramidus* picture, Bob Nicholls for his masterful recreation of ancient environments from the Carboniferous to the Jurassic, Velizar Simeonovski for his elephant bird reconstruction, April Neander for her image of the Jurassic glider *Vilevolodon diplomylos*, and the other artists whose work I have credited in these pages. Without these images, it would have been that bit harder to describe the long journey of tropical forests and their inhabitants. I would, of course, not have even got anywhere near the editing or illustration stage if it were not for my incredible agents, Joanna Swainson and Thérèse Coen at Hardman & Swainson. Thank you for having the faith in this project, for reading every page, and for negotiating contracts and payments. A first-time author of a trade book could not have asked for a better experience. Thank you also to the 'unseen editors', particularly Noel Amano, Max Findley and Rebecca Hamilton, who took the time out to provide extraordinarily helpful comments on a number of different parts of the book (and a number of different times!).

The research discussed in this book would also not have been possible without the generous funding and support I have received from various agencies throughout my career so far. They include the School of Archaeology at the University of Oxford, St Hugh's College at the University of Oxford, the Clarendon Fund, the Santander Foundation, the Natural Environmental Research Council, the Boise Fund at the University of Oxford, the PAGES and Society for African Archaeologists, the Bundesministerium für Bildung und Forschung, the Beutenburg Campus, the National Geographic Society and the European Research Council. I am also particularly grateful to the Max Planck Society for providing the funding for my position, my laboratory equipment, and my research initiatives throughout the tropics. Research funding, particularly for the Social Sciences and Humanities, is often some of the first to be cut in the face of economic crises, growing nationalism and, most recently, pandemics. The European Research Council has prominently recently faced cuts to its budget, for example. While there are undoubtedly many things more worthy of financial support in trying times, it is this funding that enables us to explore questions like those investigated in this book. Without it, none of it could have happened. I am also fully conscious of the fact that many of the same historical processes and inequalities explored in Chapters 10 and 11 have, ultimately, resulted in me having a privileged platform on which to produce this work. There is an urgent need to further support research initiatives in the tropics that are led and driven by talented researchers from the tropics, a number of whom we have heard from in the above pages.

Finally, beyond mentors, colleagues, friends and funding it is, ultimately, family that allows a book like this to be written. And I have one of the most supportive that I know. My parents, Julia and

Neil, provided the trips to museums, archaeological sites and nature reserves that sparked off my interest in the human past. My brother, Tom, put up with it. Only with their financial and loving support could I have taken the first step on the academic ladder, studying Archaeology and Anthropology at university. Every book that I asked for. No matter how boring it sounded. They bought. My grandmothers, Kay and Margaret, must also take credit here, and I get the feeling that they are still watching over me as I travel through my own 'jungles'. Kindness makes a huge difference in this life and I have never seen such kindness in a person as I have in both of them. I hope that, in some small way, I carry that irrepressible desire to help people forward. In my aunty, Ali, I also have one of the most caring, supportive and loving relatives possible. Nowadays, it is my partner, Jana, and our children, Rhys, Ida, and Livia, that are the core foundations of my tropical endeavours. They welcome the colleagues and friends that arrive on our doorstep from all over the world (though they do often get the benefits of good food . . . !). They tolerate the time I spend away on fieldwork in the tropics and at conferences with other forest-enthusiasts. They also let me find the time to write. As Livia says, '*in diesem Haus leben Chaos und Liebe*' (in this house live chaos and love). No truer words were ever spoken . . . Thank you for everything and all my love to you all. Thank you also to a wonderful acting 'mother-in-law', Bettina Ilgner, whose heart has no bounds and whose care for her family gets them through the tough, as well as the good, times more than she will ever know.

Index

Page references in *italics* indicate images.

He just wanted a decent book to read ...

Not too much to ask, is it? It was in 1935 when Allen Lane, Managing Director of Bodley Head Publishers, stood on a platform at Exeter railway station looking for something good to read on his journey back to London. His choice was limited to popular magazines and poor-quality paperbacks – the same choice faced every day by the vast majority of readers, few of whom could afford hardbacks. Lane's disappointment and subsequent anger at the range of books generally available led him to found a company – and change the world.

'We believed in the existence in this country of a vast reading public for intelligent books at a low price, and staked everything on it'
Sir Allen Lane, 1902–1970, founder of Penguin Books

The quality paperback had arrived – and not just in bookshops. Lane was adamant that his Penguins should appear in chain stores and tobacconists, and should cost no more than a packet of cigarettes.

Reading habits (and cigarette prices) have changed since 1935, but Penguin still believes in publishing the best books for everybody to enjoy. We still believe that good design costs no more than bad design, and we still believe that quality books published passionately and responsibly make the world a better place.

So wherever you see the little bird – whether it's on a piece of prize-winning literary fiction or a celebrity autobiography, political tour de force or historical masterpiece, a serial-killer thriller, reference book, world classic or a piece of pure escapism – you can bet that it represents the very best that the genre has to offer.

Whatever you like to read – trust Penguin.

read more
www.penguin.co.uk